Information Design
Design for Engagement

Editor:
Yvonne Eriksson
Professor in Information
Design and Deputy Head, Department of Health Science,
Innovation and Design Mälardalen University,
Stockholm, Sweden

CRC Press
Taylor & Francis Group
Boca Raton London New York

CRC Press is an imprint of the
Taylor & Francis Group, an **informa** business
A SCIENCE PUBLISHERS BOOK

Cover: Expanse by Ulrika Florin

First edition published 2027
by CRC Press
2385 NW Executive Center Drive, Suite 320, Boca Raton FL 33431

and by CRC Press
4 Park Square, Milton Park, Abingdon, Oxon, OX14 4RN

CRC Press is an imprint of Taylor & Francis Group, LLC

Library of Congress Cataloging-in-Publication Data (applied for)

ISBN: 978-1-032-64257-4 (hbk)
ISBN: 978-1-032-64258-1 (pbk)
ISBN: 978-1-032-64259-8 (ebk)

DOI: 10.1201/9781032642598

Typeset in Times New Roman
by Prime Publishing Services

Preface

This book is about different perspectives of Information Design and the challenges in engaging people in various communication processes. The contents span a broad spectrum of perspectives in Information Design from aesthetics, visual communication, visual management, text design, technical communication to spatial design and interaction design. Common for all chapters is a human-centered design perspective where sociocultural aspects are taken into consideration as well as perception and cognition. The majority contributions to the book comes from scholars in Information Design at Mälardalen University, where we provide education in information design in three different bachelor's level programs: Interaction Design, Spatial Design, and Visualization and Text Design. A key theme in the chapters is the role of Information Design in various contexts and how it affects individuals and groups.

The need for a systematic exploration of the role of Information Design in society is problematized by the authors and how the shape of the information affects peoples' life and understanding of the environment. Information Design is much about communication, communication in the design process and communication with the people and the environment. One of the challenges with Information Design research is to systematically evaluate the design process and its result.

The purpose of the book is to contribute to the area of Information Design and to elucidate various perspectives that the authors found in the topic. The book is directed towards a broad audience. It is also intended for students and researchers in Information Design, and design, with an aim to present the material in a form that is comprehensible for graduate, advanced undergraduate and Ph.D. students. The book is also designed to attract readers from outside of the immediate research area.

I would like to thank the authors who have contributed to the book and a special thanks to Ulrika Florin for providing the illustration ***Expanse*** to the cover of the book. Expanse means a wide continuous area of something, which highlights the aim of the book.

I would also like to thank my editor Vijay Primlani for her helpful and very important guidance through the process with the book.

Last, but not the least, I thank my partner Judit for her support, during the work with this book.

Contents

List of Contributors

Adam Palmquist
Ph.D. Senior Lecturer in Information Design and affiliated to *ExcITEd – Norwegian Centre for Excellence in IT Education (SFU)*, Department of Health Science, Innovation and Design, Mälardalen University, Box 325, 631 05 Eskilstuna, Sweden.

Anna-Lena Carlsson
Ph.D. Senior Lecturer in Information Design, Department of Health Science, Innovation, and Design Mälardalen University, Box 325, 631 05 Eskilstuna, Sweden.

Carina Söderlund
Ph.D. Associate Pofessor, Senior Lecturer in Information Design, Department of Health Science, Innovation, and Design, Mälardalen University, Box 325, 631 05 Eskilstuna, Sweden.

Helena Tobiasson
Ph.D. Senior lecture in Informations design, Department of Health Science, Innovation and Design, Mälardalen University, Box 325, 631 05 Eskilstuna, Sweden.

Jennie Schaeffer
Ph.D. Adjungt Associate Professor at Departement of Health Science, Innovation and Design, Museum director, Västmanlands Läns Museum, Karlsgatan 2, 722 14 Västerås, Sweden.

Jonatan Lundin
Ph.D Information Architect, Excosoft, Sweden. and affiliated to Department of Health Science, Innovation and Design, Department of Eskilstuna and Västerås, Mälardalen University, Box 325, 631 05 Eskilstuna, Sweden.

Karolina Uggla
Ph.D. Senior Lecturer in Information Design, Department of Health Science, Innovation and Design, Mälardalen University, Box 325, 631 05 Eskilstuna, Sweden.

Marie Sjölinder
RISE, Swedish Research Institute and affiliated to, Department of Health Science, Innovation and Design, Mälardalen University, Isafjordsgatan 22, Box 1263, 164 29 Kista.

Max Kleijberg
Ph.D. Senior Lecturer in Information Design, Department of Health Science, Innovation and Design, Mälardalen University, Box 325, 631 05 Eskilstuna, Sweden.

Peter Gärdenfors
Professor in Cognitive Science, Department of Philosophy and Cognitive Science, Lund University, Box 192, SE-22100 Lund, Sweden.

Thorbjörn Swenberg
Ph.D. Associate Professor in Audiovisual Studies, Head of Subject, Moving Image Production Chair of Swedish Media Production Academies (SMEA), Dalarna University, Högskolegatan 2, Falun, Sweden.

Ulrika Florin
Ph.D. Associate Professor and Senior lecturer and associate professor in Information design, Department of Health Science, Innovation and Design, Mälardalen University, Box 325, 631 05 Eskilstuna, Sweden.

Wibke Weber
PhD | Professor of Media Linguistics: Visual Communication, Program Director, BA Communication and Media Studies, ZHAW Zurich University of Applied Sciences, School of Applied Linguistics.

William Hagman
Ph.D. Senior Lecturer in Information Design, Department of Health Science, Innovation and Design, Mälardalen University, Box 325, 631 05 Eskilstuna, Sweden.

Yvonne Eriksson
Professor in Information Design, Department of Health Science, Innovation and Design, Mälardalen University, Box 325, 631 05 Eskilstuna, Sweden.

Introduction

Yvonne Eriksson
yvonne.eriksson@mdu.se

Today, people have access to more information than ever before. Some are connected not only to what is happening in their immediate environment but also to most of the world, by reading news and information in real-time. Information and news are often communicated speedily and without a deeper analysis of the event or situation. Widely available on digital platforms, information and news are easily shareable via social media, thus they spread fast. Society has become more complex, but much of the news and information has become simplified and is not always thoroughly formulated or analysed before it is sent out and spread further. The digital platform format, as such, encourages people to express themselves concisely and others to react with likes or dislikes.

The ongoing digital transformation involves both information challenges and extensive infrastructure change. The transformation affects *where, when,* and *how* we can (or do) access different types of information. This, in turn, poses new challenges in terms of how information and technology can and should be designed, while maintaining a focus on people's rights, privacy, and needs. The various chapters in this book discuss different aspects of information (explicit and embedded), and can be interpreted both as a unit and as parts of systems where both human needs and technological possibilities interact. For such a perspective to work, knowledge is needed about how individuals and groups interact with both each other and within organisations, as well as with different types of built environments, artefacts, interfaces, and technical solutions, both at work and in everyday life. This book contributes a deeper understanding of the need for on-site studies on various activities and within critical developmental phases in the common good, i.e., methods and approaches such as situated design, which consider that context arises between individual(s) and situation(s) and how it is constantly changing.

In Information Design, it has often been argued that, if the information is clear and concise, it will be accessible to a prospective user. In a way, it can be argued that much of the information disseminated on social media is concise and clear, and thus easy to absorb. The problem is that much information requires thought and deeper analysis in order to understand it. This applies to images of various kinds as well as text.

Information design is culturally based, and the way information is formulated and communicated differs from culture to culture. At the same time, the ability to interpret and use information varies between individuals and groups. This can be due to variations in previous experiences, education, socio economic factors, or disabilities. How people understand and use information is also partly dependent on the given context and situation. For example, in a busy environment that is experienced as stressful, an individual will likely be less receptive to information in general. Stress can also arise while looking for information and have a similar information-blocking effect.

The purpose of this book, as embodied by its title, *Information Design: Design for Engagement* is twofold. Firstly, to elucidate Information Design that is reliable, meaningful, and useful for people. Secondly, to engage students and professionals in the importance of Information Design.

In the following chapters, the authors investigate different perspectives on Information Design. Information Design is an interdisciplinary academic subject that involves scholars from various backgrounds, such as visual studies, applied linguistics, human-computer interaction, social Cognitive Science, and the humanities. All with a focus on theories and methods that provide insights for a better understanding of how people navigate and use information in various settings.

Traditionally, the aim of Information Design has been to understand how to produce information that is accessible to specific target groups. In this book, the authors have elaborated on perspectives that challenge the old school of Information Design, such as people's experience of meaningful information, aesthetics in Information Design, what is needed to bridge understanding between individuals and technologies, and how to support interaction between people, groups, and artefacts.

The different chapters are grouped roughly into four themes: Visual Communication, Text Design, Spatial Design and Information Design, and Human-Computer Interaction.

In the first chapter, Anna-Lena Carlsson explores the role of aesthetics in Information Design by discussing the variety of perspectives in the intersection of design and aesthetics. Rather than viewing aesthetics as something that is put on top of an Information Design artefact, it should be regarded as an integrated part of the design process with a focus on the user's perception, cognition, culture, and context, and the artefact itself. She emphasises that aesthetic issues should be explored in dialogue with practice.

Several chapters discuss various kinds of visualisations and visual communication in Information Design. Yvonne Eriksson discusses visual communication in Information Design and how people interpret and use various kinds of visual information. Carina Söderlund focuses on visual management as a type of Information Design, as well as how visual information can support work-related activities. Thorbjörn Swenberg brings up the challenges of using visual information in relation to uncertainties, as visuals are concrete while uncertainties and risks are not—we do not know whether they will happen at all or, if they do, when. These three chapters all discuss different aspects of visual information and what characterises visual representations.

Wibke Weber writes about the necessity of verbal information in data visualisation for understanding what is being visualised. She discusses, with the help of historical examples, how text supports visual information. Peter Gärdenfors argues that language is dynamic and that the meaning of words and language changes over time. Jonatan Lundin also discusses linguistic aspects in Information Design with a focus on technology communication.

Spatial design is a part of Information Design, and the focus of spatial design is spatial structure and how it can support navigation as well as various actions. It can be applied to workplace design, but also to activities for pleasure and various kinds of meeting spaces. Jennie Schaeffer introduces and presents a model for analysing how information is conveyed, experienced, and interacted with in both physical and digital environments. Karolina Uggla problematises different concepts that are used for designing way-finding and navigational aids in Information Design. Her chapter also gives a historical exposé on how way-showing has been designed in the past.

Meeting places can arise spontaneously or be designed for a specific purpose or type of meeting. They can also serve as a place where different activities happen. Much depends on how the meeting place is designed, and whether it is a single room, a room with several rooms, or an outdoor place. In his chapter, Max Kleijberg presents design methods for supporting meetings and interaction between individuals in various spaces, with a focus on the unspeakable. Creating space for topics such as illness, death, sexuality, gender identity, racism, discrimination, violence, and abuse opens up dialogues between people.

Ulrika Florin challenges traditional notions of presence and spatiality by exploring how presence can be understood in hybrid settings, with a focus on combining on-site work with remote work. She discusses this as a transformed workplace dynamic and contemplates what effect it will have on the understanding of presence and spatiality.

Adam Palmquist and William Hagman make a distinction between Information Design and Interaction Design, and emphasise the need to look through the lens of human cognitive limitations in both domains. They discuss aspects related to serious games. Helena Tobiasson focuses on the necessity of bodily movement in relation to interaction design and presents methods for stimulating physical literacy.

The book concludes with a chapter by Marie Sjölinder and Yvonne Eriksson, discussing the user and the perspective of different user groups and their relationships, as well as the different roles they have in different contexts and how situations affect users as aspects of complex systems.

There are some overlaps between the chapters, where authors describe the same terms and theories, but there are also variations in how the different authors define and understand Information Design. Common among all authors is that they problematise their topics in relation to Information Design and, by that, contribute with their engagement.

1 Exploring the Intersection of Aesthetics and Information Design

Anna-Lena Carlsson[1]*

Introduction

The term *aesthetics* traditionally refers to thinking related to the concepts of beauty and taste; that is, it is a focus on how the value judgement can be understood (Welsch, 1997). References to aesthetics in design today may also involve material features, and/or the sensual perception of an object (Ibid.; Nygaard Folkman, 2023). Aesthetics thus have to do with how something appears to observers and provides them with sensations. In this sense, it is easy to think that aesthetics has to do with characteristics such as shape, colour, and texture. While these characteristics of an object are not aesthetic in themselves, the ways in which they are arranged are understood to be aesthetic properties of the object, in some perspectives. In *Art*, art critic Clive Bell (1914) writes that all objects that provoke aesthetic emotion share “significant form”, referring to “lines and colours combined in a particular way, certain forms and relations of forms” (p. 8). This perspective on aesthetics and design is a formalistic outlook that focuses on the capability of the properties of an object to establish an aesthetic emotion in an observer or user. For a design student in higher education, the properties of a design are naturally crucial; however, the intersection of aesthetics and design is a more complex matter that has to do with the way a user understands a design and whether it is significant to the user. Aesthetics and design involve the sensual perceptual, the cultural, and the contextual (cf. Nygaard Folkmann, 2023).

In information design, aesthetics are often understood as supporting clearness and legibility (e.g., Tufte, 1990; Horn, 1999; Pettersson, 2002); for example, clarity in texts can be supported by following the principles of easy-to-read texts and visual hierarchies to guide users’ attention and help make the information understandable. Moreover, a balance between aesthetics and the function in design is stressed as important for asserting a communication, as in a successful infographic design (Cairo, 2012). Function should not fail, nor should aesthetics take over a design. There can also be scenarios in which the visual appeal is lacking, such that the design lacks the “potential for conveying a particular message” (Pettersson, 2002: 104).

1 Ph.D. Senior Lecturer in Information Design, Department of Health Science, Innovation, and Design Mälardalen University, Box 325, 631 05 Eskilstuna, Sweden.

* Corresponding author: anna-lena.carlsson@mdu.se

DOI: 10.1201/9781032642598-1

At the junction of design and aesthetics in information design, where users' perceptual, cognitive, cultural, and contextual settings intertwine, designers communicate with signs, through modalities in different media forms that people encounter with their senses, and designers use rules and conventions to meet users' expectations. In this chapter, different perspectives – such as views on form and function—are given as examples of how students in higher education can reflect and make use of aesthetics in design. In addition, in line with Mads Nygaard Folkmann's (2023: 1) writing on aesthetics and design in general, bringing aesthetics to the fore, can be argued to be a way of understanding the potential of information design.

The aim of this chapter is also to improve communication about aesthetics in narratives in and about design. In a higher education design programme, such as a Bachelor of the Arts, that includes the use of scientific methods, this involves including the use of theory in the students' explorations and testing. In the general literature on writing scientific essays, the value of narration is not mentioned as relevant for knowledge production; in fact, experience and narration can even be intentionally omitted as non-scientific language acts (Rienecker and Stray Jørgensen, 2017: 25). In higher education on design, however, experience and narration are crucial language acts – particularly when using ethnographic methods to understand the users, framing learning through design processes, and arguing for design choices. Design education must therefore include the literacy of a subject. This chapter thus emphasises the vocabulary linked to different perspectives; it also proposes methods for implementing students' reflections on aesthetic matters in and about design.

While this chapter is a conceptual inquiry, its pedagogy is based on a constructivist approach, underlining knowledge production through active exploration and building (e.g., Vygotsky, 1934/1999). It also has pragmatic roots in its consideration of language as gaining meaning through use (e.g., Wittgenstein, 1953/1958) and its view that education equals communication (Dewey, 1916). The references are not intended to be exhaustive; rather, they are suggested examples for students to use in taking their first step into the intersection of design and aesthetics.[#] Among these, certain philosophical approaches to design in general and to aesthetics, in particular, stands out from the last decades: for example, Glenn Parsons and Allen Carlson's *Functional Beauty* (2008), developed further by Parsons's *The Philosophy of Design* (2016), will be considered concerning function and form; and Yuriko Saito's *Everyday Aesthetics* (2007), which belongs to a cluster of philosophers embracing everyday aesthetics will be mentioned. Furthermore, Mads Nygaard Folkmann's *Design Aesthetics* (2023), which takes a phenomenological approach, is considered here as a way of implementing reflections about aesthetics within a design process by means of dialogues with prototypes. That said, these works do not cover all the aspects of design

[#] This exposé builds on the course "Design and Aesthetics", taught at Mälardalen University (MDU) in 2018 and 2020, altered and developed to focus on information design.

and aesthetics brought forth in this chapter, and none of them are considering information design.

Relevant focus areas for information design at the intersection of design and aesthetic include beauty and information clarity, the user's emotional engagement (i.e., the user experience), cultural and contextual sensitivity, narratives, and creativity (i.e., the exploration of new forms of information design). Some of these areas will be explored further in this chapter, while some will be developed in future discussions.

Information, Design, and Aesthetics

Information is pragmatically understood as a type of knowledge-based communication (*Oxford English Dictionary*, 2009) aimed at enlightening, encouraging understanding, and/or supporting choices of behaviour and action in a situation or context. Information is also understood as something that is always already designed, making "information design" a tautology. In a practical sense, it is still useful to speak about "information design" because this term emphasises the way in which a communicative artefact circulates as "information" (not, e.g., as "data", "scientific knowledge", or "fiction") and how the specific modalities and medium of a communication we relate to as information, can be altered in various ways. In some perspectives, "information" in a pragmatic sense is challenged, but the designs in these perspectives still support choices of behaviour and action in a situation or context.

Taken as a noun, *design* refers to a product; thus, it also exists among us in a practical way, and is thus separated from natural things or artworks (although the borders are blurred; e.g., Nygaard Folkmann, 2023; and cf. Kristeller, 1952 about design and art). Taken as a verb, *to design* is understood as to thoughtfully create or alter an artefact (cf. Heskett 2002: 1ff.; and Flusser 1999: 17ff.). A design can be an idea (which is also an artefact), the execution of an idea in a process, or an artifact, such as an information visualisation with graphs and texts. Additionally, it is worth noting that sometimes the user's interaction with a design continues its alteration. Thus, in some contexts, it is not possible to say that the information design is a finished thing when a user encounters the design; examples include the gamification of information in simulations, digital story-based infographics with choices (e.g., Segel and Heer, 2010) and innovating text designs that activate visitors in a museum to contribute to an exhibition (Fournial, n.d.).

Design exists in a variety of academic disciplines, and the literature devoted to design and aesthetics often presents design as something generic. This chapter attempts to introduce some of these perspectives and make them relevant for students in information design, with examples taken from visualisation and text design, interaction design, and spatial design. Nygaard Folkmann (2023) stresses that design "frames human experience" and that we "create meaning by engaging with things [we] handle, act through" (pp. x, and 9). Thus, in information design, meaning is not only about a message or a particular content; it is also about the whole artefact and our lived experience in relation to it, as will be discussed regarding critical information design, for example. In both of these understandings of meaning, aesthetics plays a role.

Finally, *aesthetics* refers to a discipline or field, as an umbrella concept (Welsch, 1997). As touched upon earlier, in the Western tradition, the term *aesthetics* refers to thinking related to the concepts of beauty and aesthetic judgements. This term was coined in 1750 to describe the discipline that focuses on knowledge related to value judgements depending on sensuous perception (Baumgarten, 1750/1988 §1). The ancient Greek concept of *aisthesis*, meaning the perception of sensuous qualities, was brought to the fore as a "science of sensitive cognition" (Ibid., Kristeller, 1952; Welsch, 1997: 79). Philosopher Immanuel Kant (1790/1952) then distinguished the aesthetic judgement of taste as "the faculty of estimating an object or a mode of representation by means of a delight or aversion apart from any interest" (§5, 1952: 479). The terms "Beauty", "pleasant" and "sublime" were used in judgements of taste and classifications in aesthetics – concepts in vogue during 18th-century Europe (e.g., Burke 1757/1823; Kant, 1952). At the time, speaking in broad terms, craft had already begun to be distinguished from the fine arts, leading to understandings of art in academies from the late 17th to the 18th centuries and disconnecting fine art from the artisans' crafts in guilds (Kristeller, 1952). During the 19th century, aesthetics became almost synonymous with the philosophy of arts (Hegel, 1835/1975; Beardsley, 1988: 234 ff.; Binkley, 1977), separated from utility and focusing on forms and relations of forms (e.g., Bell, 1914).

Today, *aesthetics* can refer to what is pleasurable, the beautiful and/or pretty on the surface; it can also refer to the aesthetic elements of economy (styled products) and reality (in simulations). The concept of *aestheticisation*—for example, of lifestyle, products on the market, human environments and media reality—has also been used since the second half of the 20th century (Welsch, 1997). As a field, aesthetics has been broadened with new objects of concern, and its focus has widened from inquiries about beauty and taste to thoughts about our "sensual engagement with the world" (Nygaard Folkmann, 2023: 7). For example, Saito (2007: 9) includes "any reactions we form toward the sensuous and/or design qualities of any object, phenomenon, or action" into the realm of aesthetics.

Formal Qualities for Perception in Design and Aesthetics

When it comes to the formal qualities for perception in design, the relevant focus areas concerning design and aesthetics are "beauty", also referred to as "visual harmony"; and "sensual perception", which is related to clarity of information. Our modern understanding of harmony of form in design aesthetics has several roots. Around 30–20 BC, Vitruvius wrote of the ancient Greek and Roman idea that some proportions are more appealing than others in his treatise *On Architecture* (Beardsley 1988: 125) which was made famous during the renaissance by Leonardo da Vinci's *Vitruvian Man* (ca. 1490) and Leon Battista Alberti's writing on perspectives and proportions, mass, depth, and harmony of form in *On Painting*

(1435). Alberti defines beauty as "Harmony of all the Parts, in whatsoever Subject it appears, fitted together with such Proportion and Connection, that nothing could be added, diminished or altered, but for the Worse" (Alberti VI, ii: 113 quoted from Beardsley 1988: 125). Kant's (1790/1952) aforementioned thinking about beauty is another source of our understanding of the pleasurable harmony of form, which he perceived as a pleasurable aesthetic experience that exists independent of interest, concept, or purpose (§5 and 6). Kant (1952) separates the concept of "free beauty" which he related to a pure aesthetic judgement (e.g., about a pattern on a wallpaper), from that of "dependent beauty", where the object has a purpose to fulfil and its beauty is related to impure judgement of taste (§16, p. 488). Based on these sources, aesthetics came to be understood as concerning a formal quality of an object—a view that persisted for a long period of time. As mentioned earlier, during the early 19th century, it was a "significant form" in art that provoked aesthetic emotions (Bell, 1914: 8). At the same time, design was influenced by modernistic thinking and the idea that form should follow function (Sullivan, 1896: 408) in terms of beauty in design.

The gestalt principles, from gestalt psychology, are fundamental to the use of formal design qualities to influence perception; they include continuity, figure ground, proximity, common region, similarity, closure and focal point (Eriksson and Göthlund, 2023: 44ff.). Writing about art and visual perception, Rudolf Arnheim (1954/1974) rejects the idea that the pleasure given by visual balance is a matter of style or a social preference (p. 19). In his writing about balance, Arnheim states that there are inherent tensions in "any percept as size, shape, location, or colour" and that there is a pleasurable and displeasurable effect related to how these forces direct our attention (p. 11f.). His idea is that a beholder of something visual is actively perceiving (p. 16), organising and creating meaningful wholes from the senses. Form carries meaning, and "[all] shape [...] is semantic" (p. 97). Accordingly, aesthetics are not emotional but cognitive. Perceptual forces then participate in making visual art both compelling and meaningful. This way of thinking – sometimes referred to as "design principles" for effective information—is reproduced in the information design literature on how to create successful visual design linked to cognition (Ware, 2022). Rune Pettersson (2002: 44) describes information design principles for clear structure, unity, clarity, simplicity and "securing a high quality of the message" which he calls the "functional principles", while harmony and proportions are the "aesthetic principles" (Ibid.). In education then, emphasising the relationship between the form and proportion of an object's properties can easily lead to the idea that aesthetics in design is related to the well-proportioned properties of an artefact creating a visual harmony for the user. Aesthetics can be viewed as a separation between the meaning of an object's message and its decorative and aesthetic characteristics.

As mentioned in the Introduction, aesthetics in information design are indeed understood as supporting clarity and legibility (Pettersson, 2002: 104), such as in easy-to-read texts, the use of visual hierarchies to guide user attention, and other efforts to make something understandable. For example, in infographics (e.g., Quispel et al., 2018), which balance aesthetics with clarity in informative visualisations. When based on design principles, aesthetics can support users' understanding (e.g., Horn, 1999; Ware, 2022). While supporting understanding is important, however, it is not the most essential part of the intersection in focus in this chapter. The use of design principles can easily become a formalistic and simplified understanding of aesthetics in information design. Pettersson (2002: 45) writes that texts and visuals should be "easy to read", and that "legibility", "readability", and "significance" for the user should be observed. It should be added that producing something that is "easy to read" is not solely and generically achieved by following design principles; it can only be realised in relation to a user's individual perceptive and cognitive abilities, as well as the shared cultural and contextual settings.

Function and Form, Meaning, and Significance

After following an initial path involving aesthetic matters that leans towards formalistic aesthetics in design, or a type of aesthetics that supports a function of design, the next step is to think more deeply about function and form while also looking at meaning and significance (i.e., the value for the user). Meaning and significance are distinguished, since "significance" is a term that can be used for both "meaning" and "importance". At the intersection of design and aesthetics, the focus areas of relevance for information design concerning function and form, meaning and significance are still beauty and the clarity of the information, along with the user's emotional engagement, cultural and contextual sensitivity.

The expression "form […] follows function" was coined by Louis Sullivan (1896: 408) in the context of architecture. It was written within a modernistic tradition of rationality, placing emphasis on the social benefits of design. Another voice in this setting was Adolf Loos (1908/2019: 87ff.), who criticised ornaments as a "plague" (p. 89) and a sign of cultural regression, and favoured simplicity in functional design. Loos criticised the social implications of aesthetics: "it is a crime that human labour, money, and material are thus wrecked in economic terms" (p. 90). In the discussion of form and function, the aesthetics in design can be understood in different ways, in attempts to embrace qualities and value from both the functional object and aesthetics. Philosopher Sven Ove Hansson (2005) writes that the most extreme functionalistic position reduces aesthetic value to its practical value, in what is called a "reduction thesis". Aesthetic considerations automatically come from judgements about function—a position held by critic Herbert Read and industrial designer Wilhelm Wagenfeld, among others. The latter's understanding was that "to be useful, also means to be beautiful" ([my translation] Wagenfeld 1990: 59–64, quoted from Hansson, 2005: 85). A design produces value through its function, rather than through its form. But what about different products with the same functions? Do they still have the same aesthetic value? (Ibid.: 86). Also, as mentioned earlier, philosopher Wolfgang Welsch (1997) refers to the aesthetic means of economy, such as that seen in styled products, where aesthetic judgements and user values can indeed come from different symbolic or aesthetic qualities. Yet another position to consider that aesthetic and practical values are independent of each other, in what is known as the "independence thesis", such as the aesthetic judgement of a design is not based on interest or purpose (cf. Kant, 1952). This perspective holds that aesthetics and function in design are independent of each other, and utility has nothing to do with beauty (Hansson, 2005: . 87, with reference to Muthesius, 1913). Donald A. Norman (1988) appears to adhere to this view when he writes that a design (a chair) with a bad function can still be appreciated but is then viewed as art and as having aesthetic qualities (1988: 153f.). As discussed below, this does not hold for what Parsons (2016) calls "functional beauty", in which an object's function is acknowledged to add to the aesthetic judgement of the object. Moreover, in line with Parsons (2016), Hansson (2005), objects to the independent thesis; as he writes, "a beautiful chair is an object that is beautiful in its capacity of being a chair" (Hansson, 2023: . 88), in what he calls the "thesis of aesthetic duality" (Ibid.). The user's satisfaction with the function contributes to an object's aesthetic value. Sometimes the aesthetic and the functional collaborate; at other times, they contradict each other (p. 89). Hansson's theses can encompass different views, as a start to reflecting on and better understanding and communicating one's own position.

Parsons' (2016) theses can also be a start to reflecting on and better understanding and communicating one's position. Parsons refers to and tests four theories about function in design and the aesthetic values. The first of these theories is that an object has certain proportions. The second is that an observer who perceives an object has an aesthetic attitude (cf. Stolnitz, 1978); that is, the observer perceives the object without considering its practical function. Upon reflecting, this means that observers perceive a design aside from its practical function or meaning. Would not basic pragmatic values then be excluded to the point that people would no longer speak of informative design? Parsons' third theory, on the topic of aesthetics and the function of design, is that to have an experience can be aesthetic. This idea has its roots in the thinking of the pragmatist John Dewey (1934/1980) and his writing on an elevated type of experience as something that is aesthetic outside of the realm of art. Parsons' (2016) fourth theory relates to "aesthetic realism"; he states that there are certain aesthetic characteristics, "certain looks or feels or impressions or appearances that emerge out of lower-order perceptual properties". What was mentioned earlier as the formalistic approach to aesthetics in design, emphasising colours, shapes, and so forth as the causes of an aesthetic experience, would then comprise lower perceptual properties that would lead to the higher perceptual properties. To express an elegance in a design, then, is an expression based on the lower perceptual properties linked to perception of certain forms in a design. Parsons discusses this idea as a "dependent (or adhesive) beauty" of what would be a homogeneous (because of the forms) world of design objects (cf. Kant, 1952: 111).

Finally, Parsons (2016: 117 ff.), using concepts from philosopher Kendall Walton (1970), then brings to the fore a "functional beauty" that acknowledges function within the aesthetic experiences. For every functional object, there are: (1) characteristics that are standard for its type, based on users' expectations; (2) characteristics that are contra-standard (very rare) for its type, which are experienced as unpredicted in the design; and (3) characteristics that are variations of its type (Ibid.). Following this, then, there are a multitude of ways in which form and function can be realised: first, a functional beauty can relate to the characteristics that are (in themselves) non-aesthetic (e.g., size, number of operational steps) and that vary, such as in technical instructions, but relate to the object's function and can be turned into an "aesthetic appearance" (p. 119). Second, the standard characteristics can give a "sense of stability", a correctness and a "unity to the work as a whole" (Ibid.), which is also an aesthetic quality. Third, the aesthetic quality called "looking fit" for a function (Ibid.) also depends on the expectations of the user. Fourth, characteristics that are contra-standard can create "visual tension" because the object does not "look fit" (p. 119f.). Functional beauty can be related to the modernist's idea of beauty in terms of form following function, but it is also a matter of how things function and how they look. Aesthetics can contribute to a design's indication of an object's function (which may be hidden), not just be associated with an object having a function. Functional beauty is not just about simplicity and elegance; it is more complex and can be expressed in a multitude of ways (Ibid.: 120). Using this reasoning requires experiences and understandings of users' expectations of a designer.

An informative illustration can then be functionally appealing in a multitude of ways, depending on users' perceptual, cognitive, cultural and contextual situation; that is, if the form works with the meaning in its culture and context, the illustration becomes understandable and significant to the user. But an informative illustration can also challenge the user, through visual tensions in the atypical or the contra-standard characteristics of the design. In information design, intertextuality and the dialogical element become important when discussing standards, contra-standards, and variations. The designer, therefore, needs to be aware of the design history, traditions, and conventions related to the user, as well as the informative situation, both during a design process and when communicating the design choices. This shifts the focus to culture and context related to aesthetics. Thus, functional beauty in information design is not just a matter of following design principles to arrange formal properties in an artefact that will then appear as visual harmony leading to or supporting the clarity of the information for a user. Functional beauty would come from applying the specific legibility, readability, and significance (Pettersson, 2002: 45) needed for the users in each informative event. This would create a variety of ways in which information can be designed to be perceived as appealing, understood in its meaning, and be meaningful for users. It is indeed possible for a contra-standard design to be the best design. In information design, this way of thinking goes beyond the discussion of function and form in the intersection of design and aesthetics. Rather than form and function (although different design choices have functions in information design), form, meaning, and significance are brought to the fore (cf. Heskett, 2002 on significance). As a concept, significance "explains how forms assume meaning in the way they are used, or the roles and meaning assigned to them, often becoming powerful symbols or icons in patterns of habit and ritual" (p. 27). Therefore, it is necessary to embrace aesthetics and design also related to the cultural and contextual factors if an information design is to be of value to a user.

Jonas Löwgren's (2009) writing on interaction design and aesthetics can be taken as an example here. Löwgren understands beauty as something more than experiencing a static object. A design must take into account the temporal aspects of interaction, which increase the quality of the interaction. Löwgren writes about four concepts that are relevant for interaction design and aesthetics: pliability (flexibility, bendiness, the responsivity of the responsive), rhythm (recognising the value of rhythmical interaction), dramaturgical structure (a story/play over time), and fluency (the blending of everyday norms and culture). In this way, function is included in the aesthetic experience through use, as is discussed further below, drawing upon both the perceptual and the cultural.

Culture and Norms in Aesthetics and Design

Aesthetic matters are now also to be understood as an aestheticisation process through culture (cf. Nygaard Folkmann, 2023: 13). The focus areas of relevance for information design include the cultural and contextual sensitivity, as well as narratives, and creativity – that is, the exploration of new forms of information design. Essential concepts related to the theme of culture at the intersection of design and aesthetics include "norms", "semiotics", "narration", "critical design", the "poetic", and "everyday aesthetics".

Information designers participate in forming representations and "re-presentations" of humans in the world using visualisations and narratives, for example. It is essential for a designer to reflect and speak about norms in taste when it comes to culture and aesthetics. Paying attention to culture and conventions in relation to aesthetics and design is to consider society's standards and norms existing as they do in systems and structures. They are produced through distinctions among phenomena: establishing of dichotomies and of a hierarchy. Historian Yvonne Hirdman (1988, with reference to Habermas, 1984) writes about how gender norms are created on three levels: norms involving our ideas about things in a cultural overlay; norms at the social integration level, where ideas are concretised in institutions and through the division of labour; and norms within socialisation on the individual level (Ibid.: . 54). People dwell within such structures of norms, which are not just about gender: there are norms in the intersection of design and aesthetic as well (e.g., Ehrnberger et al., 2012; Borgerson and Schroeder, 2002).

As will be discussed later, Nygaard Folkmann (2023) writes about aesthetic disposition; more specifically, about what role the aesthetic expressions can play in design. The disposition sets up "the space of cultural, social, institutional, political, and economic factors that set the scene for what is regarded as aesthetic in a given context and that, in relation to designers, frame the space of aesthetic parameters to act within" (p. 157). For example, design students in Sweden must consider Sweden's *Discrimination Act (2008: 567)*, which calls for "combating discrimination and in other ways promote[ing] equal rights and opportunities regardless of sex, transgender identity or expression, ethnicity, religion or other belief, disability, sexual orientation, or age" (§1). An information designer working for an organisation or a municipality must be aware of how a design is inclusive, both according to the law and regarding what type of society the organisation or town wants. Visually or verbally creating images and verbal narrations around work situations, for example, from a city to its citizens, requires an awareness of how to communicate within the existing cultural conventions, without re-representing unwanted stereotypes. This requirement is in addition to required knowledge about perception and cognition in design, knowledge about graphical profiles and context analysis, for example. The designer requires a cultural and contextual sensitivity. Standards of taste, which are viewed as cultural and contextual rather than centred on properties of objects, are entwined in society's norms. When it comes to design, an awareness of culture and taste can be a way of making information design significant and understandable for a user. Additionally, such awareness can allow a designer to participate in changing the norms.

From a socio-cultural view, according to Pierre Bourdieu (2010), norms in taste are linked to "cultural capital" in groups and the social belonging. From a semiotic perspective, focusing on how signs operate, norms can be found in the myths and ideologies we bring into communication (Barthes, 1957/1969; Chandler, 2007). In information design, deciding that a design is "good looking" is not just a Kantian subjective judgement; it is also connected to cultures, social groups, conventions, and norms. In Scandinavia, a functionalism that emphasises light, air, simplicity in form, and clean perfect surfaces is considered a standard of good taste (Skou and Munch, 2016, cf. Ahl and Olsson, 2002). For example, design artefacts can communicate a "Scandinavian design" (Skou and Munch, 2016: 1)—an ideal inspired by modernist and functionalistic ideas. This can be seen in graphic design, in simplicity, and in minimalism (Rynning, 2018). Nils Peter Skou and Anders V. Munch (2016) debate the "new" in contemporary Scandinavian design, arguing that the values and aesthetics involved are still closely linked to modernism (cf. Rynning, 2018). They also point out that norms in design objects can exist, not only *of* narratives (i.e., content) but also *in relation* to a narrative (i.e., that accompanies an object and affects the vocabulary used to describe it). Some of the key concepts found in Skou and Munch (2016, with reference to Munch, 2015) include "minimalistic", "timeless, "authentic", "harmony", "balance", "simplicity", "purity", "freshness", "simplicity", "elegance", "affordable luxury", and "tweaked functionalism" of beautiful everyday objects. This standard is contrasted with regional tastes, such as *kurbits*, Swedish folk art from Dalarna region (Ibid., with reference to Murphy, 2015). As discussed later in this chapter, timelessness and simplicity can also be contrasted with aesthetics and design from a different culture. Without reflection, such concepts may appear to be written in stone, as it were, guiding an information design process to create successful information design and telling stories about it that embed the design in norms of taste. However, a taste can itself be a myth (Ahl and Olsson, 2002).

Aesthetic Atmosphere and Poetic Disruption in Design

From an everyday aesthetic perspective, the practical side of design does not overshadow its aesthetic side, as the focus is on the user's experience of the design. Everyday aesthetics are associated with the ordinary, but an emphasis on the aesthetic is sometimes considered to be a focus on the extraordinary within the ordinary (Leddy, 2012) or a defamiliarisation of the familiar (Haapala, 2004). In the former sense, an aesthetic quality can be referred to as an aura or atmosphere (Leddy, 2012; Böhme, 2017); in the latter sense, the poetic can be actualised. The influences behind the everyday aesthetics can be found in Dewey's *Art as Experience* (1934/1980), in the work of philosopher Arnold Berleant's *The Aesthetic of Environment* (1992), and in philosopher Martin Heidegger's phenomenological thinking, for example, in "…Poetically Man Dwells…" (1951/1971).

Atmosphere: In this section, the focus areas of relevance for information design at the intersection of design and aesthetics are the user's emotional engagement, and cultural and contextual sensitivity. Philosopher Thomas Leddy (2012) separates concepts that are deep and complex, such as "beauty" and the "sublime" from more superficial concepts (cf. Welsch, 1997 on deep and superficial levels in aesthetics). As mentioned earlier, Kant (1952) separates the dependent beauty from the free beauty, favouring the pure judgement of taste, free of interest and purposes, from the impure. As there is traditionally a hierarchy in aesthetic judgements, philosophers of aesthetics have tried to separate "high" from "low" aesthetics over time, such as privileging "higher sense" (Saito, 2007: 21). In Leddy's (2012) thinking, this distinction involves the complexity within the concepts. However, he argues that the aesthetics of everyday experiences have been neglected. "Fun" can be an aesthetic term (p. 152f.), as well as "delicious" (p. 159), "stillness, "relaxing", and "soothing" (p. 172). Negative qualities can also be included. In "A Bestiary of Aesthetic Terms for Everyday Contexts", Leddy (2012) refers to "dull, "boring", and "ordinary", mostly as non-aesthetics, since they are not accompanied by pleasure (p. 152); however, "tragic", "shocking", "frightening", and "sad" can be aesthetic if existing within a narrative. For example, a Halloween mask can be "frightening", which is praiseworthy within the Halloween narrative (p. 157). The aesthetic terms, then, is profoundly linked to culture and context. If a viewer does not understand what Halloween is, or if a Halloween mask is used in a robbery, there is nothing aesthetical about the frightening experience. In gamification of information within interaction design, it is possible to think of the negative aesthetic concepts and experiences in this sense, in relation to make-believe aesthetics such as Walton's *Mimesis as Make-Believe* (1990).

Leddy (2012) refers to "aura" amid the everyday encounters with phenomena. Aura is an elevated, intense and pleasurable experience related to significance (i.e., meaning and importance) (p. 132). Aura "surrounds [the thing], something fits the idea of that a thing going beyond itself" (p. 133). It is closely related to the idea of beauty, to saying that a beautiful object "shimmers", "sparkles", or "shines" (p. 134). Rather than being an aesthetic property of a thing, aura is "an intensification of [...] a thing or its qualities" (p. 135). According to Leddy (2012: 133), aura as a positive aesthetic quality covers more than just the beautiful; there can be an aura of "elegance", "luxury", "daring", "success", or "uniqueness".

For a phenomenon in everyday life to be a positive aesthetic experience, it should be experienced as having an aura (Leddy, 2012) or as having "ambience" or "atmosphere" (Böhme, 2017). According to philosopher Gernot Böhme (2017: 2), atmosphere is a "*tuned space*" and relates to emotions; it is relevant in scenography (i.e., staging of a space in any context, not just theatrically), commodity aesthetics, advertising, architecture, and design. Instead of focusing on the functionality of a design, Böhme (2017: 5f.) suggests incorporating experiential, emotional and spatial dimensions. According to Böhme, these can be produced by "sound and illumination, but also by the geometry of a room, by signs, pictures, etc." (p. 3).

Instead of discussing information design in terms of clarity, structure and usability, it can be thought of in terms of mood, tone and sensory engagement. Böhme's (2017) ambience has been used for interaction design in a museum context, as well as focusing on navigation (Wang, 2024) and for analysing urban and digital environments (Grant, 2013). It can be useful to reflect on aura or atmosphere in information design by using semiotics, rhetorics, and narratives, thereby building a story world for information. However, Janet L. Borgerson and Jonathan Schroeder (2002: 579ff.) write about ethical and aesthetical issues in global marketing, for example, how our understanding of the "Hawaiian", as the "exotic other" has become a common understanding, designed through music, visuals, and films during the 20th century (Ibid., cf. Ehrnberger et al., 2012) – which can be viewed as an aesthetic atmosphere. In design, atmosphere can be used to create an emotional engagement and to produce an intensification of stereotypes.

Critical Design: There are a few focus areas of relevance for design and aesthetics concerning critical design: the user's emotional engagement, the cultural and contextual sensitivity, narratives and creativity (i.e., the exploration of new forms of information design). Depending on the context, it is possible to view an information design as a critique of how things were prior to its existence – that is, a critique of people being unable to understand and act on something because of a lack of legible, readable and significant information. Ensuring equitable access to information can be a way of challenging the existing power dynamics. However, according to Daniela Sangiorgi and Kakee Scott (2014: 116), "critical design" is "a methodological use of designed objects and systems to elicit reflection among users, observers and the designers themselves". "Speculative design" is a concept related to critical design that can also be thought-provoking; for example, Anthony Dunne and Fiona Raby's (2009) *Foragers* imagines uncanny tools that allow humans to eat non-traditional foods. Another related concept is "design fiction", which focuses on possible futures through narratives; for example, Tobias Revell's fictional documentary *New Mumbai* (2012) proposes a sustainable city powered by energy-generating mushrooms. While Parsons (2016) refers to contra-standard and visual tensions within functional beauty, critical design can be seen as violating expectations through design in a more drastic way, such that a disruption in design is now associated with the "poetic".

The concept of critical design came into vogue through Dunne's *Hertzian Tales* (2005), which focused on a product design that emphasises creative artistic expressions. It is critical to technical optimisation and interactive user-friendliness, arguing that these can be enslaving (p. 21). Instead, design should embrace estrangement in order to reveal hidden ideologies (pp. 22f. and 41ff., cf. semiotic analyses revealing hidden ideologies; Barthes, 1969) and nourishing independent thinking. Dunne also refers to artefacts such as electronic products that embody narratives that shape us (p. 69ff). Post-optimal and post-efficiency design encourage thinking (Dunne, 2005: 20f.; see also Ionascu, 2010; Malpass, 2016: 474, and Hallnäns and Redström, 2001 on slow technology). Dunne also discusses "the aesthetics of use" as a para-functionality of design (2005: 43ff.) and uses the concept of "poetic" to refer to situations in which a design's function is "to criticise the limits that products impose on our actions" (pp. 22ff. and 42f.). For an information designer, it is important to remember that the design can make something visible; examples include norm creativity work with inclusions (e.g., Manberger et al., 2021 on norm creative inclusions in public spaces).

The influences behind the concept of a "poetic" quality at the intersection of aesthetics and design comes from ancient Greece, including the concepts *aesthesis*, referring to the perception of sensuous qualities, and the *poietic*, meaning to bring something new forth (Heidegger, 1954/1977: 10; Welsch, 1997: 13).[#] More recently, a changed perception of avantgarde art during modernism has emerged using the concept "poetic". Philosopher Robert H. Jauss (1982) writes that aesthetics can be more than just a focus on the pleasurable and beautiful; the poetic can transcend traditional aesthetic theory. Jauss refers to the poet and critic Paul Valéry's aesthetic thinking, pointing out that art forces a change—including shock value—upon the beholder such that the latter cannot stay passive and disinterested but must cocreate in the reception of the art (Ibid.). According to Paul Martin Lester (2006), a scholar of communication, the same change is visible in the graphical designs presenting informative events in the Dada movement. Dada critically examined the social structure during the terrors of World War I, as seen in its use of asymmetrical and absurd designs that both confused observers and captured their attention (Ibid.). In 2004 in Sweden, strange texts appeared on normal road signs in various cities, with phrases such as "It hurts" and "There are no children anymore" (Carlsson, 2010). In fact, this event was advertising for Lars Norén's play *War*; the texts on the signs were quotes from the play, as well as being an information campaign about war crimes against civilians, from the Living History Forum, a Swedish authority. Like Dunne's (2005) critical design, these artefacts, produced by the artist collaboration FA+ (Ingrid Falk and Gustavo Aguerre), interrupted and shocked the beholder by

[#] This section of the poetic builds on Carlsson's (2010) "The Aesthetic and the Poietic Elements of Information Design" and Carlsson and Schaeffer's (2017) "The Poetics of Workplaces".

poetically imposing representations of the lived experiences of civilians during war, by means of everyday signs.

These examples of designs can be placed along the spectra of design that permits experimentation with artistic expressions. They bring forth information about the world and engage viewers, while disturbing the ways in which information usually circulates in society. Thus, in relation to information design and aesthetics, a poetic use of narrative and extended artistic expression can tell of our being in the world.

In Heidegger's (1951/1971) understanding of the poetic, in "…Poetically Man Dwells…", the poetic is essential to the way people dwell in the world, through relationships to our being. Our building—that is, our being in the world—can then either negate poetic dwelling or affirm building as growing out of itself, making "visual inclusions" like poetry (Ibid.: . 226) in design. As a phenomenology, this emphasises our lived experiences rather than anything objective. For example, when a spatial designer is not only focusing on the mathematical and effective but also on the experienced space, the designer is affirming users' being in the world.

In acknowledging a broader understanding of what the discipline of philosophical aesthetics can offer information design, there is a possibility of discussing a wider variance of information design artifacts. Aesthetician Arto Haapala (2004) uses the concepts of "familiar" and "unfamiliar" when writing about everyday objects and aesthetics. Certain everyday objects give us "pleasure through a kind of comforting stability, through the feeling of being at home" (p. 50). Together with Heidegger's (1971) thinking, this perspective is applied to information design relating the familiar and unfamiliar to visual-spatial artefacts found in different manufacturing plants (Carlsson and Schaeffer, 2017). This perspective can challenge the idea that visual-spatial information artefacts in the contemporary manufacturing industry only have aesthetic concerns related to what is utilitarian and functional. Pleasing aesthetics are indeed understood as being essential for the users' satisfaction during work; interior styling and branding emphasize the aesthetic experience of the workspace (Ibid.: 57, with reference to Hackston, 2015 and Venkatesh, 2015). Some informative artefacts, however, depict being as working in a manufacturing plant as a poetic dwelling, in their inclusion of the unfamiliar within the familiar in everyday workplaces, making "visible inclusions" about being at the site. For example, a donkey-dog figure in a roundabout (at a time when there was a roundabout dog phenomenon in Sweden, following a vandalization of an artist's installation of a sculpture dog in the city of Linköping), created by workers and placed in a crossing for trucks, to organise space and showing directions in a familiar way. However, the unfamiliar was seen in the workers' dwelling with the figure. It was a continuous and collective negotiation concerning its identity, the figure was reshaped and maintained (also burned and resurrected) and for a while the figure even made it to the company's web site for information on how to get to the plant (Ibid.: 60ff). The visual-spatial figure reveals information about directions for trucks at the plant, it is setting up a space for behaviour, but it also discloses visual inclusions concerning collaboration and negotiation about meaning among workers at the site. The workers' experiences become visible in the figure.

Sara Mc Laughlin (2009) writes, with reference to Heidegger (1971) that information designers are "constantly involved in searching for appropriate forms to express aspects of the world as they show up from different perspectives". Referring to Edward Tufte's writing (1990) on the isometric map of midtown Manhattan by Constantine Anderson, McLaughlin (2009: 311) comments, "[f]rom a Heideggerian perspective the widening of the streets show up as an important strategy for organising the information, allowing the viewer to activate their own understanding of the world referred to in the map, allowing the viewer to bring that world close". The mapping is thus done for the purpose of clarity from the perspective of lived experience, rather than to reveal exact data. The perspective of aesthetic matters, the poetic, is not linked to Kantian disinterestedness here, but is rather linked to the user of information design as a participator in the creation of meaning. In terms of lived experiences, this involves placing the poetic at the heart of an information designer's way of handling perception and cognition, as well as cultural and contextual factors for users. According to McLaughlin (2009: 316), this is an ethical issue. Passing on information is never neutral.

A Japanese Perspective on Aesthetics and Design: Wabi-Sabi

The relevant focus areas for information design at the intersection of design and aesthetics are the user's emotional engagement, the user experience (which is linked to the understanding of beauty), the cultural and contextual sensitivity, as well as narratives, and creativity.

Concerning the optimistic and progressive view of design in the Western worldview, Nygaard Folkmann (2023) writes that design history has become part of our "history of waste and wastefulness, of unsustainability" (p. 9). Today, there is a need for thoughtfulness about the superficial aestheticisation of the artefacts operating in our economic machine of product and service innovation and consumption. Rethinking the intersection of aesthetics and design can be part of this reflection. A first step in higher education on design could be to increase the awareness of other perspectives, both through critical design and by acknowledging other cultures' understandings of aesthetics and design.

The Western paradigm of looking at information design in terms of efficiency and as optimising a situation for some purpose (e.g., working faster and better), with aesthetics focusing on function and form, can be considered as a cultural phenomenon (cf. Ahl and Olsson, 2002), in contrast to the Japanese perspective of wabi-sabi, for example (Saito, 2007; Koren, 2008), a philosophical and spiritual world view that affects the intersection of design and aesthetics. Wabi-sabi has its roots in Buddhism and the Buddhist notion of coming to peace with our temporary, imperfect lives; it is a way of life and action (Saito, 2007: 4) with moral underpinnings (Koren, 2008: 59ff).

The status of wabi-sabi in Japan can be compared with the status of antiquity (i.e., perfection and beauty in the Western world). The Western world had the art system that emerged during the 18th century, including painting, sculpture, architecture, poetry, and music. In Japan, there was the *iemoto* system, which was already in existence in the 18th century, involving flower arrangements, calligraphy, singing, dancing, theatre, and tea ceremonies. Tea ceremonies and the flower arrangements were connected to the aesthetic and philosophical worldview of wabi-sabi (e.g., Koren, 2008: 15ff: , and Saito, 2007: 174, 194ff.).

Wabi stands for "a way of life" and *sabi* for "material objects" and "the outward" (Koren, 2008: 23). The wabi–sabi universe, then, on a metaphysical level, sees things as developing to or evolving from "nothingness" (Ibid.: p. 40). On a spiritual level, observing nature is essential for gaining truth, "greatness" is found in the overlooked details, and beauty can be found in ugliness (Ibid.). This perspective leads to a level state of mind, with acceptance and appreciation of the way of the universe. Thus, the moral percept of wabi-sabi is to get rid of everything that is not necessary and to focus on the intrinsic. Correspondingly, the material qualities in wabi-sabi design stem from natural processes, which are "irregular, intimate, unpretentious, earthy, murky", and "simple" (p. 41). For example, cracks in ceramics are preferred to perfection. This perspective brings with it a sense of melancholy and respect for what passes. To affirm this simplicity of the everyday is to take a spiritual path. Wabi-sabi, then, is also understood as a feeling in its highest expression.

The tea ceremony emphasises rustic ceramics, with defects in the glazing. According to the tea masters, one must use a thing in order to perceive it as beautiful. Tea masters not only used their eyes to enjoy beauty in contemplation (cf. Kant, 1952), but also use design through activity (Weimarck, 2003: 12 f. see also Saito, 2007: 127).

Leonard Koren (2008: 26ff.), artist and scholar, compares wabi-sabi with modernism. Wabi-sabi does not recognise progression; it is now-oriented and believes in the uncontrollability of nature. It uses an organic design language (soft, geometric organisation, and vague shapes and edges) and natural materials that are seemingly rough/simple and can be worn out. In wabi-sabi, corrosion is viewed as something that makes expression richer. Wabi-sabi offers an expansion of sensory information and is tolerant of ambiguity and contradictions; it can also be dark and vague (see also Tanizaki 2001on the praise of shadows). Modernism expresses a general, rational and logical worldview: it is absolute, with prototypical solutions; it believes in progress, is future-oriented and believes in controlling nature. Modernism romanticises technology, and its shapes are sharp, with precise edges. It makes use of created materials, is seemingly sophisticated or elegant, requires maintenance, believes that purity makes expression richer, offers a reduction of sensory information and is intolerant concerning ambiguity and contradictions (Koren 2008: 26ff.). Simplicity gains a different meaning through this comparison of wabi-sabi and modernism. Also, in modernism, beauty is not gained through use over time. In wabi-sabi, temporal aspects are recognised as being involved both in use and expression; the cracks or rust gained through times or even in mistakes are valued. Thus, wabi-sabi constitutes a form of dematerialisation.

Wabi-sabi involves ethical thinking: starting from a metaphysical base, the idea is that things come from nothing and go to nothing. The material qualities in wabi-sabi involve natural processes, irregularities, familiarity, unpretentiousness, earthiness, darkness, and simplicity. The state of mind concerns acceptance of the inevitable and appreciation of the cosmic order. Finally, the spiritual values in wabi-sabi are that truth is gained through studying nature (everything is impermanent, imperfect, and incomplete), there is grandeur in the details (Koren, 2008). For a detailed idea of the modern perspective on aesthetics in Japan, see Saito (2007: 226ff.) regarding packaging, flower arrangements, food preparation, and the tea ceremony, along with ideas on care about material, process, and others. Saito (2007) also links the reasoning on aesthetic ambience with a Japanese aesthetic perspective on the distinct character of each season in nature.

In an information design context, in terms of material qualities, a minimalism, a use of natural textures, and somewhat unsymmetrical and unpolished designs could be a signum of aesthetics pointing to an organic balance. However, embracing imperfection and the non-standard must align with the culture and context in order to work. It can be fruitful for an information designer to focus on another understanding of what simplicity is, while highlighting local material and the time-sensitiveness, in relation to local legibility, readability, and significance within a sustainability context. For example, Yixuan Du and Soong Joo (2021) emphasise cultural aspects of spatial information design and look at the role of wabi-sabi in tea houses, where it guides behaviour and spiritual values. For example, the Hojo Garden, at Ryoanji Temple, a manufactured landscape disguised as a natural environment, uses material (rocks, aged clay, and moss) to guide people to "reflect on life and practice Zen" (p. 34). Wabi-sabi also focuses on user experiences. For example, in Vasiliki Tsaknaki and Ylva Fernaeus' (2016) writing on wabi-sabi and the idea of impermanence in interaction design.

Aesthetics in Dialogue with Design Practice and Writing

This chapter closes with proposals for how thinking and concepts can be integrated into practical design work and written text. Donald Schön's *The Reflective Practitioner* (1983) has frequently been used to reflect on designing (e.g., Roozenburg and Dorst, 1998; Tan et al., 2023). Schön discusses problem-solving, in dialogue with a situation, known as the *reflection-in-action*, which refers to how emotions and previous experiences can be integrated into practical and experimental work. He also discusses *reflection-on-action*, which refers to situations distanced from activity, emotions, and reflection-in-action, when there is time for documentation within the continual creation. This entails a "good linguistic description" of the work (Schön 1983: 365). However, Nygaard Folkmann (2023) goes further in a structural approach for aesthetics in design practice, even stressing that aesthetics are important in order to understand complexity in design. Thus, for Nygaard Folkmann, aesthetics are far from being a glazing or styling in a design. He uses a negotiation of aesthetics in design practice (p. 151ff.). Although design dialogue and negotiation are already present in design practices, such as in collaborative processes (Florin 2022: 108–140), Nygaard Folkmann brings aesthetics to the fore in a dialogical method.

Nygaard Folkmann (2023: 1) writes of aesthetics as a "methodological tool" for inquiry into design and how design can appeal to users. Both sensual materials and conceptual ideas are considered in this structured method for prototypes in a design process. Nygaard Folkmann's focus is on how the "sensual, conceptual, and contextual dimensions" of aesthetics are negotiated in the design process (p. 152). In this regard, prototypes are seen as material expressions of immaterial processes. It is the temporariness of a prototype that makes it useful for negotiating aesthetic inquiries during design. Nygaard Folkmann introduces conditioning factors—namely, aesthetic latitude and aesthetic disposition—and a conceptualisation (in his case, following a modernistic tradition, "functional realism" and "formal simplicity") linked to prototypes. His epistemology (with reference to Murphy, 2017) is: "knowledge of how aesthetics, in acts of designing, is a central means to constructing meaning at the intersection of sensual, conceptual, and contextual concerns" ([italics in org.] Nygaard Folkmann 2023: 154).

Aesthetic Latitude

Aesthetic latitude is about "the range and scope of aesthetic concerns at play" in the design tasks, referring to what is sanctioned and how free a designer can be (Nygaard Folkmann 2023: 155). For example, the aesthetic latitude can limit how innovative a design can be. It can comprise formal or symbolic limits and may involve technology, purpose, and acceptability concerning visibility, for example (p. 156, with reference to Lawson, 2005). For an information designer, the aesthetic latitude must relate to legibility, readability, and significance. For each inventiveness, students must ask themselves how the users are affected. For example, if information is intended for a Swedish municipality, principles for plain language must be considered and artistic experimentation might be problematic.

Aesthetic Disposition

Aesthetic disposition is about what role the aesthetic expressions can play in a design. The disposition sets up "the space of cultural, social, institutional, political, and economic factors that set the scene for what is regarded as aesthetic in a given context and that, in relation to designers, frame the space of aesthetic parameters to act within" (p. 157). For example, it is clear how norms in aesthetic taste and thoughts about efficiency play a role here, as information design is rooted in Western modern ideas, influencing the institutional setting and the designer's "habitus" (how designer acts when designing, in relation to perception, cognition, and culture). It is about how a designer acts as an individual within an institution but also differs in relation to others. Designers are shaped in relation to their education and the job market (p. 158, with reference to Bourdieu, 1994). Also, "field" is used to represent the relations among positions (Bourdieu 1994: 68, in Nygaard Folkmann, 2023: 160). Nygaard Folkmann's (2023) uses two opposites in a field: the "heteronomous principle", which relates design to what is going on in the market, and the "autonomous principle", which relates design to the realm of art. He also set up two opposites for the design process "design as 'rational' planning"(2023: 162), which is dominant in engineering, and a "'romantic' idea of artistic creativity", which exists in artistic design education (Ibid.). These polarities can be a way of opening a space for a design process in relation to aesthetic concerns (Ibid.: 162f.).

Students are socialised into human- and user-centred designing that differs from user-centred design in a commercial setting. They are educated in thoughts about legibility, readability, and significance for the user in designing information, along with ideas of modernist functional traditions (cf. Nygaard Folkmann 2023 about the Danish culture in his cases, p. 151ff.). Aesthetic disposition can support students in orienting themselves with the design literature and finding ways of thinking related to their field, especially regarding how cultural and contextual aspects relate to the significance of an information, which is understood as both the meaning and the importance for users. This is far from "just" applying design principles to create appealing designs. Design principles are indeed relevant, but they are not sufficient for understanding aesthetics and design. During design education, artistic and poetic experimentation might be needed to promote students' reflections on users and/or human conditions in relation to sustainability and democracy.

Aesthetic Negotiation with Prototypes and Concepts

Once the preconditions of aesthetic latitude and disposition are set, aesthetic negotiation takes place in relation to the transitional prototypes concerning the sensual, the conceptual and the context (p. 164). Schön's (1983) understanding of an object that "talks back" is surpassed; instead, a prototype is at "a stage, where matters of the world are tested through scenarios as a 'rehearsal of new practices'" (Nygaard Folkmann 2023: 166). This negotiation can also be a collaborative process (cf. Florin, 2022) and can thereby touch upon a political power involved in aesthetics, concerning the "distribution of the sensible" (Ranciére, 2004, in Nygaard Folkmann 2023: 166).

In this negotiation, Nygaard Folkmann (2023) uses concepts that are suitable for the latitude and disposition in his cases: namely, "formal simplicity" and "functional realism". The questions asked in the dialogue then concern form, function, sign, and creation. These topics correspond to aesthetic categories of Nygaard Folkmann's choice based on culture and context: formal simplicity (related to form), functional realism (related to function), semantic references (related to sign), and artistic experimentation (related to creation). The aesthetic dimensions are connected to thoughts of the forthcoming final product: sensual appearance and the conceptual idea context (related to both form, formal simplicity, function and functional realism). Moreover, the aesthetic dimension linked to a forthcoming product includes conceptual reflection and contextual concerns (related to signs, semantic reference, creation, and artistic experimentation. p. 166; Figure 5.3).

The idea here is to establish a reciprocity between the material and concepts in a process that acknowledges perception, cognition, culture and context as relevant for design. This allows for different ways of understanding aesthetics in design to be understood in all their complexity, such as modernist ideas of form and function, and the poetic in critical design linked to artistic experimentation. However, this process does require some knowledge of design history, design studies, and external analysis in order to outline the aesthetic latitude, disposition and concepts (as well as the thinking behind them) and to be able to make conceptual reflections about contexts related to semiotics, semantics, and creative experimentation. Again, in information design, questions about legibility, readability, and significance must be considered in all instances, keeping the user in mind.

Negotiation in Writing

Writing can also be used to try out aesthetic matters in design, especially when using an essayistic form during a process. A non-scientific essay format that incorporates both data and personal reflections is frequently used in higher education when making practical knowledge visible (Alsterdal, 2014). Such essays are analytical, interpretative, and/or critical compositions, less systematic than a thesis (*Encyclopaedia Britannica*, 2025). This format is suitable for students aiming to try out their thinking, since it has a "trying, conceivable, presumptive relationship to modern realities" (Isenberg, 2006: 98), using empirical data linked to students' own experiences and thinking and answering questions about how to understand and create something (Ibid.).

Graham Badley (2009), a researcher in academic writing, writes that the form makes it possible for the essayist to focus on the concrete and in relation to the experiences and reflections on action; on the dialectical process; and on shaping and reshaping. That is, it is a constructivist view of knowledge. Badley comments: "Essays [...] allow us, flexibility and variously, to be analytical, autobiographical, connective, descriptive, dialogic, interpretative, meditative, reflective, and speculative according to our critical purposes" (p. 103). Reflection in writing means, according to Badley, "that we construct new thoughts as part of the essaying process. A reflective essay helps us interpret our experiences as we write" (p. 107). A practical and an essayistic approach to aesthetics in design can both enforce the exploration and reflection. Essay writing can be done in relation to Nygaard Folkmann's (2023) method, for example, starting by trying out the aesthetic latitude and disposition in the context of information design, then entering the dialogue on the thinking about aesthetic concepts, while taking users' perception, cognition, culture, and context into account.

Conclusion

This chapter introduced a variety of perspectives concerning the intersection of design and aesthetics, including views on form and function (e.g., formal qualities for aesthetic perception and functional beauty); cultural aspects (e.g., norms in taste); the everyday aesthetics involving the familiar (e.g., a familiar atmosphere), and the poetic unfamiliar in critical design; and a phenomenological view on lived experience altering understandings of what "good" information can be in a workplace. It also presented a contrasting cultural perspective—the Japanese wabi-sabi, and in contrast to modernism—and discussed how crucial concepts such as simplicity are filled with other meanings. All these outlooks can be linked, in various ways, to focus areas at the intersection between design and aesthetics: beauty and the clarity of information, the user's emotional engagement, the cultural and contextual sensitivity, narratives, and the exploration of new creative forms. Finally, based on the work of Nygaard Folkmann (2023), this chapter suggested a way of exploring aesthetics in dialogue with practice and negotiating in essayistic writing during a higher education.

References

Ahl, Z. and Olsson, E. 2002. *Svenska smak. Myter om den moderna formen.* Ordfront förlag, Stockholm.

Alsterdal, L. Essäskrivning som utforskning. In: A. Burman [Ed.] 2014. *Konst och lärande. Essäer om estetiska lärprocesser* (pp. 47–73). Södertörns högskola, Huddinge.

Arnheim, R. 1954/1974. *Art and Visual Perception.* University of California Press, Berkeley.

Badley, G. 2009, May. Valuing essays: Essaying values. *Research in Post-Compulsory Education, 15*(4): 103–115. DOI: 10.1080/13596740903565434.

Barthes, R. 1957/1969. *Mytologier*. Bo Cavefors Bokförlag, Lund.

Baumgarten, A. 1750/1988. *Theoretische Ästhetik.* Felix Meiner Verlag, Hamburg.

Beardsley, M.C. 1988. *Aesthetics: From Classical Greece to the Present.* The University of Alabama Press, Tuscaloosa.

Bell, C. 1914. *Art.* Frederick A. Stokes Company, New York.

Berleant, A. 1992. *The Aesthetic of Environment.* Temple University Press, Philadelphia.

Binkley, T. 1977, Spring. Piece: Contra Aesthetics. *The Journal of Aesthetics and Art Criticism, 35*(3): 265–277. URL: http://www.jstor.org/stable/430287.

Böhme, G. 2017. *The Aesthetics of Atmospheres*. Routledge, London.

Bourdieu, P. 2010. *Distinctions: A Social Critique of the Judgement of Taste*. Routledge, London.

Borgerson, J. and Schroeder, J. 2002 October. Ethical Issues of Global Marketing: Avoiding Bad Faith in Visual Representation. *European Journal of Marketing, 36*(5/6): 570–594. DOI: 10.1108/03090560210422399.

Burke, E. 1757/1823. *A Philosophical Inquiry into the Origin of Our Ideas of the Sublime and Beautiful*. Thomas M'Lean, Haymarket, London. Internet Archive. URL: https://archive.org.

Cairo, A. 2012. *The Functional Art*. New Riders Publishing, Berkely, California.

Carlsson, A.-L. 2010. The aesthetic and the poietic elements of information design. In: Banissi, E. et al. [Eds.] *The Proceedings of Information Visualization Conference IV London* (pp. 450–454). IEEE Computer Society.

Carlsson, A.-L. and Schaeffer, J. 2017. The Poetics of Workplaces In: C. Friberg and R. Vasquez [Eds.] 2017. *Experiencing the Everyday* (pp. 53–66). NSU Press, Copenhagen.

Chandler, D. 2007. *Semiotics*. The Basics, Routledge, London.

Dewey, J. 1934/1980. *Art as Experience*. PERIGEE, New York.

Dewey, J. 1916. *Democracy and Education*. The MacMillan Company, New York. Internet Archive. URL: https://archive.org.

Discrimination Act. 2008:567. *Swedish Law*. Retrieved at the Equality Ombudsman. URL: https://www.do.se/choose-language/english/discrimination-act-2008567.

Du, X. and Joo, A. 2021. The Influence of Cultural Information Design and the Classification of Spiritual Guidance, Based on the Cases of Japanese Wabi-Sabi. *International Journal of Contents*, *17*(1): 27–36.

Dunne, A. 2005. *Hertzian Tales: Electronic Products, Aesthetic Experience, and Critical Design*. The MIT Press, Cambridge.

Dunne, A. and Raby, F. 2009. Designs for an overpopulated planet. URL: https://dunneandraby.co.uk/content/projects/510/.

Ehrnberger, K., Räsänen, M. and Ilstedt, S. 2012. Visualising Gender Norms in Design. Meet the Mega Hurricane Mixer and the Drill Dolphia. *International Journal of Design*, *6*(3): 85–98.

Encyclopaedia Britannica. 2025. Essay. URL: https://www.britannica.com/art/essay.

Eriksson, Y. and Göthlund, A. 2023. *Foundations of Visual Communication: How Visuals Appear in Daily Life*. Routledge, New York.

Florin, U. 2022. On Design Dialogues: Their Roots, Features, and Usage. In: Y. Eriksson. [Ed.], *Different Perspectives in Design Thinking* (pp. 108–130). CRC Press, Boca Raton.

Flusser, V. 1999. *The Shape of Things: A Philosophy of Design*. Reaktion Books, London.

Fournial, M. n.d.. Effective Text Design in Museum Exhibitions. URL: https://www.academia.edu.

Grant, S. 2013. Performing an Aesthetics of Atmosphere. *Aesthetics*, *23*(1): 12–32.

Hackston, J. 2015. Type and Work Environment. A Research Study from OPP. Oxford, UK. URL: https://eu.themyersbriggs.com/-/media/Files/PDFs/White_papers/Type-and-environment-survey-findings.pdf.

Haapala, A. 2004. On the Aesthetics of the Everyday: Familiarity, Strangeness, and the Meaning of Place. In: A. Light and J.M. Smith. [Eds.], *The Aesthetics of Everyday Life* (pp. 39–55). Columbia University Press, New York.

Hallnäs, L., and Redström, J. 2001 August. Slow Technology: Designing for Reflection. *Personal and Ubiquitous Computing*, *5*: 201–212. DOI: 10.1007/PL00000019.

Hansson, S.O. 2005. Estetik och funktion. In: K. Edvardsson, S.O. Hansson and J. Nihlén Fahlquist [Eds.], *Filosofins nya möten* (pp. 83–89). Gidlunds förlag, Möklinta.

Hegel, G.W. 1835/1975. *Aesthetics. Lectures on Fine Art*. (Volume 1). Clarendon Press, Oxford.

Heidegger, M. 1951/1971. *…Poetically Man Dwells…. Poetry, Language, Thought*. Harper and Row, New York.

Heidegger, M. 1954/1977. *The Question Concerning Technology and Other Essays*. Harper Perennial, New York.

Heskett, J. 2002. *Design: A Very Short Introduction*. Oxford University Press, Oxford.

Hirdman, Y. 1988. Genussystemet: Reflektioner kring Kvinnors Sociala Underordning. *Tidskrift för Genusvetenskap*, *9*(3): 49–63.

Horn, R.E. Information Design: Emergence of a New Profession. In: R. Jacobson [Ed.] 1999. *Information Design* (pp. 15–33). The MIT Press, Cambridge, Massachusetts.

Ionascu, A. 2010. *Design Poetics, Poetic Design: A Theory of Everyday Practice*. Diss, Loughborough University, Ann Arbor.

Isenberg, B. 2006, March. Sociologisk Essäism: Essäistisk Sociologi. Om tankestils Utveckling oich Aktualitet I en Postdisciplinär Tid. *Dansk Sociologi.*, *17*(1): 87–110. DOI: 10.22439/dansoc.v17i1.806.

Jauss, H.R. 1982. Poiesis. *In: Critical Inquiry*, *8*(3): 591–608.

Kant, I. 1790/1952. *The Critique of Judgement*. William Benton, Encyclopaedia Britannica Inc., Chicago.

Koren, L. 2008. *Wabi-Sabi for Artists, Designers, Poets and Philosophers.* Imperfect Publishing, Point Reyes.

Kristeller, P.O. 1952. The Modern System of the Arts: A Study in the History of Aesthetics, part II. *Journal of the History of Ideas*, *13*: 17–46.

Leddy, T. 2012. *The Extraordinary in the Ordinary: The Aesthetics of Everyday Life*. Broadview Press, Peterborough.

Lester, P.M. 2006. *Visual Communication: Images with Messages*. Thomason Wadsworth, Belmont, California.

Loos, A. 1908/2019. *Ornament and Crime: Thoughts on Design and Materials*. Penguin Classics, London.

Löwgren, J. 2009. Toward an articulation of interaction aesthetics. *The New Review of Hypermedia and Multimedia*, *15*(2): 129–146.

Malpass, M. 2016. Critical design practice. Theoretical perspectives and Methods of engagement. *The Design Journal*, 19(3): 473–489.

Manberger, K., Sarakinis, D. and Gunnarsson, Östling, U. 2021. *Guide till normkreativ gestaltningsprocess:. Tengbom in collaboration with KTH.* URL: https://tengbom.se.

McLaughlin, S. 2009. Information design and the world that comes before us. *Visual Communication*, *8*(3): 303–316. DOI: 10.1177/1470357209106471.

Murphy, K.M. 2015. *Swedish Design: An Ethnography*. Cornell University Press, Ithaca.

Norman, D.A. 1988. *The Design of Everyday Things*. Basic Books, New York.

Nygaard Folkmann, M. 2023. *Design Aesthetics: Theoretical Basics and Studies in Implication*. The MIT Press, Cambridge, Massachusetts.

Oxford English Dictionary. 2009. Information. URL: https://www.oed.com/dictionary/information_n?tab=meaning_and_use#416073.

Parsons, G. 2016. *The Philosophy of Design*. Polity Press, Cambridge.

Parsons G. and Carlson, A. 2008. *Functional Beauty*. Clarendon Press, Oxford.

Pettersson, R. 2002. *Information Design.* John Benjamins Publishing Company, Amsterdam.

Quispel, A., Maes, A. and Schilperoord, J. 2018. Aesthetics and Clarity in Information Visualization: The Designer's Perspective. *Arts*, *7* (4): 1–16. DOI: 10.3390/arts7040072.

Revell, T. 2012. *New Mumbai*. URL: https://www.youtube.com/watch?v=4VgP-M5JMW8.

Rienecker, L. and Stray Jørgensen, P. 2017. *Att skriva en bra uppsats*. Liber, Stockholm.

Roozenburg and Dorst. 1998. Describing Design as a Reflective Practice: Observations on Schön's Theory of Practice. *In:* E. Frankenberger, H. Birkhofer, P. Badke-Schaub [Eds.], *Designers* (pp. 29–41). Springer, London. DOI: 10.1007/978-1-4471-1268-6_3.

Rynning, M. 2018. New Nordic Graphic Design: The Balance Between Scandinavian Traditional Crafting and Globalization 3.0. In: *Conference: Cumulus Wuxi 2018. Diffuse Transition & Design Opportunitets* (pp. 1–19). URL: http://cumuluswuxi2018.org/proceedings.

Saito, Y. 2007. *Everyday Aesthetics*. Oxford University Press, Oxford.

Sangiorgi, D. and Scott, K. 2014. Conducting Design Research in and for a Complex World. *In:* P. Rodgers and J. Yee. [Eds.], 2014. The Routledge Companion to Design Research (pp. 114–131).

Schön, D. 1983. *The Reflective Practitioner: How Professionals Think in Action.* Basic Books, New York.

Segel, E. and Heer, J. 2010. October. Narrative Visualization: Telling Stories with Data. *IEEE Transactions on Visualization and Computer Graphics*, *16*(6): 1139–1148. DOI: 10.1109/TVCG.210.179.

Skou, N.P. and Munch, A.V. 2016. New Nordic and Scandinavian Retro: reassessment of values and aesthetics in contemporary Nordic design. *Journal of Aesthetics and Culture*, *8*(1): 1–12. DOI: 10.3402/jac.v8.32573.

Stolnitz, J. 1978 Summer. The Aesthetic Attitude. *The Journal of Aesthetics and Art Criticism*, *36*(4): 409–422. URL: https://doi.org/10.2307/430481.

Sullivan, L. 1896. *The Tall Office Building Artistically Considered.* Lippincott's Magazine. Internet Archive. URL: https://archive.org.

Tan, L., Kocsis, A. and Burry, J. 2023. Advancing Donald Schön's Reflective Practitioner: Where to Next? *Design Issues*, *39*(3): 3–18. URL: https://doi.org/10.1162/desi_a_00722.

Tanizaki, J. 2001. *In Praise of Shadows.* Vintage Books, London.

Tsaknaki, V. and Fernaeus, Y. 2016. Expanding on Wabi-Sabi as a Design Resource in HCI. *In: CHI'16.* San Jose, CA, USA. 5970–5983. DOI: 10.1145/2858036.2858459.

Tufte, E. 1990. *Envisioning Information.* Graphic Press, Cheshire.

Venkatesh, N. 2015, January. Conceptualizing Employer Branding as an Effective Technique for Talent Acquisition and Retention: A Framework Based Analysis. *GJRA-Global Journal for Research Analysis*, *4*(1): 5–7. URL: https://ssrn.com/abstract=2652965.

Vygothsky, L.S. 1978. *Mind in Society. The Development of Higher Psychological Processes.* : Harvard University Press, Cambridge, Massachusetts.

Walton, K. 1990. *Mimesis as Make-Believe.* Harvard University Press, Cambridge, Massachusetts.

Wang, W. 2024. Research and Practice of Human-Computer Interaction Design in Museums from the Perspective of Atmospheric Aesthetics. 331–338. *In:* C. Stephanidis, M. Antona, S. Ntoa, G. Salvendy [Eds.], *HCI International 2024: Communications in Computer and Information Science, 2321* (pp. 331–338). Springer, Cham. URL: https://doi.org/10.1007/978-3-031-78561-0_33.

Ware, C. 2012. *Information Visualization. Perception for Design.* Morgan Kaufmann, Boston.

Ware, C. 2022. *Visual Thinking for Information Design.* Morgan Kaufmann, Cambridge, USA.

Weimarck, T. 2006. Introduktion. *In:* P. Cornell et al. [Eds.], *Design och konst, Skriftserien Kairos nr*, *8*(1), 11–40). Raster förlag, Helsingborg.

Welsch, W. 1997. *Undoing Aesthetics.* SAGE, London.

Wittgenstein, L. 1953/1958. *Philosophische Untersuchungen/Philosophical Investigations.* Basil Blackwell, Oxford Harvard University Press, Cambridge, Mass. USA.

2 The Role of Visual Communication in Information Design

Yvonne Eriksson[1]*

Introduction

"Love is in the Air" is a disco song from 1977, sung by John Paul Young, with lyrics that refer to a feeling of love and romance that is present in the atmosphere. Here, we can even find an analogy about information, although it is less romantic. Just as those who are in love tend to see it everywhere, those who are looking for information tend to find it in almost all places and situations—whether it is designed information or information embedded in nature itself.

We can observe a lot of graphic information simply by walking around in a city. It is literally on the streets, in the form of lines and supporting signs, alerting us to the presence of walking paths, indicating where to bicycle and in what direction, informing motorists when they are prohibited from changing lanes or obligated to stop and give precedence. The city is additionally filled with ATMs, bus stops, graffiti, signs advertising shops and companies, advertisements, and more. The built environment, as such, is part of the visual culture that surrounds us, and this visual culture is filled with information we can use to navigate it and learn what is expected from us. In this sense, information is "in the air". However, to feel love in the air, one needs to be in love, and to see all the information, one must be motivated to find it. Beyond falling in love, when building a relationship, two individuals need to develop trust in one another. Likewise, to use information as it was intended to be used, one must trust the information and be able to evaluate its content. This has to do with literacy, which UNESCO defines as follows:

"Acquiring literacy is not a one-off act. Beyond its conventional concept as a set of reading, writing, and counting skills, literacy is now understood as a means of identification, understanding, interpretation, creation, and communication in an increasingly digital, text-mediated, information-rich and fast-changing world. Literacy skills themselves are expanding and evolving as people engage more and more with information and learning through digital technology. (UNESCO, n.d)"

In Information Design, it is therefore important to be aware of the meta level and to understand what is required, in terms of literacy, for a person to be able to use

[1] Professor in Information Design, Department of Health Science, Innovation and Design, Mälardalen University, Box 325, 631 05 Eskilstuna, Sweden.

* Corresponding author: Yvonne.eriksson@mdu.se

DOI: 10.1201/9781032642598-2

the information. It is also possible to distinguish information literacy from visual literacy. Visual literacy can mean, among other things, that one understands and can use visual representations (Eriksson and Göthlund, 2023). Information literacy, on the other hand, means that one is able to use information in an intended way. However, it is also important to define the intention. As an Information Designer or communicator, one must understand the content of what is to be communicated, even if it is an assignment from an organisation and the brief is to create a readable information campaign.

This chapter will address questions regarding information and problematise what can be considered as information, with a focus on visual communication. It will investigate various dynamic aspects of information and explore the relationship between misinformation and disinformation. Misinformation is false or inaccurate information, where the communicator gets the facts wrong. Disinformation is false information that is deliberately intended to mislead, where the communicator intentionally misstates the facts.

Information as a Kind of Communication

Communication, *communicare*, means to share or to make common and includes a social process, while information, *informate*, means to give form, shape, or instruct. Communication is fundamental for human beings. People communicate with each other, with artefacts, and with pets and working animals. Communication occurs when people talk, write, gesture, and mimic (Tversky, 2019). An extensive part of communication is mediated via language, visuals, and various artefacts, but also via the built environment. The literature on communication is extensive and discusses topics such as spoken and written communication, gestures for communication, and visual communication.

Mediated communication is often based on conventions, that is, traditions and agreements within a culture. Some conventions function globally and over time, while others are limited to a specific culture and/or time span. Communication can be directed towards larger groups of people, sometimes even a whole population, or even globally, or only towards a limited target group. Different subcultures develop their own "signals" to communicate that they belong to a specific group; it could be a hairstyle, style of makeup, clothing details, or accessories. Another example can be found in sports, where supporters wear a specific colour and emblem to signal which team they support.

The way information is communicated depends on the intention of the information and to whom the information is directed. Information communicated between specialised professionals, such as medical doctors, needs to be in medical language, while information communicated from a doctor to a patient must be in everyday language. The same goes for visual information: a plan drawing made by architects and engineers, which includes the infrastructure for water, electricity, etc., is typically not readable (or interesting) to a potential buyer of the house who would, instead, prefer a clean plan drawing showing the layout of the rooms.

Information is not a container of a message that is sustainable over time. It is dynamic and can change over time, even if the content remains the same. This does not mean that information is inherently unreliable, however. Information is grounded in facts or knowledge about something that is often built upon research in specific areas. Consequently, information based on research results will no longer be correct once new knowledge is developed and replaces the old. Research results only last until new research is conducted with new findings. Over the last decades, information has spread to a high degree on social media, the internet, etc., which has led to a new behaviour among producers/designers of information and consumers/recipients.

Framing of Messages

The term framing has been used to describe the process of organising or structuring the meaning of messages (Goffman, 1974). The most common use of framing can be seen in the framework that the news or media places on the information they communicate. Framing is two fold: the "sender" of the message frames the target audience, but so do the recipients. For instance, some target groups may decide to ignore a message from the government, while others may misinterpret the information. In this way, information can be transformed into misinformation by the perceiver when they communicate it further. There are many explanations for why information is misunderstood.

People tend to orient themselves in their environment by identifying themselves with groups that conform to their own convictions. By describing their own group, family, and colleagues using specific properties, they distinguish them from others. This distinction is often based on socio economic and cultural factors, as well as political and ideological aspects. Gender, ethnicity, age, and sexuality influence people's sense of belonging to a group or not. All of these aspects are crucial for the norms and values that an individual or group embraces. On top of this, there are aspects that create exclusion; if you do not share certain values, norms, and ideologies with a group, it can lead to alienation. The delineation of what a group accepts or not, the values it shares, and how it distinguishes itself from other groups are all forms of framing, where the framing is either influenced from the outside, through grouping aspects of individuals, or by one's own group (Klintman, 2019).

This is a challenge for Information Design, since it is not always the information itself that triggers a behaviour change or the willingness to accept societal information; sometimes it is the sender. This has been discussed by sociologist Mikael Klintman (2024) in relation to the COVID-19 pandemic, but is also relevant on a more general level. He discusses information and communication in relation to framing and Apollonia trust, which emphasises scientific principles like source criticism that is distinguished from Dionysian trust, which is built on sociocultural narratives and identity. Since the perception of information is, to a high degree, based on trust, and trust is something that, for most people, is created in relation to friends, family, and colleagues, Dionysian trust plays a dominant role.

From a socio-semiotic perspective, humans' willingness to accept information that is not in line with their personal convictions is very low.

When people discuss and value the information being communicated, they do so with people they trust and know, and with whom they share similar values and experiences. Since the Dionysian trust built on sociocultural narratives is strong, the Apollonian trust demands a lot of individuals who lack self-confidence in relation to a group with strong core values (Klintman, 2024). Simplified messages are often posted on social media platforms. If a message is posted by someone who is part of one's own network, it is often read very briefly without the use of critical thinking. In many cases, the message will be forwarded to larger groups and receive a lot of likes. In this way, unselected information is spread and can lead to both misinformation and intentional disinformation.

Information Society

The term "information society" was established in the 1980s when personal computers first came into use, but the term really grew in impact in the 1990s with the rise of the internet. There are different opinions regarding the meaning of information society. Is it a new era or only a new form of capitalism? What do people gain from it, and what are the drawbacks when it comes to the exclusion of those who lack access to the digital tools that are the conditions for the so-called information society? There is therefore a drive to analyse what can be defined as the information landscape.

The information landscape includes all kinds of information and things that can be used as information. The latter, however, depends on the viewer's ability to interpret an artefact or some kind of event. Information can be analysed and interpreted with various focal points, such as text and its content, visualisations, or text and visuals in combination. What and how information is analysed and interpreted depends on the context, situation, and purpose. It could be information on the internet with a specific goal, and if you find relevant information via a search engine, for example, it is likely that you will explore further, reading more via the suggested links from the first hit. This is different from being exposed to information that you have not searched or looked for; it is just there. There are many different examples of this, where the information can be explicit or embedded. Information can be analysed by using several theories and methods. As mentioned before, several theories about communication and information are based on the viewers/recipients' socio cultural background and belonging (Dervin, 2000). But when we talk about what and how people analyse and understand various kinds of information, we also need to take cognitive and perceptual processes into consideration (Arnheim, 1969).

One method that can be suitable when interpreting information, both explicit and embedded, is semiotics. Semiotics is the study of signs and how signs are given meaning. This brings us back to communication theory, which also plays a major role in how we understand the function and use of pictures of various kinds.

Pictures and visual expressions, in all their forms, are important communication media. Pictures are representations of either objects, people, environments, or symbols, or representations of such as ideas. Visual expressions, on the other hand, may be pictures but also letters that are used to emphasise a word or to play with different forms. But people also use visual expressions to communicate with each other. Semiotics is useful when studying and analysing different kinds of visuals (expressions or visual representations). A simplified explanation of semiotics is that we interpret our surroundings by identifying objects and artefacts as icons, indexes, and symbols. An icon is a representation that looks like the object it represents, like in a photograph or naturalistic drawing. An index has a logical, causal relationship with its object; common examples are smoke indicating fire, or tracks in the snow showing that someone has walked there. A symbol is built on social conventions (Kjørup, 2004). A symbol can also be a representation of a real object, but have a specific meaning in a specific context; this is not commonly used when information is communicated.

Semiotics

From a semiotic perspective, it is possible to identify an index of what has happened or is happening in the surrounding environment and interpret it as information. An umbrella, as such, can be interpreted as an index for rain ☂, or it can be interpreted as an icon being used to tell a friend via a text message to bring an umbrella in case of rain. However, the icon itself does not convey the meaning; for that, a context and a situation are required. What is needed is either an additional text that says, "bring the ☂", or an additional icon such as ☔. From this, we understand that we should bring an umbrella in case of rain. Still, in order to draw that conclusion, we need to be familiar with that kind of communication.

While taking an evening walk, one discovers the street is wet, which is an index for rain (the street can also be wet after it has been cleaned, but since it is not only the street that is wet, it is obviously the result of rain), and the trash bins are full of broken umbrellas (Figures 1, and 2). The broken umbrella functions as an index for heavy wind. Since people don't usually throw away their umbrellas in bins, we can assume that something extraordinary has taken place. From that, it is possible to draw the conclusion that it has rained heavily (heavily because it is very wet on the streets) in combination with strong winds. Of course, this is not explicit information, but an example of the information landscape. In one's surroundings, it is possible to find a lot of information that can be defined as indexical.

More explicit information can be found in a pictogram which, traditionally, is described as a graphic symbol that conveys meaning through its visual resemblance to a physical object. In one's surroundings, it is possible to find various kinds of pictograms. Some have high iconicity, meaning they very much look like what they represent, while others are more abstract. Some of them are easy to find since they pop up in the environment, while others are more discreet. Even though a pictogram is very simple, effort can be required to analyse and interpret it as

Fig. 1–2 After heavy rain and wind. (Photograph: the author.)

something meaningful (Eriksson, 2017; Eriksson and Carlsson, 2022; Eriksson and Göthlund, 2023), like in the example above with the umbrella emojis. One needs to understand what is represented in the pictogram, what it is intended for, and how to use it.

In a public lavatory in Portugal, a small symbol or pictogram (about 50 mm wide) was placed above a hook (Figure 3).

The small pictogram of a bag indicates that the hook is intended for bags. The fact that the symbol is placed on the wall beside the basin in a public toilet suggests that one can hang the bag on the hook while washing one's hands. The interpretation of the pictogram needs to be done in relation to the hook. It seems like it would be a simple thing to discover the pictogram and understand its meaning, but since the picture is small, it could easily be ignored. One needs to discover both the picture of the bag and the hook to be able to draw the conclusion that it offers the possibility to hang your bag. But one needs to have some knowledge of bag design, that is, to be able to interpret the represented bag as a handbag or portfolio. The hook is not meant for big, heavy bags.

Information is Everywhere

So, information is everywhere, but we need to discover it, and we must also interpret it as information. Some places are very information-dense, such as a platform at a train station. At one of Stockholm's train station platforms, one can

Fig. 3 Hook for handbag. (Photograph: the author.)

spot a lot of information in addition to information about the track numbers, the train's direction and departure time, and the way out. There is also a screen with information about arrival and departure times for the entire station, and different kinds of other information. There is a sign with a symbol for a camera and a text explaining that the platform is under surveillance. Smoking is prohibited, which is indicated by a sign showing a cigarette marked with a red cross. There is also a sign with a symbol for high voltage accompanied by the text, "Deadly wire over tracks. Do not go too close to fallen wire". Most people do not notice all of this information since, for a traveller, the most important thing is to confirm that one is at the correct platform and that the train will depart on time, or how to find the way out for those arriving with the train. This kind of information is regulated by law and is required to be there.

Interestingly, many of the most iconic symbols are outdated in many senses. Cigarettes, envelopes, alarm clocks, telephone receivers, and floppy disks are some examples that belong to this group. Vaping is common, but the sign for "smoking forbidden" is a cigarette. On computer and smartphone interfaces, an envelope indicates a mailbox, an icon of a round alarm clock with hands and numbers indicates where to set an alarm, and, to make a phone call, you tap an icon of a telephone receiver. The icons remain in the form of symbols for things that are no longer common or in existence, such as a mechanical alarm clock or a telephone with a receiver that you can lift and hang up. The icon for saving in Word is a floppy disk; although it no longer exists, we accept it as a symbol for saving. The same applies to the envelope, which is used as a symbol for e-mail. These icons function as metaphors for activities that are now carried out digitally, whether we are setting the phone's alarm to wake us up in the morning, saving texts and documents, or sending an e-mail. A visual can look the same but take on different meanings over time. However, the meaning of a visual is not fixed, but flexible. That is a challenge that needs to be taken into consideration.

Sometimes, it is not enough for a representation to resemble or match the real object. For example, in many grocery shops, one must scan the barcode of their receipt in order to exit. Information about this has been limited, and the customer is expected to understand how to perform the operation intuitively. If you know how it works, you will look for the scanner, which is a small square where you align the barcode of your receipt to make the doors open. Observing an elderly man try to figure out how to open the doors, it was obvious that the visual information was not clear enough. A picture of a barcode with an arrow pointing down at the scanner below did not give him the information he needed (Figure 4). What he did was to put the barcode in front of the picture of a barcode. His logic was to match the receipt's barcode with the picture of the barcode. Since he was not familiar with the scanner, he did not notice the small scanner for the receipt's barcode.

Fig. 4 A scanner for the receipt's barcode. (Photograph: the author.)

Visual Information

Images are tangible and may be perceived as compelling at the same time if their content can be perceived arbitrarily. How we see and perceive an image, that is, what it communicates, depends on how we read and interpret it. The way an image is perceived and interpreted, in turn, depends on culture, context, and situation. It is about the viewer's preconception, but also their interest. Images have both a prominent place today while, in many cases, they have an overlooked role. In the following section, we will discuss different aspects of visual communication; that is, where images and visual representations are expected to explain, display, influence, and convince people in different ways.

Pictures are powerful—something we see is incredibly difficult to shake off and affects how we absorb new visual information. On the other hand, as a sender, it is difficult to penetrate the enormous amount of noise created by all the images surrounding us, that is, to attract attention. People have learned how to protect themselves from being "attacked" by information, especially visual information. We learn to navigate between what we are interested in or not.

However, what some people perceive as information can easily be ignored by others for several reasons. It can be because of the content, how it is formulated, or when it is addressed. Within the field of Information Design, there is a tradition of simplifying information in order to make a message more easily accessible (Simlinger, 2007). However, it is more complex, as making information "easily accessible" does not necessarily mean that it will fulfil the original purpose or intention of the message. Sometimes the content of information is complex and therefore needs to be presented in a way that signals such complexity.

The line between information and advertisement is a fine one, and sometimes we see a combination or collaboration between the two. For example, a healthy lifestyle that includes exercise and foods like fruits, vegetables, and legumes is beneficial for one's well-being. This is factual information based on research, but it is also used by companies to advertise vitamins or pharmaceutical products.

We can notice something—that means we see it—but in order to analyse and interpret it as a meaningful whole, we need to look at it more carefully. To look carefully means to describe what is in front of us. By verbalising the content, we can begin to analyse what is there. From this point, it is possible to interpret the message (Eriksson and Göthlund, 2023). There is an interwoven relationship between visuals and verbal language. Visuals embody language built on symbols. But to interpret and use visuals, a verbal language is needed. Different parts and units of a visual representation need to be verbally defined in order to interpret and use the information that is being communicated (Eriksson and Göthlund, 2023). How a visual is interpreted and understood depends on the native language as well as the cultural context. The way a description is formulated will affect how a visual information artefact is interpreted. Using only visual information, such as pictures for information, can be tricky if the recipients or users speak different languages. The terminology for some visual details can vary between languages (Strömquist, 2009).

Making people pay attention to something specific is a challenge in Information Design. To make information more convincing, pictures—especially photographs—are often used. Visuals and particularly photographs are pregnant with information and have a realistic effect on the viewer (Sandström, 1995). Therefore, the photographic or realistic picture is efficient when it comes to information. It can work as a stand-in for parts of the environment that are not physically accessible. On the other hand, a photograph can be problematic in an Information Design context because it does not show a selection that is possible to recreate in a drawing; an aspect that is characteristic of photography, with the exception of studio photography. Meanwhile, drawings, maps, diagrams, and sketches can be used to clarify complex physical relationships and spatiality.

Information does not exist in a vacuum. It is related to previous information, facts, and people's interpretation and use of information. It does not matter if it is verbal or visual information. How information is perceived is dependent on the context and situation. The context is the environment, while the situation is what one is expected to do or what is happening at a specific moment.

Visual Communication

Visual communication, in all its aspects, can also be considered to enact what the professor of Media, Culture, and Communication, Nicholas Mirzoeff (1999), calls a visual event. This event always includes at least three actors: the visual artefact/phenomenon, the spectator, and the medium or technique through which the visual appears (Eriksson and Göthlund, 2023). This is extremely relevant when it comes to visual Information Design, that is, when using different types of visualisations within Information Design. What has an impact and influence on how individuals absorb and understand the information is not only how the visualisation is designed, but also in what form it appears as a moving image, in an advertisement or on a website.

Although Information Design is a young discipline, it is crucial to learn from how people have communicated throughout history, as it can offer us a better understanding of how various conventions have developed over time. Today, we connect with others by using different kinds of social media. With a few clicks or taps on a device, we can reach people in other parts of the world. Via communication apps such as FaceTime, WhatsApp, and Snapchat, not only can we talk to people or send text messages, but we can also immediately share photos from where we are presently located. The use of visual communication dates back about 50,000 years and can be seen in rock art or cave art, which exists on almost every continent.

Of course, visual communication likely looked very different in prehistoric times. Still, the common thread throughout history is that people use visual communication to create a connection. Pictures hold a special position when it comes to mediated communications because they can be perceived and sometimes even interpreted and understood intuitively. This is because a picture often resembles what it represents and therefore, people easily recognise its contents. A picture can also arouse various emotions, which often dominate the perception of a visual and additionally influence how we interpret and understand an accompanying text. In Sweden, tobacco advertisements are not allowed. Wine may be advertised, although not with people, but only the bottle or box with a glass of wine. This can be explained by the fact that pictures are mimetic and easy to imitate, and by imitating, one can internalise various behaviours (Eriksson, 2017). So, the motivation for not allowing advertisements that show people drinking alcohol is that pictures can function as role models.

In Sweden and many other countries, a cigarette package is labelled with a photograph and text providing information on the negative effects of smoking. The motif of the photograph varies from package to package, but all of them show the different kinds of diseases smoking can cause, as well as tooth loss. The problem here is that smoking *can* cause the various diseases represented in the photographs, but it is a warning and not something that will necessarily happen. The motifs of the photographs are disgusting and likely do not have the intended effect, instead causing smokers to avoid looking at them. When looking at an advertisement for wine in Sweden, there is no visual information about the negative effects of alcohol, but only a text stating, "Starting to drink at an early age increases the risk of alcohol problems" (Att börja dricka i tidig ålder ökar risken för alkoholproblem).

For another example of such information, we can look to the pharmacy (Apoteket). They used to offer free brochures containing various kinds of information. The brochures were thematic and informed readers about topics such as insect bites, skin and sun protection, advice about the genitals, and various diseases. It started in the early 2000s, and the brochures were illustrated with photographs featuring metaphoric symbols such as a rose for female genitals, and a picture depicting a crab to illustrate the expression, "red as a crab", referring to being in the sun for so long that one's skin has turned red (Eriksson and Göthlund, 2004). Ten years later, Apotek replaced these photographic illustrations with drawings that can be defined as pictograms (Eriksson and Göthlund, 2012). Today, the brochures no longer exist, but have been replaced by information on Apotek's website. While the brochures were something one happened to see while visiting Apotek, visiting the website is an action one takes when seeking out specific information. At the same time, Apotek, like many other companies, posts information, but mostly advertisements for various cosmetic products, on social media.

Looking for information about wounds, bites, and stings, which is part of the summer experience in Sweden, you really need to know where to look on the Apotek website. It is possible to search via the search box, but you will not get the same information, except for the products that could help. The start page of the website Apoteket.se features a menu containing the following: Products, Campaigns, Wise advice, Customer club, Services, Pharmacy live, Brands, Dose customers, and Healthcare providers and companies. To find what you are looking for, you must know that information about wounds, bites, and stings can be found under the menu item titled Wise advice.

If you click on Wise advice, you will find topics of very different kinds, ranging from Pollen allergy to Sunless tanning guide, Face cream with SPF, the different phases of the common cold, and Wounds, bites, and stings. When you click on the picture for Wounds, bites, and stings portraying a child from behind as they place on a plaster on their knee, you are met with the following text:

> "Wounds, bites, and stings
> Nothing is more part of life, and especially the summer period, than stinging abrasions and throbbing bites from insects that choose to emerge after a long, Swedish winter. As the variation in injuries can vary from large to small, we at Apoteket would like to advise you to seek medical attention if you feel that the injury is extensive."

By scrolling further down on the web page, you reach the next level of information:

> "Everything for a wonderful summer!
> The summer heat is here and it's time to get ready for a wonderful season filled with vacations, sun, swimming and adventure. Let Apoteket be your best friend when it comes to taking care of your health during the summer months. Here we have collected everything you need this summer, from tips and advice to products. Enjoy a carefree summer!"

Fig. 5 Is it advice to stand outside in a rapeseed field and enjoy the summer or a picture that illustrates when you can get insect bites? (Photograph from www.apoteket.se.)

The text is illustrated with a photograph of a woman standing in a rapeseed field, raising her arms above her head and looking out at the horizon. She wears a dress or blouse that is open in the back. If one wants to avoid stings and bites from insects, walking in a rapeseed field is not recommended (Figure 5). The next step of information is:

> "Tips for mosquito bites and stings
> You usually don't need treatment for insect bites or stings. However, it is important not to scratch the bite, as it can easily become infected. One tip if it itches unbearably is to put a patch over the bite and scratch the patch instead."

Fig. 6 A harmless crane fly. (Photograph from www.apotekct.se.)

The text is illustrated with a photograph of a crane fly. Crane flies are insects that belong to the order Diptera and are characterised by their long, slender legs and relatively large wings. They are not dangerous and cannot bite or sting (Figure 6). Their larvae live in moist soil and can sometimes cause damage to lawns. This means that the photograph has not been selected by a person who has knowledge of insects, and in the context of "wise advice" from a group of pharmacists who are represented in a group photo on the web page, people might get the wrong ideas about crane flies since they are portrayed in the context of bites and stings from insects. This is an example of misinformation that, for sure, is not intentional. However, the designer of the webpage has not conducted an analysis of the content.

There is a separate section for ticks related to the potential consequences of getting a tick bite.

> "Lyme disease – the bacterial disease spread by ticks
> Lyme disease is caused by bacteria found in the tick's digestive tract. The most common symptom of a Lyme disease infection is redness that is larger than a nickel and spreads around the bite. Therefore, check the bite area for at least 4 weeks."

This is followed by a description of the symptoms of Lyme disease and what to do if you suspect you have it. Everyone who has had a tick on their body knows that they are very small and difficult to discover; you often only find them because the skin at the bite site becomes itchy. The illustration for this section of the web page shows a large tick as they often appear on a dog or a cat (Figure 7).

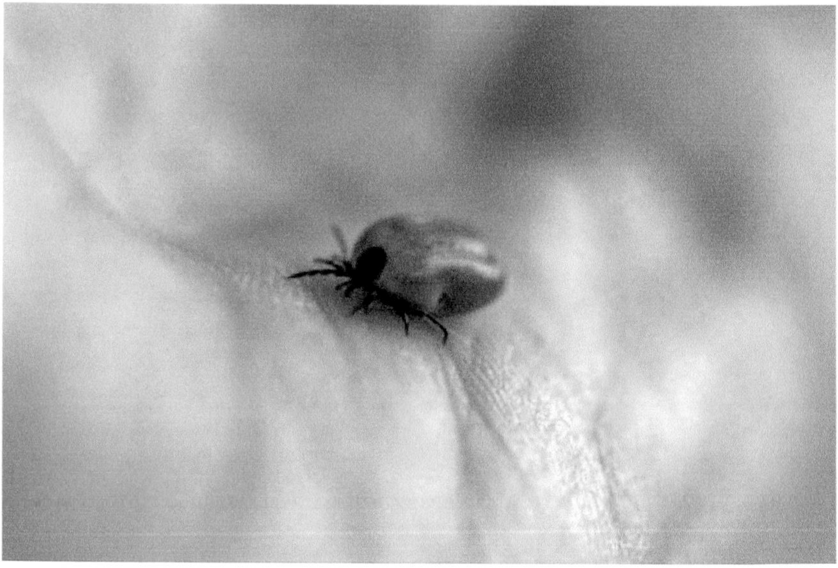

Fig. 7 A large tick. (Photograph from www.apoteket.se.)

The photo that has been chosen to illustrate the information in this section of the webpage does not provide further information or deepen one's understanding of what has been communicated. It has not been carefully selected, and the web editor has not conducted a deep analysis of the picture. The picture is not only a mismatch with this text, but it is also misleading.

The Importance of Information

As mentioned above, there is a difference between specific information that one searches for and information that is sent out or spread from organisations and governments. Whether someone is actively searching for information or is randomly exposed to it will influence their perception or vice versa: their perception will influence how they perceive the information. Perception is an active process where cognition, that is, our thinking, is present, and usually controls perception as we look for recognisable patterns. Here, cultural aspects also come into play as part of what we see and how we perceive it.

It is often stated that information is needed to raise people's awareness, but the question is, how is the information defined, formulated, and perceived? As previously mentioned in this chapter, we rarely keep information to ourselves; it is something that we communicate with others. We digest what is communicated and weigh it against our own experiences and beliefs, and confirm it with others. From there, we internalise the message, and we communicate the information further.

Reliable and credible information is crucial for a democratic society. It not only affects people's ability to participate in society's debates, but it also affects decision-making in everyday life. Not all countries have a political climate that makes it possible for people to openly produce and take part in information. Disinformation is one of the biggest challenges to trust today. When false or misleading information is spread, it affects people's ability to trust the media, governments, and even one another. Disinformation undermines trust, especially trust in institutions, and makes it difficult for people to navigate within the flow of information.

A phenomenon that has been discussed in the context of information and social media is trolling, as well as troll factory. Trolling refers to a troll, and the troll has its roots in Scandinavian mythology. From the mythology, we learn that these creatures are found underneath bridges where they wait for travellers. Until the travellers provide the correct answer to a riddle, they are refused the freedom to cross the bridge. Troller has been defined as follows:

> "A troller is a CMC [Computer Mediated Communication] user who constructs the identity of sincerely wishing to be a part of the group in question, including professing, or conveying pseudo-sincere intentions, but whose real intention(s) is/are to cause disruption and/or to trigger or exacerbate conflict for the purposes of their own amusement" (Dynell, 2016).

A troll is a person who posts or engages in inflammatory, misleading, and rambling actions on the internet. Their aim is to provoke emotional responses (Marcondes et al., 2024). Trolls often organise themselves in a troll factory, where members share the same goal and actions on social media (Marcondes et al., 2024). Some troll factories are professional while others are not. The spread of misinformation is a complex ecosystem. Troll factories can use both real people and automated programmes (bots) to spread their messages, which may include memes and deepfakes. Much of the trolling is formulated in such a way that makes the message clear and easy to read and understand, helping it become, most importantly, meaningful for the readers/recipients of the "information".

Disinformation is often spread and comes to life on social media, especially in closed groups. Members show loyalty to each other and, instead of exercising a critical analysis of statements, they confirm them. This kind of community can be very hard to move or penetrate, since the culture of a subgroup is often very strong, as is the loyalty within the group.

When people are repeatedly exposed to disinformation, it can create scepticism towards traditional news outlets, researchers, and authorities. This leads to a fragmented public debate where different groups struggle to agree on the basic facts. Disinformation can be used to create division in society. Narratives built on false or distorted information reinforce contradictions between different groups, which reduces the willingness to listen to other perspectives and instead leads to hostility. When people do not know what is true, their ability to make informed decisions is affected. This applies to everything from personal relationships to democratic elections. Mistrust spreads, and in the long run, it can reduce the fundamental trust that holds society together.

In a digital world where information spreads at lightning speed, it is difficult to determine what is accurate or not. Disinformation plays on emotions and often spreads faster than facts, making it difficult for people to build a solid foundation of reliable knowledge.

Building trust is about transparency, source criticism, and awareness. Training in digital literacy and critical thinking helps people identify false narratives and strengthen their trust in well-founded information.

Different societies react to disinformation in different ways, depending on factors such as political culture, media landscape, and education level. In many societies, disinformation leads to increased distrust of authorities and the media. When false narratives are spread, they can create division and intensify existing conflicts between different groups. There is a debate going on regarding restrictions to combat disinformation, for example, requiring transparency from social media platforms or introducing penalties for the intentional spread of false information.

Technology companies and researchers are developing AI-based tools to detect and flag false information, which can help reduce the spread of disinformation. In some societies, there are organisations and initiatives dedicated to fact-checking and disseminating accurate information to counter disinformation.

Troll, trolling, and troll factories are a challenge for Information Design since an Information Designer needs to get through disinformation and support Apollonia trust, which was previously referred to in this chapter. Creating awareness and stimulating information literacy is necessary.

The Visual as An Eyewitness and Confirmation

Images can evoke emotions, which can either be problematic, magical, or effective. By that, they often attract attention, if they are noticed. There is a saying: “I won’t believe it until I see it”, and often, “see it” includes a photograph. Even though we know that a photograph can be manipulated, it still functions as a kind of eyewitness. How a picture is interpreted depends on the text that goes with it, as well as the context. In an information context, we are more likely to trust the photograph. Especially if the content of the picture confirms or shows something we sympathise with. In a research study, participants who were exposed to a photograph that had been manipulated and later received information about how the picture was manipulated did not change their opinion about the picture. They stuck with their first impression and interpretation of the content. Although it is a limited study, it indicates how strong one's perception is, as well as how powerful a first impression is—when experiencing something visually, one tends to hold on to it.

A drawing, painting, or cartoon is rarely discussed from the point of reliability or authenticity since expectations are different for them compared to a photograph. One can question whether caricatures are fair or not, too rude, or just funny. When it comes to photography, there has been a discussion about its authenticity since its very beginning in 1830, as different techniques have made it possible to retouch and later manipulate them to various degrees. Even though we know that a photograph is not a blueprint, we tend to perceive it as such. The power of images is huge, but still, we need to observe them and take them into consideration: that is, in order to make use of them.

As mentioned in the introduction of this section, whether a given example functions as information or visual communication depends on how we perceive and interpret it.

We suppose that we see what is in front of us, but many times we only see what we think we see. This is because the relationship between the mental image (the imagination) and the visual surroundings is an intricate one. Mental images have a substantial impact on visual perception. People who witness a crime or an accident interpret the event based on previous experiences, which often come from movies. Or when they start to talk about what they have witnessed, they create a narrative which, in turn, creates new perceptions about the event.

When different types of visual representations are used for informational purposes to communicate something specific, it is common that the information is absorbed to various degrees depending on the recipient, but also on the group level. The quote, "When I read that drinking was bad for you, I gave up reading," is taken from an ironic picture showing a woman wearing a white dress and sunglasses, with her hair up. In her hand, she holds a bottle of what appears to be vodka (Figure 8). The picture and the text may be something to laugh at or to be horrified by. Regardless, it can be regarded as a metaphor for our behaviour, our tendency to avoid things that do not match our own thoughts or preferred behaviour.

Perception is a process that involves what we notice in the environment and our perception of something. The perception can partly be based on knowledge of something, i.e., previous experiences, and partly on what we have received from others. Much of what we acquire in the form of information occurs through what is usually called the visual image—this especially applies to visual information. However, the visual image is not a message but a structure of possible connections; it lacks a given structure of meaning.

Our ability to see or certify something visible often works as a means of justifying that something has taken place. The question to raise is: what do we see?

Fig. 8 People often make decisions that align with their own interests.

The fact that there is something in front of our eyes does not necessarily mean that we are noticing its details or really looking at what is there; it might be that we are only looking at what we think is in front of us. The expectation could be stronger than the perception. Most importantly, when it comes to perceiving information, people need to feel pointed. That is, to a big extent, dependent on culture and previous experience, but also on living situation (Eriksson & Carlsson 2022). In the parking lot of a beach area, there is a gate in front of which parking is prohibited. There is a sign stating this, with the text "Do not park in front of the gate"; despite this, a car has parked in front of the gate in an otherwise empty parking lot. Is it a matter of arrogance, or did the driver simply not notice the information? (Figure 9.)

News, whether on television, in the paper, or online, is often presented using photography to verify that an event, accident, or incident has taken place. It could be scenes from a war, demonstration, flood, fire, or from a more harmless event such as the first sunny spring day when people are eating ice cream outside. It is hard to verify whether a picture was actually taken on site in the referred context or situation, or sourced from a picture stock or AI-generated. A trained eye or a person with visual literacy and the required competencies for the specific situation may be able to recognise details in the photograph that verify its authenticity. Do

Fig. 9 A car has parked in front of the gate in an otherwise empty parking lot, despite a sign stating this, with the text "Do not park in front of the gate" when the gate must be clear to allow emergency vehicles through. (Photograph: the author.)

the environmental details in the picture showing scenes from a war confirm the photograph was taken in the actual place, or in another part of the country or city? Is the picture of the fire from the same place as the one described by the reporter or not? Does the type of clothing people are wearing or the variety of ice cream they are eating reveal whether the spring day event is authentic or not? Perhaps the ice cream is no longer being produced.

Conclusion

The motivation to absorb information and to be convinced by it is based on socio cultural affiliation. The saying, "As you shout, so shall you be answered," is relevant to information seeking. There are many "truths" that abound in the media noise, posing a challenge to well-founded factual information. At the same time, it is important to recognise that not all misinformation is deliberate disinformation; some of it may very well be based on misunderstandings that have been amplified through social networks. The use of images in different types of information artefacts requires that the designer of an information artefact has knowledge of what and how images communicate, but above all, that they choose visualisations with care. For a user or viewer, it is about acquiring sufficient visual literacy, as seeing and using images is something that is constantly learned and developed over time.

The aim of this chapter was to problematise information and its relationship to communication, since communication is the interaction between what is being communicated (the content) and a receiver's way of interpreting the message. This chapter has emphasised visual communication as a source of information. On the one hand, visual information is an efficient tool. On the other hand, it is problematic since people need to have visual literacy and a language to interpret pictorial symbols and information. Having visual literacy means that one can interpret, understand, and know how to use visual information. As the examples in this chapter have shown, visual literacy is also required among people who produce various kinds of information.

As an Information Designer, one must deal with various aspects of information, such as what the aim of a given piece of information is that should be spread, what the core message is, and who the intended recipients are. An Information Designer needs to ensure that the text and the associated picture communicate the same thing, and that the most effective media is chosen.

When placing information in an environment—indoor or outdoor—it must be easy for people to notice, see, and look at. It is not until we have had the chance to interpret information as meaningful that it will make sense to us and become useful. Therefore, it is important to involve users or at least listen to a user's perspective. Try to walk beside the intended users and involve various perspectives of what could be useful for them in different situations.

References

Arnheim, R. 1969. *Visual Thinking*. University of California Press, Berkeley.

Dervin, B. 2000. Chaos, Order and Sense-Making: A Proposed Theory of Information Design In: R. Jacobsen [Ed.], *Information Design* (pp. 35–58). MIT Press, MA Cambridge.

Dynell, M. 2016. "Trolling is not stupid": Internet trolling as the art of deception serving entertainment. *Intercultural Pragmatics*, *13*(3): 353–381. DOI 10.1515/ip-2016-0015.

Eriksson, Y. and Carlsson, A-L., 2022. The Challenge of Designing Meaningful Information. In: Y. Eriksson [Ed.], *Different Perspectives on Design Thinking.* CRC Press, Boca Raton.

Eriksson, Y. and Göthlund, A. 2023. *Foundations of Visual Communication. How Visuals Appear in Daily Life*. Routledge, New York and London.

Eriksson, Y. and Göthlund, A. 2004. Möten med bilder. Att tolka visuella uttryck. Studentlitteratur, Lund.

Eriksson, Y. and Göthlund, A. 2012. Möten med bilder. Att tolka visuella uttryck. Second edition Studentlitteratur, Lund.

Eriksson, Y. 2017. *Bildens tysta budskap. Interaktion mellan bild och text*. Studentlitteratur, Lund.

Goffman, E. 1974. *Frame Analysis: An Essay on the Organization of Experience*. Northeastern University Press, Boston MA.

Kjørup, S. 2004. *Semiotik*. Studentlitteratur, Lund.

Klintman, M. 2019. *Knowledge Resistance: How We Avoid Insight from Others*. Manchester University Press, Manchester.

Klintman, M. 2024. 2 Apollonian and Dionysian trust in vaccination. In: L. Borin et al. [Ed.], *Vaccine Hesitancy in the Nordic Countries*. Routledge, London.

Marcondes, F., Almedia, J.J. and Novais, P. 2024. An exploratory design research on troll factories. *Integrated Computer-Aided Engineering* , *31*: 95–115. doi: 10.3233/ICA-230720.

Mirzoeff, N. 1999. *An Introduction to Visual Culture*. Routledge, London.

Sandström, S. 1995. *Intuition och Åskådlighet*. Carlssons Bokförlag, Stockholm.

Simlinger, P. 2007. *idX Development of International Core Competencies and Student and Faculty Exchange in Information Design within the EU/ US Cooperation Program in Higher Education and Vocational Education and Training*. International Institute for Information Design, Wien.

Strömqvist, S., Nordqvist, Å. and Wengelin, Å. 2004. *Writing the Frog Story: Developmental and Cross-Modal Perspective*. Psychology Press, Taylor and Francis.

Tversky, B. 2019. *Mind in Motion. How Action Shapes Thought*. Basic Books, New York.

3 The Complexity of Effective Communication

Visual Management and Visual Information Design in Industrial Workplaces

Carina Söderlund[1*]

Introduction

This chapter critically examines the concept of visual management (VM) as an information and communication strategy aimed at achieving effective workplace communication. VM is employed across a variety of workplace settings, such as hospitals (Williamsson and Poksinska, 2019) and construction sites (Valente et al., 2019). This chapter focuses specifically on industrial contexts, with particular emphasis on the manufacturing industry and the potential role of VM within this domain.

VM is rooted in visual communication and involves presenting work-related information using visual devices. (Terms such as *visual tools* or *visual systems* are also commonly used in literature.) The information conveyed visually using these devices is referred to as *visual information*. The concept of information is discussed in greater detail later in the chapter: in general, it refers here to visual elements present in the work environment; ranging from pathways and furniture arrangements to mediated information such as various types of drawings and diagrams, colour codes, graphical markings, tables, and signage (cf. Davison et al., 2012; Meyer et al., 2013). Information that has been intentionally designed by someone or something is referred to in this chapter as visual information design (visual ID), whether created by technical communicators, consultants, workplace staff, or software such as Microsoft Word.

The chapter offers a description and analysis of various visual devices, drawing on both original studies and those of other researchers, in conjunction with relevant theoretical perspectives. The primary focus is on analogue and printed media, as well as light-based media and digital displays, given their prevalence in today's workplace settings. Technologies such as VR (virtual reality), AR (augmented reality), MR (mixed reality) and AI (artificial intelligence) are not addressed in this chapter, although they offer new possibilities for VM.

The chapter begins with a historical reflection on visual ID in industrial work environments, followed by a brief discussion of the concept of information

[1] Ph.D. Senior Lecturer in Information Design, Department of Health Science, Innovation, and Design, Mälardalen University, Box 325, 631 05 Eskilstuna, Sweden.
* Corresponding author: carina.soderlund@mdu.se

DOI: 10.1201/9781032642598-3

and how it can be approached in a workplace context. Subsequently, various visual devices and their information design are described and analysed by means of diverse theoretical concepts, perspectives, and disciplinary lenses. In this context, the chapter also problematises how these devices and their information design contribute to effective communication in workplaces. Finally, the chapter summarises the theoretical approaches applied, emphasising the importance of adopting interdisciplinary and transdisciplinary perspectives in future VM research.

Visual Information Design in Industrial Contexts: A Brief Historical Overview

To grasp the purpose of visual ID in industrial contexts, it can be helpful to adopt a historical perspective. This section does not present a comprehensive historical account or an in-depth analysis, but instead offers a concise overview highlighting selected technological advancements, societal developments, and design approaches relevant to the subject.

According to Stiff (2005), the term "information design" was scarcely mentioned prior to the early 1980s. That period, and even earlier, was marked by significant technological innovations. It is reasonable to assume that information design as a concept gained momentum with the rise of digital technologies, particularly as personal computers and mobile phones became both consumer products and tools of production across industries, especially in Western societies. During this era, concepts such as graphical user interfaces and human-computer interaction began to emerge as commonplace at industrial work sites. By then, engineer Mike Cooley had advocated a human-centred approach to technology development, while American author and researcher Don Norman introduced the concept of user-centred design; to name just two of the scholars who have emphasised human needs, prerequisites, and conditions in the development of products, systems, and technologies.

As early as the 1950s, Theodor Larson highlighted the growing significance of "visual information design" in the American construction industry (Larson, 1952: 210). He emphasised the importance of designing information from the user's perspective; to make it not only accessible, but also clear, comprehensible, and easy to recall. Larson refers to Sweets' catalogue files (first published in 1906), which contained construction materials and products, and to Sweets' educational initiatives aimed at facilitating users' understanding of them. Thus, in 1950 they supported the publication *Catalog Design Progress: Advancing Standards in Visual Communication.*[#]

Even earlier, around 1940, the concept of VM and visual communication in industry emerged with the development of Toyota's production system and the introduction of what is known as the Lean philosophy, a set of practices and principles rooted in Japanese industrial traditions. Central to this approach

[#] Authors Knud Lönberg-Holm and Ladislav Sutnar.

were waste reduction, process optimisation, customer value, and continuous improvement, all of which influenced Western manufacturing industries after World War II. This influence brought with it several Japanese concepts and terms—notably *muda, mura, and muri*—which can broadly be understood as principles for achieving efficient workflows and organisational processes in which improvement forms part of everyday operations. Principles related to the Toyota production system and the Lean philosophy also introduced a visual approach to workplace communication, aimed at effectively distributing work-related information through visual devices such as signage and colour codes (Liker, 2004). It is likely that technological developments played a key role here, as printing techniques became more accessible and affordable; and the automatic photocopier was introduced by the late 1950s.

However, this visual approach to workplace information and communication could be traced even further back in time. It was introduced as early as during 19th-century industrialism and further developed in the early 20th century with Taylorism. Joanne Yates, in her book *Control Through Communication: The Rise of System in American Management* (1993/1989), describes how American companies during the industrial era were influenced by Taylorism and the principles of scientific management when developing communication systems, including visual devices, to increase efficiency in processes and workflows. Charles Kostelnick and Michael Hassett (2003) also discuss the impact of industrialism and Taylorism in their book *Shaping Information: The Rhetoric of Visual Conventions*. Among other things, they explored how visual ID through conventional drawings, tables, charts, and manuals was assumed to contribute to an organised workplace, and thus efficient processes and workflows. Within this historical framework, *Graphic Methods for Presenting Facts* (1919) by Wizard Brinton emerges as a noteworthy contribution. He demonstrates the communicative potential of visual ID in early industrial contexts, and provides, for instance, an illustration of how pictorial or linear perspective renders material flows and transport between factory floors visible.

Against this historical backdrop and the rise of industrialism, it may be noted that visual ID entered and began to shape industrial work and the workplace more than a century ago.

The Visual Factory and Workspace, and Information in the Work Environment

VM is associated with the concepts of "visual factory" and "visual workspace". In the late 1980s and early 1990s, these concepts began to appear in various publications in relation to Computer-Supported Cooperative Work (CSCW). In 1991, Michel Grief linked the concepts to the manufacturing industry in his book on the visual factory and visual workspace. Here, Richard Schonenberger (1990) explains that Grief contributes a new approach to the visual workspace; one that does not concern computer-supported work, but instead highlights the role of the physical environment, the industrial built space. Grief (1991) argues that visual communication aims to distribute and share work-related information to increase efficiency in the workplace. This form of communication enables information flows that are not only hierarchical and vertical, from management to staff, but also horizontal flows between the physical industrial space and staff, and among the staff themselves. Grief explains that horizontal flows allow employees to gain an overview, understand, and be included in whatever is happening in the workplace. This can be seen as a shift towards a more inclusive and democratic work environment, where access to information and involvement in ongoing activities and processes are no longer reserved for management alone.

Given the significance attributed to physical and built space in VM and to visual communication in workplace settings, it is relevant to reflect on the concept of information, and visual information in particular. Although the concept of information appears central to VM, it is rarely discussed in the literature concerning (Lean) manufacturing, production, and operations management. However, the concept of information is well debated within Information and Library Science. Influential information scholar Birger Hjørland describes, in simplified terms, two main perspectives: information as something subjective or objective (2007). When the concept of information is viewed as subjective, it is understood as being the result of an interpretive process in which individuals assign meaning to what they perceive in their environment, shaped by context and situation. Conversely, when information is viewed as objective, it refers to something that exists independently of human interpretation and meaning-making. Hjørland (2009) argues that information cannot be both at the same time, and therefore we must choose a perspective.

Considering the close relationship between visual management, visual communication, and the physical work environment, an environmental psychology perspective on the concept of information is introduced. The view of environmental psychologists Rachel and Stephen Kaplan on the concept of information may be suitable in this regard. In their book, *With People in Mind: Design and Management of Everyday Nature* (1998), Rachel and Stephen Kaplan, together with landscape architect Robert Ryan, provide examples of information as visual and spatial elements in green outdoor environments. Here, information as a concept encompasses both elements derived from nature and those created by humans (that is, artefacts), as well as their arrangement. The visual and spatial appearance and arrangement of these elements affect how individuals gain an overview and understanding of a place, as well as how they interact with its constituent features.

In summary, if information is understood as visual and spatial elements within a physical industrial environment in which people interact with one another, with space, its technologies and various objects, then how such information is designed becomes crucial. Its influence lies in how various designs may shape the ways in which staff interpret, understand, and respond to work-related information and apply it in practice.

A Few Common Visual Devices

In a VM context, various visual devices are used to make work-related information—such as deviations, problems, or status updates—visible to staff. Certain visual devices are subject to legal regulation, including signage and indicators related to fire safety, first aid, and other warnings or instructions intended to alert staff to potential risks, mandatory actions, and operational guidelines. In the Swedish context, such regulations are issued by authorities including the Swedish Work Environment Authority (*Arbetsmiljöverket)*, which provides directives concerning placement, accessibility, and symbolic representation, for instance.

In industrial workplaces, however, there are also visual devices that are not directly subject to legal regulation: these are covered in the following section. It draws on Michel Grief's book on the visual factory (1991), Gwendolyn Galsworth's, *The Visual Workplace* (2017) and research by Tezel, Koskela, and Tzortzopoulos (2016), who are among the most established scholars in the field of VM and its application in production and construction environments. The section also builds on visual devices observed and documented in original studies (Andersson, 2010; Lundin, Söderlund, and Eriksson, 2017; Söderlund and Hansson, 2021; Söderlund, 2023).

Visual Devices for Visual Management and Visual Control

The literature on VM within (Lean) production, manufacturing, and operations management rarely addresses the information design of visual devices, particularly regarding how such design is interpreted and understood by staff, and how it influences their responses (Söderlund and Hansson, 2021). In *The Visual Workplace*, however, Gwendolyn Galsworth offers a comprehensive review of various visual devices and their information design. She categorises these devices into four groups – indicators, signals, controls and guarantees – based on the extent to which they control and regulate staff behaviour, actions, and interactions, and constrain interpretations of work-related information.

"Indicators" are visual elements which staff may choose to respond to or ignore, such as a text label indicating the contents of a package. "Signals" include devices such as light indicators and wayfinding systems, designed to allow for a somewhat narrower interpretive scope and to regulate behaviour to some extent. Examples include red, yellow, and green lights indicating operational status, or graphic floor markings guiding staff movement on the factory floor. "Controls" are visual devices designed to further reduce interpretive scope and behaviour. One example is a shadow board designed to indicate the designated placement of tools, often using silhouettes to guide correct positioning and ensure order and accessibility. Finally, "Guarantees" are devices whose design requires little to no interpretation and fully determines behaviour; for instance, the glass cover of an emergency stop button that must be broken to activate the stop function.

Since VM can be seen as synonymous with visual control, it is reasonable to categorise visual devices on the basis of their potential to guide and control the interpretive space available to staff, including their actions, interactions, and behaviours. However, categorisation is not without challenges: categories may be difficult to distinguish from one another, for instance. In this case, a single visual device may stimulate multiple types of actions, interactions, and behaviours which refer to several categories.

Whether and how visual devices—categorised as indicators, signals, controls, and guarantees—contribute to effective workplace communication requires critical examination.

Prior knowledge and experience play a significant role in this regard. For example, what may be an obvious interpretation for an experienced machine operator might appear unclear to a newly hired employee. A red-light signal may elicit varying responses depending on the workplace context; between a manufacturing workshop producing individual components and a process industry such as a paper mill, for instance, where a deviation or fault can rapidly propagate throughout the entire system. Moreover, visual devices in VM often, though not always, convey relatively brief messages. These are simplifications of more complex situations and concepts, which in turn may lead to simplified interpretations rather than a broader understanding of workplace phenomena.

Studies on information design for industrial maintenance at paper mills show how interpretations of, and responses to, condition indicators (green, yellow, and red) on handheld tools and control room displays can vary (Andersson, 2010). Here, red was intended to indicate poor machine condition (that is, operational status or machine health), while green signified good condition. The study revealed that yellow was sometimes difficult to interpret, and that colour coding was perceived as a generalisation of machine condition. To interpret the colour semantics, staff also relied on prior experience, such as continuous observation on the factory floor and documentation of machine condition over time (that is, machine history). This was done to avoid the potential risk of rigid interpretation of the red signal, which could lead to unnecessary and costly maintenance. Thus, an experienced maintenance technician might treat the red colour as a cautionary indicator and postpone maintenance.

From a socio-materialist perspective, visual devices can be approached as visual actors. They actively participate in guiding, directing, and controlling actions, interactions, behaviours, and interpretations, as well as in shaping what occurs, and is expected to occur, in the workplace. For example, a directional sign legitimises movement in a certain direction, graphic markings regulate where objects should be placed, and a flashing light could indicate desired or undesired interactions.

However, the example of colour coding and machine condition also illustrates the fact that visual devices do not always function as visual actors, nor are they necessarily equal actors in the workplace in relation to staff. This may depend on the other types and sources of information on which staff rely in the workplace, and whether they operate within a team that collectively interprets workplace events and the information conveyed by visual devices, thereby creating new conventions (Ibid).

Making the Work Environment Tangible, Concealed, Delimited, and Expandable

Visual devices and their associated information design can be understood in terms of how they render concepts and phenomena in the work environment; not merely as visible, but also as tangible, concealed, delimited, and expandable. This is illustrated by commonly used devices in industrial contexts such as instructions, guides, and various procedural documents such as standardised work instructions (Standard Operating Sheets, SOSs) and one-point lessons (OPLs).

Such documents require, for instance, clear and structured presentation of information to support the operator in understanding the sequence of a procedure; to switch off a machine before cleaning it, for example. If the document design is a combination of pictures, text, and graphic markers, it can support the operator in identifying relevant components such as oil reservoirs and gauges, and indicate where oil should be replenished. In this context, visual ID helps to delimit the physical work environment and reduces the need for visual attention across the workspace in order to complete the task.

Fig. 1 Mark-ups and labels. (AI-generated illustration, Microsoft Copilot, 2025.)

Another common application is the use of markings (*mark-ups*), for instance, on materials and packaging in storage areas or on shelves or containers. Here, a black marking may indicate combustible materials, while green signifies items designated for recycling. In this case, visual ID serves to make complex concepts such as combustibility and recyclability visible and concrete to staff.

Labels are another kind of marking. Although labels may appear visually similar, their contributions to the work environment can vary significantly. For instance, a label can draw attention to hazardous or unpleasant substances, including those that are colourless and odourless and therefore imperceptible to the senses. Conversely, labels on storage shelves delimit the space in the workwear supply room by spatially distinguishing work gloves from safety goggles, for instance, and eliminate the need to search the entire room visually in order to locate specific items. In such cases, effective communication involves unfolding and concretising otherwise abstract and complex concepts or phenomena and delimiting the built industrial space, depending on what is considered meaningful in a given context.

As a further example, consider the use of andon lights—visual light signals that employ colour combinations and sometimes form part of light towers. They utilise green, yellow, and red lights in a traffic-light metaphor to indicate the operational status of a production line. Green signifies normal operation, yellow denotes a deviation (such as rising temperature), and red indicates a malfunction or stoppage. Additional colours, such as blue, may be used to signal a request for assistance and are typically activated by an operator or assembler.

In some cases, the operator or assembler may already be aware of the condition of the production line. Here, the visual device and its information design serve to externalise and validate this knowledge, making it accessible to others. When andon lights are employed, their illumination and colour-coding function as communicative signals to other staff, indicating the need for support or maintenance. For example, maintenance staff can prepare appropriate maintenance actions based on aspects such as verbal communication and the computerised maintenance system.

Here, effective workplace communication relies on visual devices that render otherwise abstract concepts—such as operational status and system conditions—concrete and tangible, not only at the immediate production site but also across broader areas of the industrial environment.

Spatial/Environmental Graphic Design and Light Design

Other commonly used visual devices in industrial settings are grounded in "environmental graphic design" or "spatial graphic design". This is particularly relevant considering the communicative role attributed by Michel Grief to the physical industrial space, as discussed earlier in the chapter.

Graphical markings are one example of practical applications of spatial graphic design within industrial environments. These may include yellow (or other coloured) dashed lines indicating where objects may or may not be placed, or yellow-and-black striped barrier tape indicating non-accessible areas on the factory floor. Additional examples include white or coloured floor markings that indicate designated walking areas for staff. This type of visual ID helps, for instance, to separate staff from forklift traffic, thereby reducing the risk of collisions. It may also delimit the area of machine stations and clarify the placement of mechanical equipment, thereby minimising exposure to potential accidents. Such visual information can be compared to elements commonly found in traffic environments, such as parking spaces, dashed lines along road edges and directional arrows at roundabouts and junctions.

Environmental or spatial graphic design is also employed to make various (material) flows visible. This may involve graphically marked areas indicating where empty bottles should be placed before refilling, conveyor belts with shaped compartments, or zones marked with white tape to indicate where a component should be positioned. Barbara Tversky's theories on visual thinking (2011) are relevant in this context, suggesting that the graphically designed workplace can be perceived as a diagram. It constitutes a visual structure that influences the ability of staff to perceive and understand otherwise invisible phenomena within the work environment. These may include imperceptible material flows and expected movement patterns, or desirable and prohibited activities and interactions.

One aspect that is rarely addressed in VM literature concerning production, manufacturing, and operations management is light design. In the construction industry, light design could be used to visualize aspects of time, warnings, status, conditions etcetera (Tezel, 2011). In the context of VM and spatial graphic design in built space, light can also be used to make spatiality visible and to define spatial zones on the factory floor. Additionally, when a seemingly ordinary door with a green illuminated exit sign is opened and light pours out from the adjacent space, this visually signals a designated and active evacuation route, a safe route out.

An additional example of light design in this context is the use of lighting to create distinct spatial zones and "light rooms" that form room-within-room configurations on the factory floor. These illuminated areas may be perceived as places—that is, spaces—imbued with meaning and associated with specific functions and purposes. Conversely, areas shrouded in darkness may be interpreted as "non-places" or closed-off zones that are either inaccessible or should be avoided.

These two cases demonstrate how visual devices and information in the form of light design can create, conceal, and delimit spaces on the factory floor, as well as open transitions between rooms and places that form part of a larger whole. Effective communication in this context involves guiding and directing staff regarding where they may go and where they should remain, while also visually communicating the flexibility of the workshop or factory floor and its potential to develop places or conceal spaces as needed.

Visual Devices as Boards

Other commonly used visual devices include various types of boards, both digital and analogue. One example is shadow boards, which display black silhouette outlines of tools to visually indicate which tools are associated with a specific workstation, as well as where they are expected to be stored. An empty shadow board may raise questions about whether the workstation is no longer active, or whether cleaning discipline at the workstation is lacking.

Fig. 2 A shadow board for cleaning tools and a wet floor-sign. (AI-generated illustration, Microsoft Copilot, 2025.)

Other types of boards include what are known as "andon boards", which graphically present the status and condition of the production line by means of colours and markings, for example. There are also "kanban boards", which are used to visually present details related to work activities and tasks. These may be connected to "kanban cards"; that is, coloured notes containing brief texts indicating work tasks, such as when something needs to be moved or replenished.

In the above cases, the boards and their visual ID help to make otherwise abstract concepts – such as expected workstation order, work planning and structure, or material flows – concrete and tangible in the workplace.

Various types of meeting boards are a final example of visual devices as boards. These are referred to here as VM boards; although in practice they are known by several names, such as daily management boards, team boards, Lean boards, or planning boards. Such boards are distributed and located throughout an organisation where staff meet regularly. During what are known as board meetings, work is organised and coordinated, and activities and performance are created, planned and followed up (Zarbo et al., 2015). The boards are also used by staff to support development work and continuous improvement efforts in the workplace (Jaca et al., 2014).

These VM boards present various types of work-related information with differing visual designs. This may include documents containing general information from trade unions, as well as lists, graphs, and charts presenting measurement results related to key performance indicators such as safety, delivery, and financial performance. It may also include handwritten notes concerning problems, solutions, and prioritisation of tasks. In addition, the VM boards may include names and photographs indicating who is responsible for what, when, and how, as well as who is available on a given day.

Effective communication in this context is partly achieved through the boards' ability to provide a visual gestalt to a department's results, performance, and activities. The VM boards then make complex concepts such as safety and finance more tangible and make the available resources and competencies more concrete. In turn, this can facilitate the planning, follow-up and improvement of work activities, performance, and the workplace. Furthermore, the graphic design and information structure of a VM board—through the arrangement of tables, charts, and notes, for example—make the meeting agenda visible and guide the structure of the meeting and the order of speakers. Additionally, the VM boards assist with the creation of meaningful spaces—meeting places—where staff, for instance, share and exchange information, experiences, and knowledge with one another (Söderlund and Hansson, 2021).

Designing Information for Sustainability

It is common to associate visual management (VM) and visual devices with the 5S (or 6S) work methodology, aiming at creating order and clarity for a safe and efficient workplace (cf. Glasworth, 2017). The letter 'S' originates from Japanese terms, which in this chapter are translated into English with less reliance on alliteration in order to enhance clarity: *Seiri*: select and sort; *Seiton*: structure and systematise; *Seiso*: maintain, care for and improve; *Seiketsu*: promote uniformity, standardise; and *Shitsuke*: establish and uphold routines.

Through their information design, visual devices related to the 5S framework make visible not only the expected orderliness of the workplace, but also a sustainable approach to resource management. This is exemplified by the previously mentioned shadow boards with tool silhouettes, as well as shapes and "imprints" of tools in foam or wooden materials, placed in toolboxes or similar containers.

When tools such as oil filter spanners or drive belt tools are sorted and placed near the corresponding machinery or machine components, staff can easily recognise which tools are intended for specific machines and quickly locate the relevant tools when needed. This reduces stress and irritation and saves time for staff. Shadow boards or storage containers also protect tools by allowing them to be stored appropriately and preventing accidental injury to staff.

In the context of 5S, this chapter highlights the potential of visual devices to reveal various spatial relationships between objects and their functions, provided that central parts of the object are depicted; referring to Biederman's (1987) theory of recognition-by-components. Returning to the previously mentioned shadow boards, this example features rendered with contoured lines depicting cleaning tools. When the outlines of a floor mop and a bucket are placed close together, it is possible to understand that the objects are related. If a mop wringer is also depicted near the bucket, its function becomes clearer in relation to the other objects; namely, to wring out the mop to avoid excess water on the floor, thereby preventing slippery conditions. Furthermore, the mop dries more quickly when it is wrung out, which is particularly relevant in instances where hygiene requirements are in place. This example, which includes the shadow board and cleaning tools, may be considered simple. In practice, workplaces feature far more advanced machines, components, and tools, with complex spatial relationships between them. The basic principle is the same, however, as visual ID may clarify spatial relationships between objects and illuminate their functions. When it comes to information design, it is important to understand that it carries values and norms, not just facts (Eriksson, 2017). With that in mind, what also constitutes effective communication in the examples of tool storage and shadow boards is the capacity of information design to highlight a sustainable approach to resources that emphasises care and preservation.

Analysing Visual Devices and Information Design through Muda, Mura, and Muri

How to design information to support a sustainable approach to resource use is not self-evident. For this reason, this section highlights three central Japanese concepts within Lean philosophy—*muda*, *mura* and *muri*—which are applied to further analyse visual devices and their information design. These three concepts function as both principles and standardised methods for reducing waste within organisations, aiming to create more efficient flows and processes in environments where continuous improvement is integral to daily operations. In this chapter, the concepts are examined from an information design perspective. *Muda* can then be associated with work-related information whose visual design supports the reduction of waste in the workplace, focusing on comprehension, readability and clarity. *Mura* pertains to accessibility and legibility, ensuring that employees can access and assimilate the information they need. In addition, *muri* relates to the adequacy of information, ensuring that the amount of information provided is neither insufficient nor excessive for employees' needs.

To further explore the concepts of *muda*, *mura* and *muri*, they are applied to original VM research studies based on visual rhetoric as a theoretical framework. The aim is not to generalise, but to render the concepts comprehensible while simultaneously highlighting their complexity.

In a previous project, the management of a manufacturing company expressed a desire to involve employees in ongoing workplace activities and processes. To this end, they used charts and graphs to communicate measurement results based on commonly used key performance indicators, such as financial performance, deliveries, and safety. These visualisations were displayed near the factory floor and in the operators' lunch and break areas, spaces where employees naturally gathered. However, conversations with staff revealed that they did not fully understand the content of the charts and graphs. Moreover, they were printed in black-and-white, using greyscale, which made it difficult to discern what the bars and graphs represented and to distinguish between them.

When work-related information is visually designed in a way that lacks legibility, it can lead to frustration and misunderstandings, thereby contributing to waste in line with the concept of *muda*. Furthermore, interpreting charts and graphs is a learned skill, influenced by factors such as prior knowledge of the subject matter and experience in analysing such visual data. A lack of such competencies may also result in work-related waste.

This example also draws attention to the concept of *mura* as the charts and graphs were placed in the break room, making them easily accessible to staff. However, it is worth questioning whether a space intended for rest and recovery is appropriate for displaying work-related information, such as key performance indicators and measurement results. This practice could potentially reduce opportunities for recovery and increase stress (cf. *muri*).

In another study conducted at a manufacturing company, staff were observed during their regular meetings around their VM boards (Söderlund and Hansson, 2021). These boards displayed aspects such as key performance indicators and measurement results in various graphs, charts, lists and tables. They also included handwritten notes on problems requiring resolution, lists of prioritised work tasks and documents containing general information, such as trade union communications. The VM boards and their content were studied through surveys and interviews with staff, as well as field observations.

The research findings indicated that the visual design of the content of the boards, such as the font size of charts and tables, was sometimes illegible and difficult to see. Additionally, the placement of the VM boards in cluttered and noisy work environments negatively affected the ability to hear discussions during meetings. Thus, both the unsuitable location of VM boards and the poor legibility of their content can be considered forms of waste in accordance with *muda*. Furthermore, the work-related information on the boards was occasionally outdated, and in some cases staff were already aware of the content. This may lead to waste associated with *mura*. Such accumulated inefficiencies can undermine the credibility of the VM boards and their content, resulting in underutilisation (cf. *muri*).

This section demonstrates that an in-depth analysis of visual devices and their information design can help identify design-related work waste. It also shows how poor information design may exclude staff and undermine resource sustainability.

Importantly, recognising when things do not work as intended is a way to reveal the complexity of effective communication in the context in which the visual devices and their information design operate.

Visual Devices and Visual Information through Participatory Design

The literature on VM within (Lean) production, manufacturing and operations management demonstrates strong confidence in visual modalities. It highlights the fact that various forms of visualisation can enhance transparency in workplace processes and activities and support the provision of accessible and comprehensible information.[#] This reliance on visual information such as charts and light signals reflects an underlying assumption that if information obtains a visual gestalt, it is inherently intuitive and immediately comprehensible (Söderlund, 2023). In addition, within the VM literature, a common objective of visual information is to support, guide, direct, or control employees' interactions and work activities, as well as to frame and reinforce their interpretations, decisions, and behaviours. In other words, the emphasis is on cognitive and perceptual processes, enabling individuals to make informed decisions and achieve their information-related goals efficiently and effectively. This reflects a communicative approach aimed at producing specific effects and tangible outcomes, an orientation also evident in the academic field of Information Design.[†]

In this context, a critique of standardisation may be warranted, given its roots in Taylorism and its emphasis on control and order (cf. Yates, 1989/1993), as it risks inhibiting qualities such as creativity and flexibility. It is therefore important to emphasise that parts of the VM literature argue that such visual communication and information should also promote more elusive qualities, such as enhanced morale, inclusion, and engagement among employees (Liff and Posey, 2004; Eaidgah et al., 2016). Nevertheless, the literature remains vague in its explanations as to whether or how the visual devices and the work-related information can be designed to support such elusive qualities.

The following section will now discuss these aspects based on original research (Söderlund, 2023), where municipal employees were involved in the process of designing an information visualisation representing perceived workload. Prior to the study, they used colour-coded magnets (green, yellow, red) and handwritten text on their VM boards to visualise psychosocial factors such as well-being, a visualisation method also found in industrial contexts. In this case, multiple participatory design methods were employed, involving both managers and employees in the design process. For instance, the participants conducted photo studies and heuristic evaluations of existing VM boards. Additionally, they created

---

[#] Further reading: Singh and Kumar, 2021; Murata, 2021; Galsworth, 2017; Eaidgah et al., 2016; Bititci, Cocca, and Ates, 2015; Singh and Singh (2015), Parry and Turner, 2006; Liff and Posey, 2004.

[†] Further reading: Ware, 2021; Fascara, 2019; Thornhill et al., 2017; Mollerup, 2015; Krug, 2014; Albers, 2005; Pettersson, 2002.

mood boards in a collaborative workshop to visualise and discuss metaphors associated with workload and well-being in their workplace.

The participatory methods and the findings formed the basis for two prototypes related to perceived workload, which were subsequently evaluated by the project participants. One of these was selected for further assessment by the work teams (Figure 3). During the participatory design process, the prototype was called "The Flower".

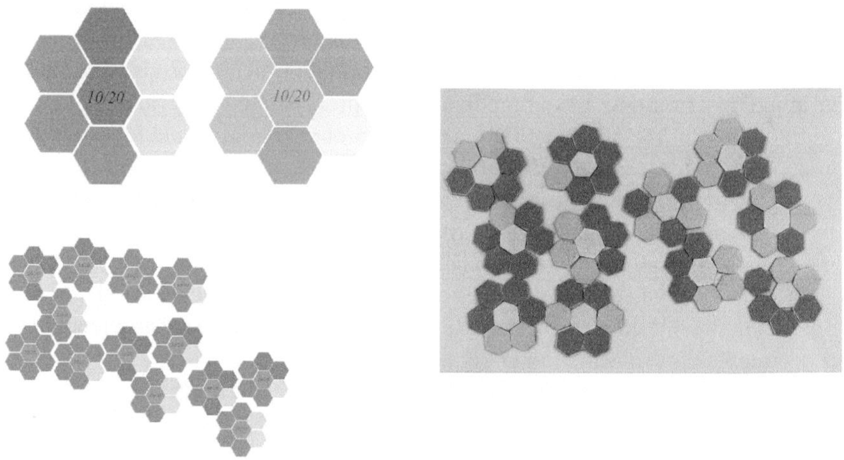

Fig. 3 "The Flower"—2019 prototypes. To the left, a digital sketch of "The Flower" created in Adobe Illustrator, designed to visually express perceived workload on VM boards. The petals are in pastel colours, green, yellow, and red. The image on the right presents a higher-fidelity version of "The Flower" using coloured magnets on a VM board.

The visual appearance of "The Flower" challenges established norms regarding how work-related information is traditionally designed in VM praxis and on VM boards. The basic idea is that when employees meet regularly at their VM boards, each places six petals in pastel shades of green, yellow, or red. Each petal corresponds to a variable or statement that the employee evaluates using the colour scale. Two examples of such variables or statements include: "The planning of my work tasks allows for unexpected tasks" and "I have received the support I need to complete a work task". At the VM board meeting, employees then created their own unique flowers to visualise a subjective phenomenon—in this case, perceived workload—and the variables related to it. In this way, psychosocial phenomena such as workload or well-being are not generalised into a single colour, such as a red magnet.

When all employees have created flowers on the VM board, a visual metaphor—"The Meadow"—emerges representing the work team's perceived workload. In this context, "The Meadow" seeks to give visual form to the relationship between the individual and the collective, acknowledging variations in psychosocial experiences of working life and across work teams.

The organisation that participated in the study has continued using flowers over the years to visualise psychosocial aspects on their VM boards (Figure 4). One of the development managers recently stated that their participation in the research project laid the foundation for employees to further create emergent forms of visualisations on their VM boards (Telephone interview, 11 October 2025). Naturally, further studies are needed to provide evidence of aspects to which collaborative and participatory design of VM board information might contribute in the long term.

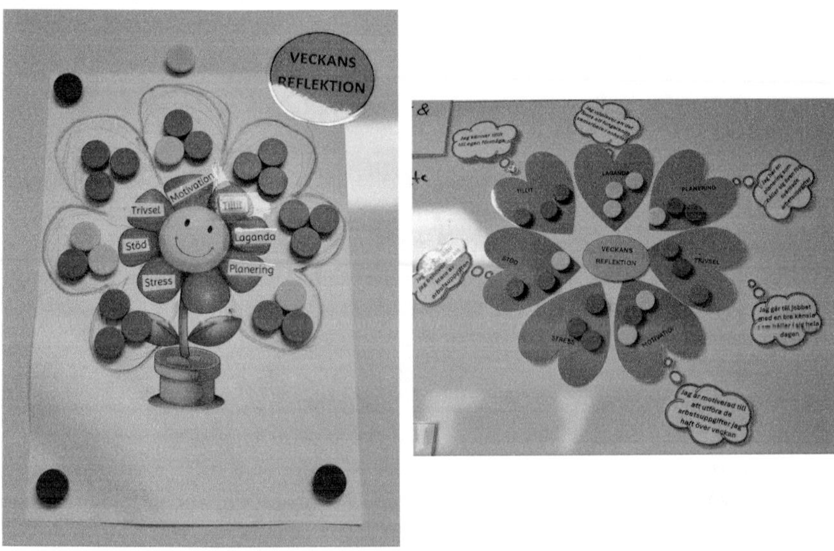

Fig. 4 Flowers on VM boards, 2025. The organisation that participated in the study currently uses these "flower visualisations" to display psychosocial aspects of work on VM boards in two of six departments. The "inner" petals in the image to the left are red, while the "outer" ones are outlined by hand in black. On the right, the petals are heart-shaped and purple. Surrounding cloud-shaped graphics explain the variable or statement represented by each petal on VM boards. Flowers are used in several departments at the case company; however, the visual design varies according to the choices made by each work team.

By involving managers and co-workers in a participatory design process and in the development of "The Flower", they became active participants in designing information, rather than relying on computer-generated charts or pre-purchased magnets in various colours. To further explain "The Flower" as a visual device, it is analysed here through the lens of performative theory, as a series of visual performative acts. "The Flower" and "The Meadow" can be approached as co-creators of how well-being is perceived at the workplace. For example, when employees create flowers representing perceived workload on a VM board, this can be seen as a collective performative act that nurtures interaction between the individual and the work team, based on the notion that it takes many different flowers to form a meadow. Additionally, through metaphors such as "cultivating one's own garden", "The Flower" and "The Meadow" visually express notions such as self-reflection, personal responsibility, and transformation on the VM board. Furthermore, this information design establishes a visual narrative in which variation in form, shape, and colour communicates diversity and change – just as in nature, the workplace too reflects the rhythms of multiplicity and transformation.

Given this background, visual information design for effective workplace communication is also understood as something beyond control and order, reinforcing qualities such as co-creation, engagement, involvement, and inclusion in workplace activities and processes.

Concluding Discussion

This chapter has explored VM as an information and communication strategy, primarily within industrial environments, with particular emphasis on visual devices and their visual information design (visual ID). The chapter describes and analyses common visual devices and their visual ID through concepts and terms associated with the Toyota Production System and Lean philosophy (*muda*, *muri*, and *mudi*). Additionally, various theoretical perspectives and fields of knowledge concerning visual thinking and visual rhetoric have been highlighted, including theoretical concepts such as "visual actor" and "visual performative act", which are linked to both performativity theory and theories of socio-materiality.

The chapter critically examines the notion of effective communication through selected dimensions of management control, as well as the support and guidance of cognitive and perceptual processes to enable informed decision-making and goal attainment. The chapter demonstrates that effective communication in a workplace context is indeed complex. It illustrates how common visual devices and their information design may both support, and fail to support employees' actions, interpretations, and behaviours. The chapter also emphasises a design-oriented perspective on visual management, its visual devices and their information design, and highlights the industrial built environment as an interface. These perspectives offer additional approaches to understanding the effectiveness of communication in this context, as visual devices and their information design not only make otherwise abstract concepts and phenomena visible, by implication, accessible, and comprehensible, something this chapter has shown is not always the case. They also render such concepts and phenomena tangible, while simultaneously concealing, delineating, and expanding aspects of the workplace environment. Effective communication in this context is not solely concerned with making information visible, rather, also encompassing the spatial capacity to embody information. Furthermore, the information design of these visual devices also contributes to effective communication by visually conveying or prevailing desired norms, values, and narratives within the workplace; in this case, a sustainable approach—not only in economic terms or as a counterpoint to a throwaway culture, but also in relation to work-related well-being.

To understand the complexity of effective communication regarding visual devices, visual ID and VM in industrial settings, further studies are needed—for instance—to explore the role that visual devices and their information design may play in fostering elusive qualities, such as engagement in workplace processes and activities. Given that visual devices constitute a multimodal experience, additional studies are also needed that address aspects of light, sound, and tactile design in the workplace. Future VM research could greatly benefit from multi-, inter- and transdisciplinary collaborations involving participants with diverse experiences and perspectives on VM as a concept, together with scholars from fields such as visual studies, visual communication, information design, and other design disciplines. Only then can VM be thoroughly understood as an information and communication strategy for effective workplace communication.

References

Albers, M.J. 2005. *Communication of Complex Information: User Goals and Information Needs for Dynamic Web Information*. Mahwah, N.J.: Lawrence Erlbaum Associates, Incorporated.

Andersson, C. 2010. *Informationsdesign i tillståndsövervakning. (Eng. Information Design in Condition Monitoring: A Study of a User Interface for Condition Monitoring and condition-based Maintenance.)* Doctoral dissertation. Mälardalen University, School of Innovation, Design and Engineering. Mälardalen University Press Dissertations No. 88.

Biederman, I. 1987. Recognition-by-components: A theory of human image understanding. *Psychol. Rev., 94* (2): 115–147.

Bititci, U., Cocca, P. and Ates, A. 2015 Impact of visual performance management systems on the performance management practices of organisations. *International Journal of Production Research, 54* (6): 1–23.

Brinton, W.C. (1919). Graphic methods for presenting facts. Engineering magazine company.

Davison, J., McLean, C. and Warren, S. 2012 Exploring the visual in organizations and management. *Qualitative Research in Organizations and Management: An International Journal, 7* (1): 5–15.

Eaidgah, Y., Maki, A.A., Kurczewski, K. and Abdekhodaee, A. 2016. Visual management, performance management and continuous improvement. A lean manufacturing approach. *International Journal of Lean Six Sigma, 72*: 187–210.

Eriksson, Y. 2017. *Bildens tysta budskap*. Studentlitteratur.

Frascara, J. 2015. What is information design? In: Jorge Frascara (Ed.), Information design as principled action: Making information accessible, relevant, understandable and usable (pp. 1–56). Common Ground Publishing.

Frascara, J. 2019. Information design from my window. *Information Design Journal, 253*: 334–347.

Galsworth, G.D. 2017. *Visual Workplace Visual Thinking: Creating Enterprise Excellence through the Technologies of the Visual Workplace*. Taylor and Francis.

Greif, M. 1991. *The Visual Factory: Building Participation through Shared Information*. Productivity Press.

Hjørland, B. 2007. Information: Objective or subjective/situational? *Journal of the American Society for Information Science and Technology, 58* (10): 1448–1456.

Hjørland, B. 2009. The controversy of the concept of "information". A rejoinder to Professor Bates. *Journal of the American Society for Information Science and Technology, 60* (3): 643.

Horn, R.E. 1999. Information design: The emergence of a new profession. In: R. Jacobson (Ed.), *Information Design* (pp. 15–33). Cambridge. Mass: MIT Press.

Jaca, C., Viles, E., Jurburg, D. and Tanco, M. 2014. Do companies with greater deployment of participation systems use visual management more extensively? An exploratory study. *International Journal of Production Research, 52* (6): 1750–1770.

Kaplan, R., Kaplan, S. and Ryan, R.L. 1998. *With People in Mind: Design and Management of Everyday Nature*. Washington, D.C.: Island Press.

Kostelnick, C. and Hassett, M. (2003). Shaping information: the rhetoric of visual conventions. Southern Illinois University Press.

Krug, S. 2014. *Don't Make Me Think. Revisited. A Common Sense Approach to Web Usability*. New Riders.

Larson, C.T. (1952). The US building industry: A case-problem in communications. *Journal of the American Society for Information Science, 3* (4): 208.

Liff, S. and Posey, P.A. 2004. *Seeing Is Believing: How the New Art of Visual Management Can Boost Performance Throughout Your Organisation*. AMACOM.

Liker, J.K. 2004. *The Toyota Way – 14 Management Principles from the World's Greatest Manufacturer*. McGraw-Hill.

Lundin, J., Söderlund, C. and Eriksson, Y. 2016. Design challenges for technical communicators. In: *Proceedings of the 34th ACM International Conference on the Design of Communication* (*SIGDOC* '16) (pp. 1–6). Silver Spring, USA. ACM Press.

Meyer, R.E., Höllerer, M.A., Jancsary, D. and van Leeuwen, T. 2013. The visual dimension in organising, organisation and organisation research core ideas, current developments and promising avenues. *Academy of Management Annals, 7* (1): 487–553.

Mollerup, P. 2015. *Data Design: Visualising Quantities, Locations, Connection*. Bloomsbury Academic.

Murata, K. 2021. Internal mechanisms framework of lean implementation using the visual management systems. *International Journal of Industrial Management, 9* (1): 1–14.

Parry, G.C. and Turner, C.E. 2006. Application of lean visual process management tools. *Production Planning and Control: The Management of Operations, 17* (1): 77–86.

Pettersson, R. 2002. *Information Design: An Introduction*. Amsterdam/Philadelphia: John Benjamins Publishing Company.

Singh, S. and Kumar, K. 2021. A study of lean construction and visual management tools through cluster analysis. *Ain Shams Engineering Journal, 12* (1): 1153–1162.

Singh, J. and Singh, H. 2015. Continuous improvement philosophy: Literature review and directions. *Benchmarking, An International Journal, 22* (1): 75–119.

Schonberger, R. 1990. *Building a Chain of Customers: Linking Business Functions to Create the World Class Company*. Simon and Schuster.

Söderlund, C. 2023. Recognizing non-designers' contribution in the process of designing information on visual management boards: A metaphorical approach. *Design Science, 9*: e18.

Söderlund, C. and Hansson, M. 2021. A visual and rhetorical perspective on management control systems. *International Journal of Lean Six Sigma, 123*: 536–552..

Stiff, P. 2005. Some documents for a history of information design. *Information Design Journal + Document Design, 133*: 216–228.

Tezel, A., Koskela, L. and Tzortzopoulos, P. 2016. Visual management in production management: A literature synthesis. *Journal of Manufacturing Technology Management, 27* (6): 766–799.

Tezel, A. 2011. *Visual Management: An Exploration of the Concept and Its Implementation in Construction*. School of the Built Environment: College of Science and Technology. University of Salford.

Thornhill, B., Potter, E., Nakarada-Kordic, I. and Reay, S. 2017. Designing clinical guidelines to support effective decision-making in emergency surgery. *Information Design Journal, 233*: 268–289.

Tversky, B. 2011. Visualizing Thought. *Topics in Cognitive Science, 3*: 499–535.

Valente, C.P., Brandalise, F.M.P. and Formoso, C.T. 2019. Model for devising visual management systems on construction sites. *Journal of Construction Engineering and Management*, *145*(2).

Ware, C. 2021. *Visual Thinking for Information Design* (2nd Edn.). Morgan Kaufmann.

Williamsson, C. and Poksinska, B. 2019. Visual management in hospitals: A systematic literature review. In: A. Galiano, J. Treinta and M. M. Cruz [Eds.], *Proceedings of the International Joint Conference on Industrial Engineering and Operations Management* (*IJCIEOM* 2020) pp. 161–168. Springer Proceedings in Mathematics & Statistics, Vol. 337. Springer.

Zarbo, R.J., Varney, R.C., Copeland, R.J., D'Angelo, R. and Sharma, G. 2015. Daily Management System of the Henry Ford Production System: QTIPS to Focus Continuous Improvements at the Level of the Work. *American Journal of Clinical Pathology*, *144*: 122–136.

Yates, J.A. 1993. *Control through Communication: The Rise of System in American Management*. The John Hopkins University Press.

4 Challenges for Comprehension of Risk and Uncertainty in Visualizations

Thorbjörn Swenberg[1*]

Introduction

This chapter is about the pros and cons of visualizing risks and the uncertainties that arise with risks. The risk of something happening is an eventuality and, as such, abstract. A heavy rainfall could cause a river to flood and a landslide to occur. These might happen. When they do, their effect is all too concrete and may be disastrous to those affected. If we want to visualize the risk, should we somehow show the potential disasters, like houses and streets under water or neighborhoods in collapse? Or should we try to explain the likelihood of things happening while interpreting the statistics? Or perhaps the possible levels of mishaps, from minor inconveniences to major catastrophes, such as the water level of a river slightly—or greatly—crossing its boundaries? This first issue revolves around what to visualize. What information would be the most useful to convey?

The second issue regards uncertainty because risk is something estimated from one or several dataset(s), a calculation. How sure can we be about the estimated risk? We might know that there will eventually be heavy rainfall: in the next 1,000 years there is a 99.9% certainty of this at least once; in 100 years, there is a 99% certainty. But how sure are we about there being a disastrously heavy rainfall the next 10 years or 25—the time during which any of us would consider living in a specific building at a particular site before moving? There are statistical methods for pointing out how certain, or uncertain, a specific risk is: for instance, the risk of a house basement flooding. But this kind of uncertainty is even more abstract than the risk itself. The question is, therefore, how can uncertainty and the level of that uncertainty be visualized?

The visualization challenge is to make two interrelated abstractions comprehensible in a concrete visual form. The strength of visualizations is that abstract figures, which might be wholly incomprehensible to the average person, are transformed into something with a concrete meaning in a context where they can be understood. But this is also challenging because a visualization is always specific, whereas risk and uncertainty are dynamic and relative to several

[1] Ph.D. Associate Professor in Audiovisual Studies, Head of Subject, Moving Image Production Chair of Swedish Media Production Academies (SMEA), Dalarna University, Högskolegatan 2, Falun, Sweden.

* Corresponding author: tsw@du.se

DOI: 10.1201/9781032642598-4

conditions. And the person who sees the visualization must interpret it, as best they can. Yet, a range of possible interpretations might still exist.

Risk and Uncertainty as Information Visualization Issues

Contemporary society has developed into a sophisticated and complex form of life that depends on organization, technology, and a wide range of general and specialized knowledge. A by-product of this society is the continuous emergence of a range of risks that threaten its persistence and us as its members (Beck, 1986).

Understanding risk is hard in itself and often a matter of seizing complexity. Complexity occurs when there are several factors or aspects involved in a situation, often expressed as "variables", which increase exponentially with more variables. Therefore, the overview of a complex situation is challenging, and understanding risk in such situations is even more so.

One inherent parameter of such complexity is the uncertainty involved in the risk. Risk may be understood in several ways, but always entails a certain likelihood that it will lead to a disadvantageous event or condition that we do not wish to be exposed to or affected by. Uncertainty is also a multifaceted concept that has several meanings in the sense that there are different ways in which we may be unsure about something. Here, risk is considered to be the statistical probability of something happening combined with its consequences, whereas uncertainty includes both the specific probability and the statistical prediction interval for the lower and upper limit of the probability.

Statistical probability is a distinct figure—a proportion or percentage of the phenomenon being accounted for. Statistical uncertainty is then a numerically defined span around that figure, presenting a possible range inside where the appointed value is certain to be found if the appointed figure differs from the actual one. A problem is that humans tend to disregard uncertainty when they estimate probability or they misinterpret the uncertainty depending on the form of visualization (Hullman, 2019), or they misunderstand or simplify the statistical concepts when they are mathematically more complex.

There are several visualization forms for statistical risk and uncertainty that are used in different contexts and that present risk and uncertainty (some common and recent examples are presented in Figures 1–8); however, statistical knowledge and training are required to understand and use them. Intervals come with ambiguity in terms of what is shown since a span can represent different measures, estimates, or observations, and may also draw more attention to central key values at the expense of what is to be found at the edges. An alternative is to try to map probability by means of density in various ways, but density may not be understood as indicating probability or it can be biased toward higher relevance of darker areas and be, quite simply, hard to read for specific values. Plotting, as another alternative, may infer occlusion of data points or imprecision if sampling is used, which induces uncertainty errors.

One attempt to bypass the intricacy of statistical misinterpretation is to avoid preciseness and instead rely on vague illustrations, which, however, tend to hide or blur uncertainties so that they become hard to grasp. Each of the following visualizations comes with strengths and weaknesses.

The boxplot (Figure 1) provides a clear abstraction of statistical values and their range; however, it does not present actual data very clearly (it is rather ambiguous), nor the interrelations between various data; nor does it contextualize the statistics with respect to possible consequences and the particular situation. Furthermore, it tends to highlight central values.

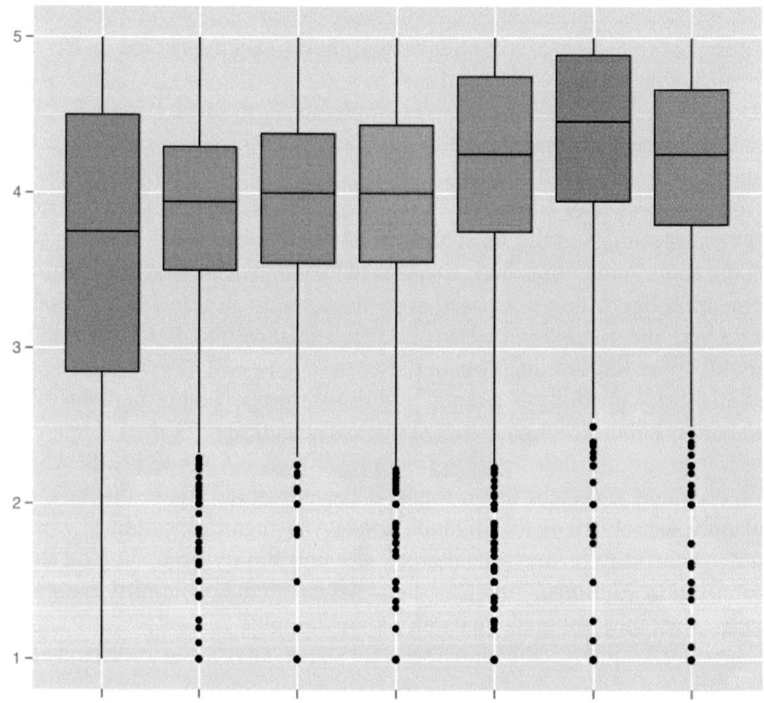

Fig. 1 Boxplot with risk as mean value (black line) and uncertainty as colored space (adapted from Proton 2012, CC BY-SA 3.0).

The heatmap diagram (Figure 2), relates probability to consequences better than the boxplot, indicates abstract relations between different data, and gives an uncertainty indication for the given values; however it presents risk only in an alarmistic way without putting it in any situational context or representing data spread.

Fig. 2 Risk heat map diagram with risk as a position with regard to probability (Y-axis) and consequence (X-axis).

The radar diagram (Figure 3), on the other hand, gives a more graspable abstract pattern of complex data relations on a risk probability scale where uncertainties may be apparent; however, the radar diagram does not monitor the data spread, situational context, or consequences.

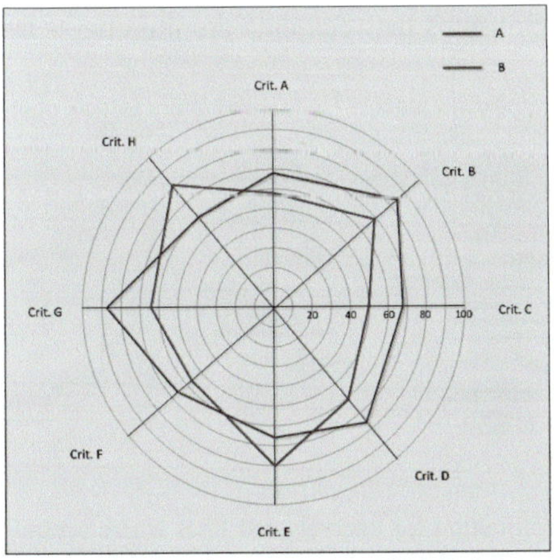

Fig. 3 Risk radar diagram (spider chart) where several dimensions of risk (criteria A-H) form a pattern (Klever, 2015, CC BY-SA 4.0).

The Figure 4 risk scatterplot matrix shows the relationship between probability and consequence using both positive and negative values. It can present uncertainty as well as the abstract relation between data, while data spread is not shown and nor is situational context.

Fig. 4 Risk scatterplot matrix where dimensions of risk have both positive and negative values of probability (y-axis) and consequence (x-axis) respectively (Datawrapper, 2022, reprinted with permission).

If we turn our attention to uncertainty as the prime data to be visualized, uncertainty distribution for intervals and lines (Figure 5) does the job well, representing risk data and its distribution, whereas specific values may be hard to read and the relationships between data are missing as are consequences and situational context.

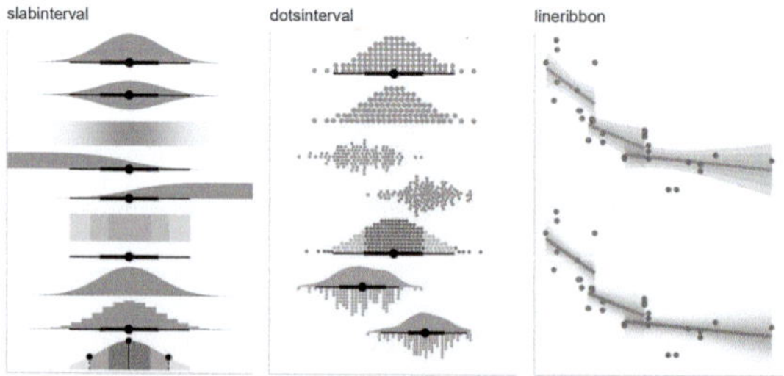

Fig. 5 Uncertainty distribution for intervals and lines where central values as well as span (x-axis) and probability within the span (y-axis) is presented (Kay, 2025; GNU General Public License v3.0).

The uncertainty bubble tree map (Figure 6) uses uneven surrounding lines as indications of uncertainty, risk correlating with size, and data spread, and data relationships presented by position in the tree structure. Yet, this visual form is very imprecise and cannot deliver situational context or consequences.

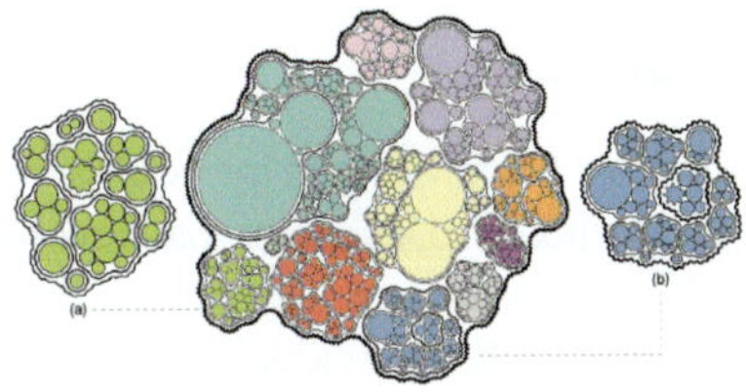

Fig. 6 Uncertainty bubble tree map presenting clusters (same color blobs) and individual values (singular blobs) within clusters as size and uncertainty via the surrounding lines (amplitude of line waveform) (Görtler et al., 2018, reprinted with permission).

In Figure 7, the hue–saturation–intensity color map of uncertainty presents risk probability and uncertainty together with interrelations of different data and can be overlaid over, e.g., a map to give a situational context, though not exposing consequences or the actual data spread.

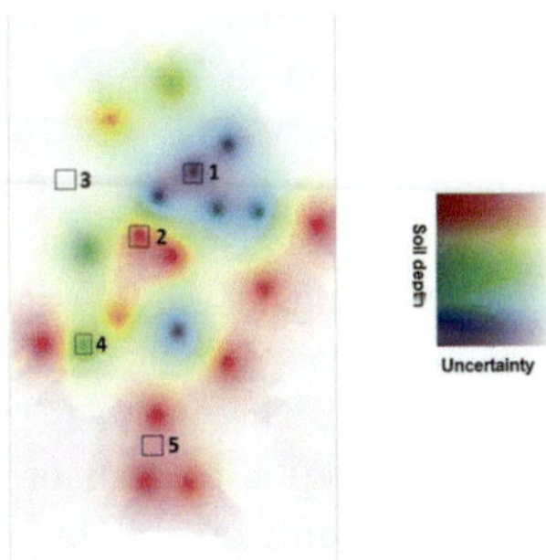

Fig. 7 Uncertainty hue–saturation–intensity color map where uncertainty is presented via saturation and intensity combined (high saturation/intensity equals low uncertainty and vice versa) (Kubíček and Šašinka, 2011, reprinted with permission).

The abstract 3D terrain visualization graphic of uncertainty (Figure 8) reveals both uncertainty, situational context, and type of data but not the risk probability itself, relationships between different data, or consequences.

Fig. 8 Uncertainty graphical abstract 3D terrain visualization where color (descrete level of uncertainty) and height (continous uncertainty value) of glyphs represent uncertainty (Dübel et al., 2017, reprinted with permission).

Risk Communication in Context

Risk and its uncertainties must be considered by experts in different fields and by public and private decision-makers. Decisions that involve risk are common at the highest societal level, such as those relating to vaccination programs or city relocations (e.g., Kiruna), and at the individual level, such as finding real estate on which to build your next home. For instance, if you want to buy property, you may want it close to water. You will then have to be aware of the risk of flooding, including uncertainties, at that particular location as well as any potential consequences: a flooded garden, a filled basement, roof-high water, undermined land, and the risk of landslide. These risks, uncertainties and consequences are not easily grasped for a lay person to whom statistics might appear abstract and confusing. Information needs to be straightforward, clear, and understandable, engaging the reader beyond the abstract, so that possible consequences of events are acknowledged before the events become real.

Furthermore, professionals need straightforward information and a clear overview. For example, in health care, epidemiologists must analyze risk factors and the possible spread of disease when developing public health strategies; risk and uncertainty must also be considered when the physician makes a diagnosis or with the development of treatment. They must also plan and prioritize resources in the event of catastrophe and threat, involving not only the statistically knowledgeable.

The insurance sector constantly deals with the risk of damage or injury, when setting insurance rates or reinsurances. This includes long-term risks on a large scale, such as in the case with construction projects or the occurrence of natural disasters. The insurance risk analyst must be able to identify and quantify such risks and create a risk profile before an insurance is agreed to, and the analyst must communicate this to their client.

Within the energy and climate sectors, risk and uncertainty are constant since energy production and consumption prognoses require repeated analyses, and these take climate factors into close consideration. Decisions on energy resources and infrastructure rely on the operations analyst's computations of complex risks; on the climate modeler's simulations of climate impact; and on the urban planner's risk assessment in urban development and infrastructural planning. Decisions are made by those at high political levels as well as by any one of us in our domestic life, and such decisions require an understanding of such risks.

Closely related to energy and climate are natural and environmental resources, which are also affected by climate factors and natural disasters, the exploitation of limited resources, and pollution: these are, therefore, subject to risk and uncertainty. These issues are managed by the environmental risk assessor, who identifies and analyzes such environmental risks, as well as by risk analysts, operations analysts, climate modelers, and urban planners, since the effects of changes in nature or other environments might have an impact on such diverse contexts as crisis management, infrastructure projects, energy supply, and insurance issues. Thus, risk in nature and the environment can potentially affect everyone and should, therefore, be understood by many.

In the construction sector in particular, assessments of climate impact and earthquakes on buildings and other facilities are risks that must be considered when calculating strength. First and foremost, the construction engineer and architect are presented with these risks when calculating material strength and dimensions and also when considering safety at the construction site. Yet, all the professionals mentioned in the previous paragraph may just as well be engaged in the risks of construction projects since credit decisions are based on repayment capacity and associated risk; insurance decisions are based on the risk profile; environmental decisions consider possible pollution or natural disasters; the impact of disasters must be accounted for; and future societal needs and risks. In this context, it cannot be expected that everyone affected will be well versed in statistics, at least not the average property buyer.

Logistics and transportation present yet another context where risk and uncertainty prevails: supply chains, traffic flows, and accident models are managed by means of statistical methods in order to optimize the flow of goods and minimize risk. The logistics specialist controls transportation and storage and supply chains with uncertainties in supply routes and demand. The production planner deals with risk and uncertainty in supply chains and production processes. In some of the more complex contexts, an operations analyst is involved. Energy supply is also a risk factor to be accounted for in this context.

Visualization of the Abstract

In the context of information design, risk and uncertainty may be presented as mere figures, which, as a form of information, is entirely abstract. The question arises as to whether such figures are comprehensible to the decision-maker, be they expert, politician, or future house owner. Or would the decision-maker benefit from some sort of visualization that places the risk and uncertainty information in context? If information is presented in some sort of context, it becomes more concrete and relational.

If future consequences were related to present conditions, there would be better understanding of how a place or situation needs to adapt to a changed climate, how different strategies for nature preservation will play out, how alternative decisions on renewable energy resources will affect the environment, and what the consequences of various emission reductions would be. Likewise, an overview of emerging risks upon the construction of roads, railways, airports, or other infrastructure and to displays of risk in traffic, logistics, or energy supplies can help make the planning processes and the assessment of alternatives more comprehensible.

The greatest challenge, however, is to give a reasonable account of the unknown, which is a factor in many predictions (Hullman, 2019). For instance, when adapting spaces and environments to climate change, the exact weather in coming years is almost impossible to forecast; therefore, we must rely on possible scenarios. Likewise, economic and political change can present unknown factors that are hard to foresee. Yet, we want to be prepared to deal with natural disasters, pandemics, and technical breakdowns in terms of minimizing risk and prioritizing resources. Much societal information, particularly when it comes to crises, employs visualizations of possible future risks.

The following section will use disasters to illustrate this theme. For instance, Geographical Information Systems (GIS) are used to visualize risk zones for flooding, earthquakes, and wild-fires. Animations are applied to simulate courses of disaster. In addition, sometimes, as consequences of natural disasters, yet other disasters follow. Environmental disasters are visualized using satellite images and interactive maps to display the spread of pollution from, for example, oil spills or chemical leakages. Food supply crises are monitored using interactive visualizations of supply chains, where risk areas and threats are indicated. Energy supply crises are visualized to represent energy flows and capacity utilization, with accompanying scenarios for power outages and their consequences.

The problem is that the uncertainties, including the probability and its upper and lower limits, are hard to convey graphically and are not always well understood by decision-makers (Andreasson and Riviero, 2014). What is not clearly understood causes confusion and a vague perception of risk, and, therefore, lessens confidence in decision-making. Skills and experience are required to interpret currently used visual presentations (Williams et al., 2023), particularly when it comes to more difficult and extreme situations.

Yet, both risk and uncertainty must be assessed based on such visualizations. From that outset, information design also concerns the cognitive efforts involved in interpreting visualizations in their various forms. The chain of reasoning (seizing event character—judging its possible impact—understanding the risk—considering its uncertainty) needs clear and understandable information that is easy to process at each step and that is comprehensible when compared and related to each other.

In the following sections, I will elaborate on these challenges, using visualizations of flood risk as examples, since these are fairly easy to relate to and address both experts as well as public and private decision-makers. Most of us have seen media reports on severe flooding or have witnessed flooded areas—we may have even experienced flooding first-hand. Therefore, we understand the nature of the risks that are at stake when flooding occurs.

Flood risks obviously require the attention of experts such as civil planners, risk assessors, and rescue forces or public decision-makers, such as politicians and high officials, and others, such as home buyers, house builders, and real estate agents. For each position in a land area, it must be understood how likely it is that an occurring flood might spread in such a way that it covers that position, how severe that coverage is likely to be in terms of depth and streams of water, and also where it is most likely to come from, and, thus, what it might be carrying with it: e.g., garbage, debris, and items from up-stream, or harmful substances if there are obvious sources for anything alike. In addition, the time and cost of restoration must be acknowledged as part of the consequences, which might be vast and hard to predict, especially for non-experts.

Worldwide, flooding is the most common natural disaster. Between the years 2000–2014, there were at least 85,000 flood-related fatalities, with floods impacting around 1.4 billion people and resulting in a total cost equivalent to USD 400 billion (de Brito and Evers, 2016). In addition, decision-making regarding flood risk management is a rather complex and uncertain matter since it requires the involvement of multiple stakeholders who have different views, knowledge, and competing objectives. What is more, the size of a flood will be unknown as will its impact.

Global climate change is expected to result in more extreme weather (Olsson and Foster, 2013). This will increase the likelihood of damage to property and infrastructure and a general increase in societal vulnerability. Thus, the work of estimating and assessing flood risk has an impact on society, from the instance of an individual citizen's housing to large-scale considerations that come with, for example, urbanization, hospital location, or road placement. The EU estimates that the cost of climate change adaptation in Europe will hit billions of Euros (European Environment Agency, 2023) or, in terms of one specific nation, 14–21 billions of Euros in Sweden alone (Government of Sweden, 2018).

Flooding will not only lead to a loss in human and animal life, but will also result in major damage to private and public property and infrastructure, such as roads, bridges, and railways, which will require extensive restoration. Houses and other buildings may collapse, be left in ruins, or, at the very minimum, require repairs. Fresh-water supplies may be contaminated by drain runoff, chemicals, and other unhealthy substances, which would cause disease and sanitation problems. Agriculture may suffer due to ruined crops and loss of livestock, which would affect the local food supply. All of these result in economic costs at both the private and public level and would mean immense expenses for insurance companies. Indeed, the breakdown in and loss of production may be huge. An expert can readily understand and calculate the financial cost of such damage; however, for the private homeowner, it would be beyond their comprehension: how much time and money would be needed to restore land, mend a road, and repair a building and other property—and the private homeowner may question whether they would manage during the time required to do so?

The long-term effects of floods extend from hazardous substance contamination and damage to eco systems and habitats for animals and plants to migration and other large-scale transformations of society with socioeconomic effects and conflicts as possible aftermaths. These are large-scale societal effects that specialists at public authorities must understand.

Challenges in Flood Visualizations

To appreciate how visual information is understood and used, I will reference research on visual perception and cognition. How we cognitively digest visual information is decisive for how we understand and use the information in terms of both how we process it and how we assess it. Saliency, for instance, is decisive for what information is extracted quickly and is highly likely to be considered—as is oddity. Vague and relational information, meanwhile, is slower to process and requires committed attention and thought.

The characteristic of visuals of various kinds is saliency in the sense that visual information is distinct and difficult to express sweepingly or vaguely. For example, the challenge of presenting possible future floods on a map is that lines or surfaces that mark areas appear concrete and are not perceived as eventualities (Cook et al., 2019); instead they are perceived as real (Eriksson, 2017). The span between low risk—low uncertainty and high risk—high uncertainty is then not understood to be a continuum that requires a deep analysis of several variables, where the momentary visualization presents only a limited set of options of all possible outcomes. It is rather understood as distinctly presenting the existing alternatives: seemingly, either low risk, which we can be sure of, or high risk, that is very uncertain. In such a case, there is a huge misunderstanding.

In visualizations of uncertainty from flood modeling, graphs and diagrams are commonly used to show changes, errors, and variations in, for example, flood stages, discharge of water, flood depths, and inundation sizes (Brandt et al., 2021). Uncertainty can also be embedded in flood maps as part of the information to provide geographic context to the areas that may be affected by flooding. The most common visual variables for representing uncertainty in flood maps are color and value, which indicate different statuses of flooding (Figures 9 and 10).

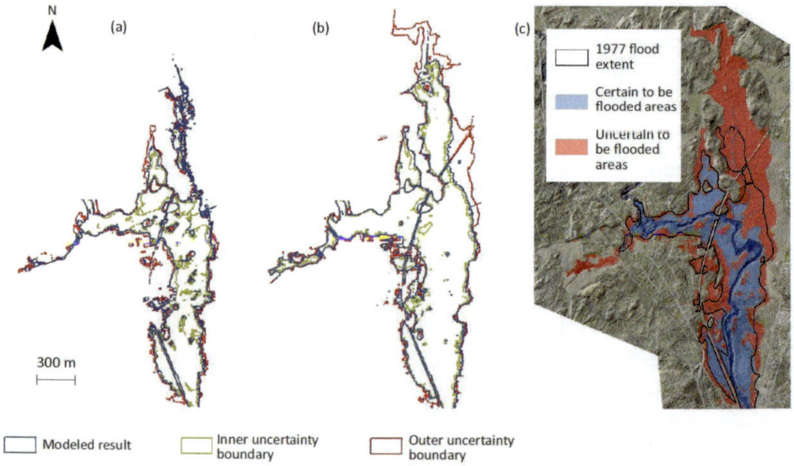

Fig. 9 Maps with uncertainty zones regarding flooding of a river section, (a) and (b), and (c) resulting in increased uncertainty zone (resilience zone) on a structured map (from Brandt et al., 2021, reprinted with permission).

Different semantic expressions, such as symbols, signs, and acronyms, are also commonly applied, along with visual variables, such as colors, patterns, and structures, to communicate the information of the maps (e.g., Horritt, 2006). Uncertainty, on the other hand, is often represented by using either a mixture of the colors in the middle of the scheme (Figure 10, Seipel and Lim, 2017) or two colors to represent the certain and uncertain conditions (Figure 9, Brandt et al., 2021; Aerts et al., 2003). Yet, several other kinds of visual flood representations exist.

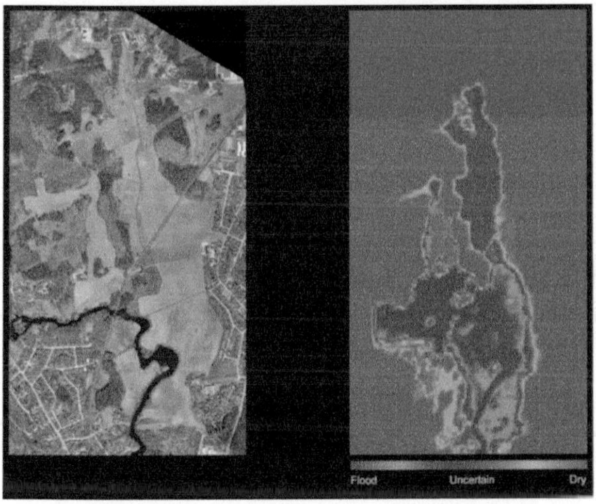

Fig. 10 Flood map visualization with an orthophoto of a river landscape (left) for geographic reference and a color map to express both expected risk and uncertainty of flooding (right) (from Seipel and Lim, 2017, reprinted with permission).

Flood level may also be represented as a profile in a simplified illustration of familiar examples of structures or objects (Figure 11). The flood can also be visualized as a 3D graphic with an angle to provide perspective on a full landscape or situation (Figure 12).

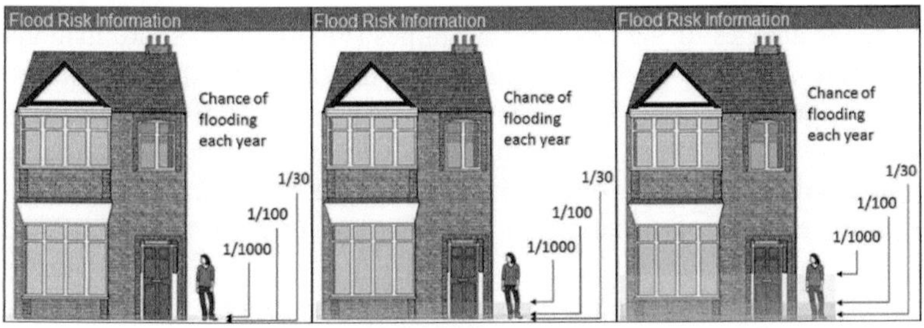

Fig. 11 From Dobson et al., 2018: Graphic illustration of flood levels with familiar structures and objects (Reprinted with permission).

Fig. 12 3D graphic with an angle onto a flooded landscape (Waser, N.D., reprinted with permission).

Yet another alternative is to present flooding as an augmented reality where an actual photograph of a specific location is used with transparent overlays of graphics to indicate the flood (Figure 13).

Fig. 13 Augmented reality with transparent overlays of graphics over a photograph (ArcGIS N.D., Copyright © Esri, Esri R&D Center Stuttgart, Esri Netherlands, Kavel10, Climate Effect Atlas, and Deltares. Reprinted with permission).

The advantages of the last three examples (Figures 11–13) are that anyone can imagine the scope of damages caused by the flood and thereby foresee the large costs and time required for restoration. Still, even though these less abstract visualizations portray possible flood risk, all fail to convey uncertainties, the complexity of variables from different factors, and the interrelation of these factors. And all three forms of visualization appear as concrete and actual instances of flooding.

Contemporary visualization tools allow for many different forms of data and information visualization, and depending on what visual design is used in each case, a particular visualization can be rather deceptive. The data used in a flood risk visualization is modeled based on scenarios where several variables are taken into account: First, the topography of the landscape is accounted for, but how fine-grained is the information? Is altitude above sea-level taken from contour lines on a map, with 5- or 10-meter altitude intervals in between, or are there satellite measures for each square meter in the whole area, and how exact and correct are the measures? Second, are ground conditions accounted for as only rough general types of vegetation, or is the information more detailed in terms of type and size of trees and bushes, and are stone blocks, levees, buildings, and other structures included, aspects that add friction to a flood? Third, how well is the variation in rainfall taken into account, estimated as general spread or calculated from varied intensity, with subsequent uncertainty regarding discharge and depth? Similarly, visualization can fail to account for fine-grained variation in a landscape, ground conditions, and a combination of uncertainty variables. Generally, uncertainty can stem already from the acquisition of data, arise in its transfer into a processing tool, or it can emerge in the reformation process into visual shape and form, which are all steps where data has to be seized and categorized in one way or another (Pang et al., 1997). Every such step includes either interpretation (possibility for misinterpretation) or adaptation to fit set categories (slight alteration of values).

The user needs to understand which data is uncertain and to what degree; they also need to understand how to draw conclusions from complex or contradictory information. Once again, there are several users and each differs in terms of their level of competence and their reason for using the visualizations. As such, their conclusions will differ. Nevertheless, faults in measurement or too rough data approximations may result in values that contradict what is visualized. Furthermore, the visualizations may also contain approximations and combined or clustered variables, which are easily cluttered and hard to understand as a complex—a complex that also contains a risk structure; this invites misperception.

Uncertainty visualization aims to make users aware of inaccuracies in acquired information, from the acquisition or transformation of data, by presenting such limitations visually to help improve data interpretation, assessment, and decision-making. This can be achieved as understated indications, emphasized information, or even overstated pointers pointing at existing uncertainties, in such ways that it is feasible for the user to include uncertainty in the analysis. Recurring techniques for such visualizations are, for example, glyphs (Figure 8), scatter-plots (Figure 4), coloration (Figure 7), geometrical properties (e.g., Figure 6), textures (Figure 9, right), or difference images (Figure 9). They differ in terms of their capacity to present discrete and/or continuous data well and also in terms of whether they are abstract or associated with human experience (Pang et al., 1997).

Data theorist Edward Tufte (1997) developed ideas about visualization to enhance the understanding of statistics, the primary goal being to make complex data comprehensible. This means showing the data and focusing on its core meaning and statistics to make it relevant through clarity, precision, and efficiency in the design. The visualization should be quick and easy to read and should use the least amount of space and the fewest graphical elements as possible (data density), while avoiding distortions, ambiguities, and unnecessary graphics. Every visual item must be in proportion to the represented data, and any visual variation must refer to data variation.

Furthermore, military researchers elaborated on principles for the use of visualizations in effective decision-making to avoid information overload or cluttering (Summers et al., 2005). Their suggestion is to present visual information in semi-transparent layers that are optional at the choice of the decision-maker, which works in digital media.

Yet, the issue of how to best visualize uncertainty remains unresolved, despite efforts to assess it. Data scientist Maria Riveiro (2007) assessed cognitive/perceptual stances in regard to different visualization methods. Color, contrast, texture, transparency, and blur have been evaluated cognitively as indicators of uncertainty as has the use of additional graphical elements, or text labels, in different positions. There are some positive results for texture, transparency, and lightness color scales (white to saturated), as well as for positional graphic indicators and text labels, if these are simple and data density is balanced.

Another research trend is toward empirical studies of particular visualization usages (Liu et al., 2014) in four major categories: empirical methodologies, with a general emphasis on visualizing research data; interactions, which focus user interest, selection, and analysis; frameworks, which provide libraries, toolkits, and possibilities for users to develop their own forms of visualizations; and applications, which are visualizations adapted for particular data and situations and/or users.

Other studies expose the challenges that come with the interpretation of uncertainty and risk when data is visualized for the purpose of decision-making. People tend to interpret data differently when it is presented in different visual forms. Each form has advantages and disadvantages and can represent some kind of idea or abstraction better than another form, simultaneously as it has its shortcomings in representing other kinds of ideas or abstractions. It is a matter of how well the form associates with the idea or abstraction it is supposed to represent (Tversky, 2019). But it also depends on how knowledgeable the user is at reading visualizations and understanding the abstract concepts the visualizations attempt to convey.

Visual Information for Risk Assessment in Decision-Making

Risk assessment in decision-making has been found depends on visual attention and whatever item that has happened to capture such attention. More salient information is likely to be adhered to and processed than less salient information, and to inform subsequent choices (Lohse, 1997). Therefore, properly chosen visualization techniques and forms of saliency can guide decision-makers toward making better decisions, particularly in the case of multi-criteria decision-making (MCDM) (Milutinovic et al., 2020). However, if there is too much and too varied information that is salient, decision-makers' attention may be drawn to information that is arbitrary in regard to the aims of their decisions (Orquin et al. 2018); meanwhile, other, more relevant information might be ignored (Glaze et al., 1992). A decision-maker may pay attention to any amount of information (Hullman, 2019)—some that is even inappropriate despite being trusted—however, salient: this can result in poor decisions (Fundel et al., 2019).

A particular scenario is when visual information is moving, such as it is in animations or VR/AR-applications. In such cases, fast movements attract the viewer's attention more so than slow movements (Pang et al., 1997). And from studies of video-gaming, we know that handling a great deal of complex visual information quickly requires cognitive effort and focus, as simultaneous, distractive information that is constantly present competes for attention (Cardoso and Carvalhais, 2014).

Further, when uncertainty is the core issue, one has to be careful when designing the visual information so that the critical information is not neglected in the presentation since an "anchoring effect" might occur when well-known data categories such as a mean or a median is also presented, which shifts attention from important extreme data values (Williams et al., 2023). Experts are found to be better at directing their attention to the information crucial for decision-making than novices are (Ibid). Consequently, time spent attending probabilities correlates with better-made decisions when the usefulness of the decision is considered (Harrison and Swarthout, 2019).

Complex decision-making, such as with regard to risk and uncertainty, needs to be managed by some sort of multi-criteria decision analysis (MCDA) (de Brito and Evers, 2016), which requires visualizations to convey such complex data in an adapted and appropriate way.

Focusing specifically on flood risk visualizations, different kinds of maps have long been used for this purpose, and it is known that users focus less attention on areas of little information and that map reading is a complex endeavor, and there are many individual approaches to it (Steinke, 1979). Again, in this more specific context, experts focus more attention on important areas of the map and garner more information from those than do lay persons, who view maps more randomly and who are somewhat guided by color (Fuchs et al., 2009).

Yet, some evidence shows that using *Gestalt* principles in the visual design of flood risk information is helpful in assessing flood risk information and in successive decision-making since such principles organize information details in ways that lead the eye in the search for what is important (Dobson et al., 2018).

Combining maps with uncertainty information in other graphical forms that convey more precise detail for the locations under scrutiny has been shown to be effective in decision-making since it "makes information processing safer and more reliable" (Lim et al., 2016). If, in addition, users know the purpose and consequences of their decisions, the implications of those decisions can be better understood (Ibid.).

As such, flood risk assessment, including uncertainties, is an intellectual endeavor that is mind-driven (thus, top-down perception dominant) and goal-directed. Attention is, therefore, strongly steered by the nature of the task (Betz et al., 2010; Hegarty, 2011). However, irrelevant information, be it salient, can also grab attention and cause disorientation when complex data is displayed. Therefore, due to working memory and attention limitations, it is cognitively costly to spend time and attention on subtracting irrelevant information from what is presented

(Hegarty, 2011). Some of those salient features even occur as distractors that draw attention away from the actual risk and uncertainty information (Sullivan et al., 2017). The critical situation occurs when decisions must be made quickly, because those decisions are biased by bottom-up perception, which is drawn primarily to saliency.

Further, the complexity of decision-making regarding flood risk encompasses several aspects, and Roth (2009) studied the relationship between the assessment of risk, the comprehension of flood maps, and the perceived assessment difficulty. He found that user confidence is affected by both domain expertise and map-use expertise, and that difficulty in risk assessment relates to map-use experience. Novices in the use of maps and risk assessment find risk assessment hard, whereas experts in risk assessment make good decisions yet find doing so difficult if they are not also experienced map users. Particularly precarious when it comes to interpretations is when decisions are made by experienced map readers who know little about hydrology since they are confident in map usage but fail to see the consequences included in the risk and thus draw hazardous conclusions.

Similarly, sociologist Jens Zinn (2019) concludes that non-experts in a field are generally more willing to take risks than experts are. However, this depends on whether the individual has previous experience of the consequences of the risk and also on what other factors must be negotiated in regard to the risk: for instance, costs or comfort. A house-owner might not want to build a wall between the house and a nearby creek for a number of reasons: it might be expensive; the creek has yet not had any severe floods; and a wall would limit the access to the creek for leisure. This may not be the case, however, if the house-owner has experience of previous flooding and therefore knows the time and cost of restoration.

Moreover, the predicted likelihood of floods does not always correspond with the actual incidence of floods, and currently used visualizations are not always valuable to and effective in the decision-making process in the sense that they convey comprehensible information. The combination of geographical data (e.g., a map) with overlays of risk and uncertainty data is hard for a user to assess and harder still for non-experts than for experts. Geospatial researchers Lim et al. (2016) and Seipel and Lim (2017) have shown that flood risk uncertainty representations affect spatial choices when uncertainty is visualized separately, yet connected to a particular position (location), with its own discernable pattern of how uncertainty is distributed for different scenarios (these researchers used color bars with different colors for certainty and uncertainty).

Furthermore, the added risk and uncertainty information, both explicitly and implicitly provided in the visualizations, may be ignored or even misunderstood depending on the decision-maker's educational or professional background. The familiarity with established forms of visualizations of risk and uncertainty is one key factor; another is the choice of language, where supporting terms such as "certainty", "confidence", and "best estimate" are to be preferred in decision-making over "uncertainty", to minimize misunderstanding (Beven et al., 2014).

These differences arise from the experts' ability to develop sophisticated mental schemas and visual patterns that guide their attention to the most pertinent areas of an image. The ability to distinguish between task-relevant and task-irrelevant information is critical to visual inspections. Experienced inspectors are better equipped to prioritize their attention toward structural elements crucial upon damage assessment. This is supported by theories of visual attention and cognitive load, which suggest that experts can manage their cognitive resources more effectively than novices. Research in visual attention indicates that experts can focus their attention on task-relevant information while ignoring distractions. Cognitive load theory further explains that experts have a higher working memory capacity for domain-specific tasks, allowing them to process relevant information more efficiently without becoming overwhelmed by irrelevant details (Saleem, Mayne, and Napolitano, 2024).

Thus, there is a wide range of aspects that differ among decision-makers, from very experienced experts to common people with no experience at all, who must still evaluate risk and understand its uncertainties. No matter what type of decision-maker, they may attend to inappropriate information, may be distracted by simultaneous moving information, may be unable to grasp the information that is crucial for the decision in making, may fail to assess the complexity of the data or may defocus their attention.

A proper visualization, on the other hand, can lead the eye in its search for what is important by limiting salient features to what is important from each decision perspective. In addition, the cognitive capacity of humans to process visual information is limited, so how visualization methods are used to support decision-making becomes a delicate issue. As such, there is a need both to make risk maps easy to interpret and to reduce the risk of bad decisions.

Future Visualizations of Risk and Uncertainty

Lastly, I will elaborate on how to find better forms of visualization and what design process to use for visualization of risk and its uncertainties, thus addressing decision-making that has to consider risk, including uncertainty. Some of these recommendations elaborate on the recommendations of Information visualization researcher Jessica Hullman et al. (2019).

As we noted earlier, contexts for risk vary a great deal. Furthermore conditions can be expected to differ greatly when risk and uncertainty are compared between different contexts; risk and uncertainty means other things in health care diagnoses or development of treatments than in assessments of risks of damage and injury when insurance companies set rates or create a risk profile. Risk is something different in logistics and transportation when supply chains, traffic flows, and accident models must be optimized for the flow of goods than it is in the energy and climate sectors where energy production and consumption need prognoses that include climate factors. Considering risks following the exploitation of limited resources, as well as pollution, with natural and environmental consequences for all of us, differs from risks in the construction sector where climate impacts buildings and other facilities.

A starting point for designing a visualization should be to consider the context in which it will be used, the point of interest in that context, the situations that belong there, and the potential specific needs. These are matters of empirical research–observations of the context and situations occurring there, possibly accompanied by interviews or other interactions with the people from the context.

- First, visualizations of risk and uncertainty may have a limited group of expert users or may aim to address a broad range of different kinds of users, including anyone from the general public. It is necessary both to know who the user of the visualization will be and their purpose—for example, decision-making—as well as to design the visualization while keeping in mind the least knowledgeable statistics user.
- Second, the aim of the design of the visualization should be to explain what the risk consists of, what uncertainty factors are included, and how uncertain the uncertainty is so that this information is possible to understand in its context and to use in decisions.
- Third, perceptual considerations must be made so that saliency is used in a balanced way, prioritizing first the crucial then the important information while diminishing the less important information and using *Gestalt* principles to guide the eye in the reading.
- Fourth, user-centered-design with testing should be applied so that visualizations are assessed during the design process that involve participants from the intended user group(s). User scenarios can be arranged with decision-making as close as possible to actual situations to provide context for the tests and include thorough understanding of the consequences involved in risks.
- Fifth, elicitation can be used to understand the extent to which the participating decision-makers are certain about their decisions and whether this certainty is well founded or not, whether there is an ambiguity involved in the certainty, and whether the consequences are well understood and considered.
- Sixth, users' interpretations of the visualizations should be captured in order to understand what meaning potentials there are in the specific designs.
- Seventh, the elicited responses can be validated with regard to the background of the participants: this includes previous knowledge, attitudes to risk-taking, experience in statistics and visual communication, comprehension of written language, and familiarity with the particular case contexts used in the test.

When testing and assessment are carried out, the combining of parts or aspects from different forms of visualization may be the best option as can finding solutions that visualize various probable situations to be compared. Hullman (2019) suggests "hypothetical outcome plots", that is, several possible outcomes animated in a fast order to avoid the possible neglect of alternative scenarios. This would make risks and uncertainty understandable to as many as possible.

Since current society has distributed its power and decision-making to many, with different degrees of influence and varied kinds of knowledge and competence (Beck, 1986), it is highly important that the communication of risk and its uncertainty is clearly understood by as many as possible so that these risks can be properly addressed. Better risk visualization is key to the improvement of this kind of communication.

References

Aerts, J.C.J.H., Clarke, K.C. and Keuper, A.D. 2003. Testing popular visualization techniques for representing model uncertainty. *Cartography and Geographic Information Science*, *30*(3): 249–261.

Andreasson, R. and Riveiro, M. 2014. Effects of visualizing missing data: an empirical evaluation. In: *2014 18th International Conference on Information Visualisation* (pp. 132–138). IEEE.

ArcGIS. N.D. Analyze flood impact. Retrieved 2025-11-07: https://doc.arcgis.com/en/3d/workflows/analysis/analyze-flood-impact.htm.

Beck, U. 1986. *Risikogesellschaft: Auf dem Weg in eine andere Moderne*. Suhrkamp Verlag, Frankfurt am Main.

Betz, T., Kietzmann, T.C., Wilming, N. and Koenig, P. 2010. Investigating task-dependent top-down effects on overt visual attention. *Journal of Vision*, *10*(3): 15.

Beven, K.J., Leedal, D.T. and McCarthy, S. 2014. *Framework for Assessing Uncertainty in Fluvial Flood Risk Mapping*. CIRIA Report C721/2014.

Brandt, S.A. et al. 2021. Mapping flood risk uncertainty zones in support of urban resilience planning. *Urban Planning*, *6*(3): 258–271.

Cardoso, P. and Carvalhais, M. 2014. Un-de-focused: variations of attention in player action in video games. *Computation Communication Aesthetics and X*: 169–183.

Cook, C.J., et al. 2019. Communicating outputs from risk assessment models: A picture paints a thousand words. *Microbial Risk Analysis*, *13*: 100084.

Datawrapper. 2022. In the climate crisis, vulnerable countries bear the least responsibility. Retrieved 2025-11-07: https://blog.datawrapper.de/climate-risk-readiness-responsibility/ .

de Brito, M.M and Evers, M. 2016. Multi-criteria decision-making for flood risk management: A survey of the current state of the art. *Natural Hazards and Earth System Sciences*, *16*(4): 1019–1033.

Dobson, B.A., Miles-Wilson, J.J., Gilchrist, I.D., Leslie, D.S. and Wagener, T. 2018. Effects of flood hazard visualization format on house purchasing decisions. *Urban Water Journal*, *15*(7): 671–681.

Dübel, S., Röhlig, M., Tominski, C. and Schumann, H. 2017. Visualizing 3D Terrain, Geo-Spatial Data and Uncertainty. *Informatics*, *4*(1), 6. https://doi.org/10.3390/informatics4010006.

Eriksson, Y. 2017. *Bildens tysta budskap: Interaktion mellan bild och text*. Studentlitteratur.

European Environment Agency. 2023. Assessing the costs and benefits of climate change adaptation [breifing]. Retrieved 2025-01-27: https://www.eea.europa.eu/publications/assesing-the-costs-and-benefits-of .

Fuchs, S., Spachinger, K., Dorner, W., Rochman, J. and Serrhini, K. 2009. Evaluating cartographic design in flood risk mapping. *Environmental Hazards*, *8*(1): 52–70.

Fundel, V.J., Fleischhut, N., Herzog, S.M., Göber, M. and Hagedorn, R. 2019. Promoting the use of probabilistic weather forecasts through a dialogue between scientists, developers and end-users. *Q. J. Roy. Meteor. Soc.*, *145*: 210–231. https://doi.org/10.1002/qj.3482.

Glaze, R., Steckel, J.H. and Winer, R.S. 1992. Locally rational decision making: the distracting effect of information on managerial performance. *Manag. Sci.*, *38*(2): 212–226.

Görtler, J., Schulz, C., Weiskopf, D. and Deussen, O. 2018. Bubble Treemaps for Uncertainty Visualization. *IEEE Transactions on Visualization and Computer Graphics*, *24*: 719–728. Retrieved 2025-11-07: http://graphics.uni-konstanz.de/publikationen/Goertler2018BubbleTreemapsUncertainty.

Harrison, G.W. and Swarthout, J.T. 2019. Eye-tracking and economic theories of choice under risk. *Journal of the Economic Science Association*, *5*(1): 26–37.

Hegarty, M. 2011. The cognitive science of visual-spatial displays: Implications for design. *Topics in Cognitive Science*, *3*(3): 446–474.

Horritt, M.S. 2006. A methodology for the validation of uncertain flood inundation models. *Journal of Hydrology*, *326*(1–4): 153–165.

Hullman, J. 2019. Confronting Unknowns. *Scientific American*, *321*(3): 80–83.

Hullman, J., Qiao, X., Correll, M., Kale, A. and Kay, M. 2019. In pursuit of error: A survey of uncertainty visualization evaluation. *IEEE Transactions on Visualization and Computer Graphics*, *25*(1): 903–913.

Kay, M. 2025. ggdist: Visualizations of Distributions and Uncertainty. R package version 3.3.3. Retrieved 2025-11-07: https://mjskay.github.io/ggdist/. DOI: 10.5281/zenodo.3879620.

Klever. 2015. Retrieved 2025-11-07: https://commons.wikimedia.org/wiki/File:Spindiagram.png.

Kubíček, P. and Šašinka, Č. 2011. Thematic uncertainty visualization usability: Comparison of basic methods. *Annals of GIS*, *17*(4), 253–263. https://doi.org/10.1080/19475683.2011.625978.

Lim, N.J, Brandt, S.A. and Seipel, S. 2016. Visualisation and evaluation of flood uncertainties based on ensemble modeling. *International Journal of Geographical Information Science*, *30*(2): 240–262.

Liu, S. et al. 2014. A survey on information visualization: recent advances and challenges. *The Visual Computer*, *30*(12): 1373–1393.

Lohse, G.L. 1997. Consumer eye movement patterns on yellow pages advertising. *Journal of Advertising*, *26*(1): 61–73.

Milutinović, G., Ahonen-Jonnarth, U. and Seipel, S. 2021. Does visual saliency affect decision-making? *Journal of Visualization*, *24*: 1267–1285.

Olsson, J. and Foster, K. 2013. Extrem korttidsnederbörd i klimatprojektioner för Sverige. *Klimatologi*, 6, SMHI.

Orquin J.L., Perkovic S. and Grunert K.G. 2018. Visual biases in decision making. *Appl. Econ. Perspect. Policy*, *40*(4): 523–553.

Pang, A.T., Witterbrink, C.M. and Lodha, S.K. 1997. Approaches to uncertainty visualization. *The Visual Computer*, *13*(8): 370–390.

Protonk. 2012. Retrieved 2025-11-07: https://commons.wikimedia.org/wiki/File:Boxplot_of_Average_Article_Feedback_ratings_by_project_rated_quality.svg.

Riveiro, M. 2007. Evaluation of uncertainty visualization techniques for information fusion. In: *10th International Conference on Information Fusion*.

Roth, R.E. 2009. The impact of user expertise on geographic risk assessment under uncertain conditions. *Cartography and Geographic Information Science*, *36*(1): 29–43.

Saleem, M.R., Mayne, R. and Napolitano, R. 2024. Evaluating Human Expert Knowledge in Damage Assessment Using Eye Tracking: A Disaster Case Study. *Buildings*, *14*(7): 2114. https://doi.org/10.3390/buildings14072114.

Seipel, S. and Lim, N.J. 2017. Color map design for visualization in flood risk assessment. *International Journal of Geographical Information Science*, *31*(11): 2286–2309.

Steinke, T.R. 1979. *An Evaluation of Map Design and Map Reading Using Eye Movement Recordings*. University of Kansas.

Sullivan, H.W., Boudewyns, V., O'Donoghue, A., Marshall, S. and Williams, P.A. 2017. Attention to and distraction from risk information in prescription drug advertising: An eye-tracking study. *Journal of Public Policy and Marketing*, *36*(2): 236–245.

Summers, V., Jones, R. and Flo, R. 2005. Increasing Situational Awareness by Visualizing Uncertainty. In: *Proceedings of the 10th International Command and Control Research and Technology Symposium*.

Swedish Government 2018. *Nationell strategi för klimatanpassning. Regeringens proposition* 2017/18:163. https://www.regeringen.se/contentassets/8c1f4fe980ec4fcb8448251acde6bd08/171816300_webb.pdf Retrieved 2025-02-12.

Tufte, E.R. 1997. *Visual Explanations: Images and Quantities, Evidence and Narrative*. Graphic Press.

Tversky, B.G. 2019. *Mind in Motion: How Action Shapes Thought*. Basic Books.

Waser, J.N.D. floodvisor by VRVis. https://www.vrvis.at/en/scenarify. Retrieved 2025-11-07: https://www.vrvis.at/en/research/research-projects/former-projects/floodvisor .

Williams, L., et al. 2023. Understanding representations of uncertainty, an eye-tracking study part II: The effect of expertise. *EGUsphere* [preprint]. https://doi.org/10.5194/egusphere-2022-929.

Zinn, J.O. 2019. The meaning of risk-taking: k Key concepts and dimensions. *Journal of Risk Research*, *22*(1): 1–15.

5 Why Text Matters in Data Visualization
Exploring the Text-Image Relationship

Wibke Weber[1*]

Introduction

Information design has traditionally been about producing information that is accessible to the intended audience and presenting it in an effective and efficient way, i.e., clear, user-friendly and attractive. An essential part of information design is data visualization. With the rapid advancement of digital technologies and big data, data visualization has gained significant popularity in a growing number of professional and academic fields. Graphs, charts and maps are used to show, explain, argue and tell stories; they help understand what is hidden in complex data and make it easier for audiences to understand patterns, trends and relationships within the data. The Covid-19 pandemic further demonstrated the societal relevance of data visualization. Data scientists analyzed massive amounts of data to model predictions about the course of the pandemic. Policymakers and government agencies relied on data cockpits and dashboards to decide what actions to take during the pandemic. News channels used data visualizations to communicate a data-driven perspective and encourage informed decision-making and critical discourse. Scientists leveraged visualizations to anchor data-driven insights. In short, data visualizations are being used more than ever because they have the power to get the message across more effectively. They have become an indispensable form of communication in our society (Kennedy and Engebretsen, 2020).

From a social semiotic perspective, data visualization can be defined as a multimodal "super-genre"[#] (Engebretsen and Weber, 2018) that encompasses several sub-genres such as diverse types of graphs, maps, dashboards and text. As a combination of text and image—sound is rarely found in data visualization—it brings together various semiotic resources such as points, lines, arrows, colors,

[#] In social semiotics, genre is defined as a set of conventionalized social practices and actions that are used to achieve a specific communicative goal. For example, a text is characterized by certain features and patterns that it shares with other texts regarding content, form, style and social function (Gnach et al., 2022).

[1] PhD | Professor of Media Linguistics: Visual Communication, Program Director, BA Communication and Media Studies, ZHAW Zurich University of Applied Sciences, School of Applied Linguistics.

* Corresponding author: webw@zhaw.ch

DOI: 10.1201/9781032642598-5

words, phrases and sentences, icons, typography and spatial arrangement. To translate data into information, the French cartographer and semiologist Jacques Bertin (1983) defined a core set of visual variables for graphs and maps. His graphic system consists of eight variables: "A visible mark expressing a pertinent correspondence can vary in relation to the TWO DIMENSIONS OF THE PLANE. It can further vary in SIZE, VALUE, TEXTURE, COLOR, ORIENTATION and SHAPE. Within the plane, this mark can represent a point (a position without area), a line (a linear position without area), or an area." (Bertin, 1983) Over the years, researchers (e.g., Andrienko and Andrienko, 2006; Carpendale, 2003; MacEachren, 1995) have extended Bertin's graphic sign system to include additional variables that enhance the visual representation of the data, such as crispness, resolution, transparency, motion and height (in 3D visualizations), as well as non-visual variables such as sound or haptic elements.

Bertin took a semiological approach in order to define the graphic system. His point of departure for his book *Semiology of Graphics: Diagrams, Networks, Maps* (1983) was the verbal language in data visualizations. He assumed that text would catch the reader's eye more than the graphic language: "Since time is of the essence, the reader simply ignores the drawings and turns to the text for the simplification necessary for comprehension and communication" (Bertin, 1983). Since then, many scholars and practitioners have focused primarily on the visuality and design space of data visualizations and text has tended to take a back seat in both research and practical guides. This chapter returns to the verbal language in data visualization. Taking a social semiotic approach, the chapter explores the verbal-visual interplay in data visualization. As recent studies have shown (Hearst, 2023; Kim, 2021; Lundgard and Satyanarayan, 2022; Sorapure, 2019; Zhu et al., 2022) text has its place in data visualization because it plays an important role in interpreting both the meaning of the data and the message conveyed by the visual elements. In this chapter, the focus is on the basic textual elements used in data visualization, i.e., headings, labels, legends, captions, tooltips, annotations, credits and other text elements. These microtexts can provide context, explanations, or specifications for reading the visual representation; they can also frame the message of a visualization by suggesting a particular interpretation. Two general questions will guide this chapter:

(i) What types of text are used in data visualizations?
(ii) What semiotic work can text do in data visualization?

This chapter contributes to the field of information design by providing insights into the different types of text used in data visualizations, the tasks that text performs and the text-image relations within data visualizations. The famous graphic of the Russian campaign designed by Charles Joseph Minard in 1869 will serve as a case study. This graphic has often been debated among scholars (e.g., Friendly, 2002; Kraak, 2014, 2021; Wainer, 1997), with discussions focusing on whether and if so, why it is considered "the best statistical graphic ever drawn" (Tufte, 2001). The chapter is structured as follows: Beginning with the definition of image-centricity, Section 2 introduces key terms and taxonomies related to text-image relations in multimodal artifacts within the broader context of social semiotics. Barthes' seminal work "Image–Music–Text" (1964/1977) will serve as a reference point here. It introduces three basic forms of text-image relations: illustration, anchorage and relay. Section 3 reviews relevant articles and studies on text in data visualizations. Section 4 describes four different levels of text types in data visualizations and the various functions of text. *Section 5* demonstrates how text appears in data visualizations, using Minard's famous map of Napoleon's Russian campaign as a case study. The chapter also explores how text can frame the message of a data visualization. Finally, Section 6 discusses the potentials and challenges of using text in data visualizations as a resource for informed citizenship.

Methodologically, the chapter includes: (i) a literature review on text-image relations and the role of text in data visualization, and, based on these insights, (ii) a text analysis of Minard's map, incorporating a frame analysis. Additionally, the chapter draws on findings from several research projects on data visualization and infographics in which the author has been involved over the past 15 years (e.g., Engebretsen and Weber, 2018; Weber, 2020, 2019 a, b, 2017; Weber, Burmester and Tille, 2013; Weber, Engebretsen and Kennedy, 2018).

Image-Centricity and Image-Text Relations

The advent of new, visually oriented digital media at the turn of the 21st century has placed visual representations at the forefront of communication. This shift from traditional text-based forms of communication to a visual culture is referred to as the visual turn, also iconic or pictorial turn (Boehm, 1994; Mitchell, 1994). The visual turn acknowledges the growing significance of images[#], visual storytelling and visual literacy. Kress (2003) describes this turn as "a move from telling the world to showing the world". Caple and Knox (2015) state that "images dominate the verbal story space and have the potential to become the story themselves". In the course of the visual turn, a concept has emerged that Stöckl, Caple and Pflaeging (2020) termed "image-centricity". Image-centricity is anchored in multimodality theory which is a broader approach to communication that extends beyond the domain of language and verbal text to include a range of non-linguistic modes, such as images, sounds and gestures—essentially, all the semiotic resources that contribute to the process of meaning-making (Gnach et al., 2022). The semiotic interrelationship of these modes is one area of research in multimodality (e.g., Bateman, 2014; Royce, 2016; Stöckl, 2017; van Leeuwen, 2005). Within this framework of multimodality, the concept of image-centricity suggests that images become the dominant superordinate mode of expression in a multimodal artifact in terms of semantic and functional weight (Engebretsen, 2020; Stöckl, Caple and Pflaeging, 2020), with text assuming the subordinate mode. Due to its perceptual

[#] The term image encompasses a range of visual representations including photographs, illustrations, visualizations, graphs, charts, maps, comics, animations, etc.

dominance, the image serves as an entry point for the reading path and meaning construction (Engebretsen, 2020). Examples of genres that reflect image-centricity include advertising (Bünzli and Eppler, 2024; Stöckl and Pflaeging, 2022), news reporting (Stöckl, 2020), social media (Caple, 2020) and data visualization (Engebretsen, 2020).

Drawing on Barthes (1964/1977), image-centricity is the result of a text-image relation of "anchorage", where text anchors the meaning of the image. Barthes describes three semantic text-image-relations: illustration, anchorage and relay. The term *illustration* refers to the text-image relation in which an image supports the text. In this case, the text is primary (text-centricity) and the image is subordinate, visually presenting information that is largely contained in the text. When text supports an image, this is called *anchorage*, as in the case of captions or titles that guide viewers on which elements of the image to focus on. The underlying assumption is that images are polysemous and "too open to a variety of possible meanings" (Kress and van Leeuwen, 1996) and that verbal text can potentially direct, or restrict, the interpretation of an image (Cheema et al., 2023). As Barthes (1964/1977) puts it, "language clearly has a function of elucidation"; it is "a control, bearing a responsibility—in the face of the projective power of pictures—for the use of the message". In this case, the image becomes the superordinate mode in a multimodal artifact and this is image-centricity. The third relation between image and text is *relay*. In relay, text and image have an equal, non-hierarchical relationship where text and image complement and enhance each other. Although Barthes makes a clear distinction between anchorage and relay, he also says that the two functions can co-exist, "but the dominance of the one or the other is of consequence for the general economy of a work" (Barthes 1964/1977/).

In social semiotic multimodality theory, Barthes' concept of anchorage and relay is reflected in the notion of information linking (van Leeuwen, 2005;

Fig. 1 Elaboration: A Labrador stands alert among the trees in a forest. (Pexels: Thijs van der Weide, https://www.pexels.com/photo/labrador-998251/)

Stöckl and Pflaeging, 2022), which is a key aspect of multimodal cohesion in a multimodal artifact. Information linking refers to how modes are linked to form a coherent and meaningful whole (e.g., through temporal or causal links). It concerns types of relations similar to anchorage and relay, but van Leeuwen (2005) uses the terms *elaboration* and *extension* instead. The term *elaboration* (Figure 1), which corresponds to Barthes' concept of anchorage, refers to instances where one mode explains, illustrates, or restates another mode, or specifies it in greater detail by exemplifying it. In other words, elaboration can take the form of explanation, clarification, exemplification, or specification. Conversely, the term *extension* (Figure 2) refers to instances where "two items—one verbal, one visual—provide different, but semantically related information" (van Leeuwen, 2005). This occurs when one mode provides new information not present in the other, or when one mode enhances the other by qualifying it with temporal or spatial details, logical features such as cause and effect, result, reason, or rhetorical figures. Examples of extension are similarity, contrast (Figure 3), cause-and-effect-relationship, addition of new aspects, or temporal and spatial information. Extension corresponds to Barthes' concept of relay (Gnach et al., 2022; Stöckl and Pflaeging, 2022; van Leeuwen, 2005). Over the years, a wide range of taxonomies have been developed to address the relationship between text and image in a variety of fields, many of which refer back to Barthes (1964/1977) and Halliday (1985)[#]. For a good overview of studies dealing with text-image-relations see, for example, Martinec and Salway (2005), Bateman (2014), Stöckl, Caple and Pflaeging (2020) and Cheema et al. (2023).

Fig. 2 Extension: Trained as a rescue dog, Sam has saved three hikers in the past year. (Pexels: Thijs van der Weide, https://www.pexels.com/photo/labrador-998251/)

[#] The linguist M.A.K. Halliday examined the various logico-semantic relations that link clauses together to form clause complexes.

Fig. 3 Contrast as a specific form of extension: Dogs are playful creatures, always full of energy and ready for adventure. (Pexels: Torsten Dettlaff, https://www.pexels.com/photo/short-coated-brown-puppy-sleeping-on-brown-mat-97863/)

In summary, the text-image relation in image-centric artifacts is characterized as an unequal status relation of anchorage (elaboration), where text is subordinate and semantically anchors the image. But how can image-centricity, with a dominant visual mode, align with the concept of multimodality, which postulates a multimodal interplay in the meaning-making process where no single mode leads? Stöckl identifies two additional defining criteria for image-centricity: "(1) the image is large and dominant in the layout—its content and design have the power of arresting viewer attention, and (2) the largest part of the accompanying text directly relates to the image so that strong intersemiotic cohesive ties emerge" (Stöckl, 2020). Considering this conceptualization of image-centricity, data visualizations are defined as image-centric multimodal artifacts with the image being elaborated through text. How text performs in data visualization will be discussed in the next two sections.

Text in Data Visualization: Related Work

Data visualizations without text are rare and difficult to interpret (Sorapure, 2019). Text is essential for properly reading and understanding the data. Although often small and subordinate, text plays a critical role in shaping a viewer's interpretation of a visualization. Titles, captions and annotations can direct readers toward the intended meaning, thereby influencing how they interpret the data. Borkin et al. (2016) found that participants spent the most time focusing on the text in a visualization. Their study also showed that titles significantly impact what viewers take away from the visualization and what they remember about its message. In recent years, other researchers have begun exploring the relationship between text and images in data visualization (Engebretsen, 2020; Hearst, 2023; Kim, Setlur and Agrawala, 2021; Lundgard and Satyanarayan, 2022; Stokes et al., 2023; Zhu, Cheng and Wu, 2022). Based on studies (Stokes et al., 2023), Hearst (2023) argues that "the field of data visualization should treat textual content as a primary design element", emphasizing that text and visual elements should receive equal attention in the design process. She also advocates for more research on text in data visualization and "the need for better cognitive models that combine reading and understanding visualizations".

To promote accessible visualization through natural language[#] descriptions, Lundgard and Satyanarayan (2022) studied which type of text for describing visualizations is preferred by blind readers versus sighted readers. They developed a model that spans four levels of semantic content:

Level 1: enumerating visualization construction properties (e.g., marks and encodings);

Level 2: reporting statistical concepts and relationships (e.g., outliers and correlations);

Level 3: identifying perceptual and cognitive phenomena (e.g., explanations, complex trends, patterns);

Level 4: domain-specific insights (e.g. social and political context).

Their study revealed significant differences in the types of content preferred by the two groups of readers. While sighted readers ranked higher levels of semantic content (Levels 3 and 4) as more useful, blind readers considered Levels 2 and 3 to be most useful (Lundgard and Satyanarayan, 2022). The authors conclude that the divergent reader preferences for semantic content suggest that natural language should be viewed "not only as an interface for constructing and exploring visualizations", but also as "an interface into visualization, for *understanding* the semantic content they convey". Stokes et al. (2023) utilized the four-level framework developed by Lundgard and Satyanarayan (2022) for their study on the amount of text in data visualization, exploring the question: How much text is too much? The texts used as stimuli in univariate line charts ranged from no text (except for the axes) to a written paragraph with no visuals. A key finding was that "participants preferred the charts with the largest number of textual annotations over charts with fewer annotations or text alone" (Stokes et al., 2023).

Kim, Setlur and Agrawala (2021) also investigated text in line charts, namely, how captions affect the people's takeaways. They found that when both the graph and the caption highlight a prominent feature, readers treat the double-emphasized feature as the takeaway. However, when the chart and caption emphasize different features, readers are less likely to incorporate information from the caption into their takeaways. They also found "that external information that provides context, helps further convey the caption's message to the reader" (Kim, Setlur and Agrawala, 2021).

Engebretsen (2020) took a social semiotic approach to qualitatively analyze the role of text in data visualization. As a case study, he used two stand-alone graphics[†] published on the Norwegian news site "VG.no". He concludes that

[#] Natural language means ordinary human language people naturally use in everyday speech or writing (like English, Spanish, or Mandarin).

[†] A stand-alone graphic is a visualization that works independently to convey its message without requiring additional context or being embedded in a textual environment.

despite the strong image-centric focus where the text anchors the interpretation of the visualization, the text plays an important role in fulfilling the communicative purpose of the data visualization. He argues that in data visualizations, the abstract forms (e.g., circles, bars) may indicate internal proportions and relations between visual units, but they carry no ideational meaning until they are linked to a legend or similar explanatory co-text (Engebretsen, 2020). The verbal legend thus has the performative, defining function of establishing the semiotic relationship between form and content.

To sum up: The findings of the studies suggest that, although data visualizations function as image-centric artifacts with the visual as the dominant mode of representation, text is needed as a co-active verbal mode. The verbal mode—the various types of text and their functions in data visualization—will be the focus of the next section.

Text Types and Their Functions

Due to their different semiotic natures, the visual and verbal modes offer distinct affordances; that is, they differ in terms of their communicative potentials and limitations for meaning-making (Bezemer 2023; Kress and van Leeuwen, 1996). Therefore, they can be used "to do specific semiotic work" (Kress, 2010). The visual mode can convey concrete details, such as the surface of an object, but it cannot communicate how much it weighs or who owns it—that requires words or numbers. Text can express abstract ideas and arguments, while images create meaning through the positioning of elements, as well as through color, shape, size, etc. In a data visualization, geometric forms such as squares, triangles, or lines can easily show proportions, quantitative relationships, or trends over time, but text is needed to explain what these forms represent. Data visualizations often contain uncertainties, margins of error, or assumptions. Here, text can clarify these aspects for the audience and prevent misinterpretation. The verbal mode offers a range of affordances, such as labeling, specifying, identifying, emphasizing, highlighting, describing, explaining, elucidating, facilitating the visual narrative, interpreting and contextualizing—all of which are distinct types of verbal anchoring. In the following, the types of text in data visualizations will be presented according to the four semantic levels developed by Lundgard and Satyanarayan (2022).

At the first level, text types are encountered that refer to the fundamental properties of the data visualization, such as *axis titles, axis labels, axis scales, data labels a*nd *legends*. These types of text are "words and fragments anchored to quantifiers that provide context to the number within the visualization" (Setlur and Cogley, 2022). They help clarify, specify, or refine the meaning of the data. Without these qualifiers, data can become abstract or ambiguous, leaving viewers wondering what the bars or lines represent. *Axis titles* describe what the data show along the entire axes, e.g., sales, months; *axis labels* are the specific names or numbers for the categories or values plotted on the horizontal and vertical axes, e.g., January, February, March, etc. They give meaning to the tick marks[#] by describing what the ticks represent, such as units of measurement, categories, or time. *Data labels* specify the value of individual data points, for example, showing exact values in a chart. *Axis scales* determine the numerical range and intervals, i.e., how data is displayed on the axes, helping the viewer understand the relationships between variables. Without labels, the data points are meaningless and open to interpretation. *Legends* tell the audience what each visual element (e.g., color, shape, or size) corresponds to in terms of the data in a chart or map.

Headings (titles and subtitles) convey the main message, summarize the key insight or conclusion, or indicate what the reader should learn from the data. They generate interest and attract attention with the aim of drawing the reader into the data visualization. They also place the data in context. *Captions* are short pieces of written information, usually placed inside or below the chart. They provide context or explain the main point.

Data sources, authorship, credits, copyrights, time and *date stamps* and even the *logo* of the producer or sender can be characterized as "functional text" (Setlur and Cogley, 2022), i.e., they do not refer directly to the data displayed but are essential textual elements that should be included in any data visualization. *Footnotes* and *disclaimers* can explain uncertainties or data sources to substantiate the content of the visualization. In addition, interactive elements such as navigation, filters, or drop-down menus are also considered functional text elements.

At the second level, text types are encountered that refer to the dataset displayed in the visualization, specifically descriptive statistics and relational content, such as extrema, outliers, range, mean and standard deviation. Annotations and callouts help readers quickly grasp the main points and better understand the content by clarifying specific data, such as notable shifts in the data or the presence of outliers. These elements highlight key data points, trends, or anomalies. For example, a callout might emphasize a peak in exports during a particular period, with an arrow pointing to the relevant data point, e.g., "peak in sales". Tooltips serve a similar purpose but are intended for use in interactive data visualizations. They are useful for displaying the exact underlying values of data points. Tooltips appear when hovering over relevant data points, ideally accompanied by a short instruction such as "Hover over a circle to see details."

At the third level, text types are encountered that still refer to the data displayed in the visualization but now focus on "perceptual and cognitive phenomena appearing in the visual representation" (Lundgard and Satyanarayan, 2022), such as trends, changes in trends, patterns, comparisons, or gaps. In other words, readers are shown something more than what is presented at Level 2. For example, "Sales figures steadily increased throughout 2021, with a significant peak in Q4". Here, too, the explanatory function of text can be fulfilled by *annotations* and *tooltips*, as well as by longer texts such as *explanatory notes*. To ensure that

[#] Ticks are the small marks or lines along the axis. They represent specific values or intervals on the axis.

the audience fully understands the data visualization, text can clarify specific terms, variables, or metrics used in the visualization. *Explanatory notes* can break down complex visualization into digestible steps or point to specific interactive features, thus helping users navigate through complex, interactive data stories. *How-to-read texts* guide readers on how to correctly interpret the data and prevent misinterpretation or incorrect assumptions, especially for those unfamiliar with the visualization. Other types of explanatory text include *longer captions* and *parallax scrolling text*.

At the fourth level, text types are encountered that refer to the context and domain-specific knowledge that is not visible in the visualization. Text types at this level no longer anchor the meaning of the data visualization, but extend it, i.e., the text mode provides additional information that is not present in the visual mode. This extension corresponds to Barthes' text-image relation of relay. Extension does not imply questioning the image-centricity of a data visualization overall, but at Level 4, the textual and visual modes begin to establish an equal, non-hierarchical relationship, complementing and enhancing each other. Text types at this level include, for instance, *comments* or *introductions* that provide the reason for or relevance of the visualization (Kirk, 2016), or context about related past or current events (e.g., social, political, environmental, cultural, economic). Comments can be expressed in the title, as Simon Scarr did in his iconic graphic "Iraq's Bloody Toll", published in the *"South China Morning Post"* (Scarr, 2011; Scarr, n.d.). The data visualization depicts the death toll of the Iraq war from 2003 to 2011. The graphic is called iconic because Scarr used the visual metaphor of blood to visualize the data, which can be interpreted as a comment on the war. Scarr explains, "One deliberate design choice with this graphic was the visual metaphor of blood. This striking visual would hopefully draw the reader into the graphic." (Scarr, n.d.) The effect of the visualization in the shape of a pool of blood makes the graphic more emotional and dramatic and less objective or even neutral. The provoking title "Iraq's Bloody Toll" anchors the interpretation of the visualization and frames the interpretation of the graphic in a negative way. Titles, as well as other types of text that stand out in a graph or map, therefore have the potential to frame the message of a visualization and make it appear biased. *Methodological explanations* and *How-we-did-it stories* provide insight into data integrity, how and which data was collected or analyzed. In contrast to the visual mode, the textual mode has the communicative potential or affordance to foster transparency, which, in turn, enables the audience to assess the trustworthiness and credibility of the data visualization. Referring to van Leeuwen (2005), Engebretsen suggests "that such elements of background information serve the function of 'legitimizing', that is, attaching the social practice of DV production to the context in which it appears" (Engebretsen, 2020).

Last but not least, the *alt-text descriptions*[#] of digital data visualizations should be mentioned here as another type of text that serves the function of ensuring access

[#] Alt-text: short for alternative text.

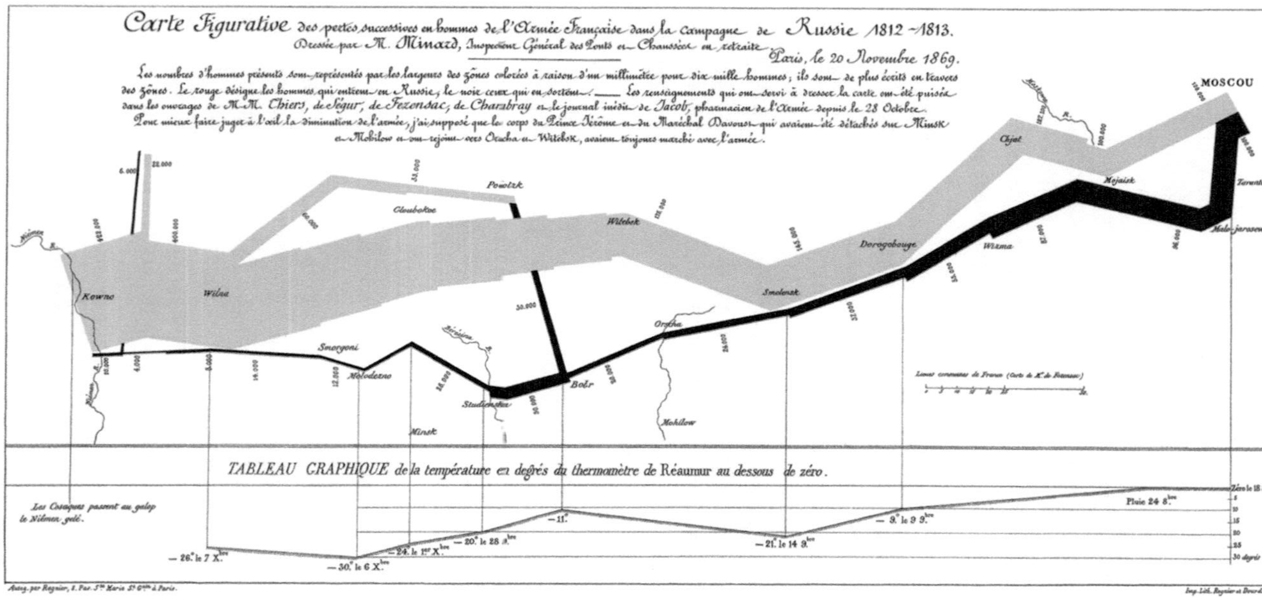

Fig. 4 Minard's map of the Russian campaign. Source: Public domain taken from https://en.wikipedia.org/wiki/File:Minard.png.

Note: Title and introductory text in English (Tufte, 2001): Figurative Map of the successive losses in men of the French Army during the Russian campaign 1812–1813. Drawn up by M. Minard, Inspector General of Bridges and Roads in retirement. Paris, 20 November 1869. The numbers of men present are represented by the widths of the colored zones at a rate of one millimeter for every ten thousand men; they are further written across the zones. The red [here rendered as gray, W.W.] designates the men who enter into Russia, the black those who leave it.—The information which has served to draw up the map has been extracted from the works of M. M. Thiers, of Ségur, of Fezensac, of Chambray and the unpublished diary of Jacob, pharmacist of the army since October 28th. In order to better judge with the eye the diminution of the army, I have assumed that the troops of Prince Jérôme and of Marshal Davout who had been detached at Minsk and Mogilev and have rejoined around Orcha and Vitebsk, had always marched with the army.

to the data for visually impaired people. As Lundgard and Satyanarayan (2022) have shown, *alt text* encompasses all four levels of semantic content.

In summary, the affordances of text in data visualization are manifold, ranging from axis labeling that provides meaning to data points, to placing the data in a broader context to emphasize its social relevance. Text is therefore a critical element in data visualization, ensuring that the content is meaningful and that readers, especially those unfamiliar with data visualization, can fully comprehend it and better understand what the visualization represents. The next section illustrates the semiotic role of text in Minard's historical data visualization of Napoleon's Russian campaign.

The Semiotic Work That Text Can Do: A Case Study

The text types described in Section 4 are not novel; they have already been used in historical data visualizations. The iconic map of Napoleon's Russian campaign in 1812, created by Minard in 1869 (Figure 4), serves as a case study for the analysis of text integrated into a data visualization. The map was printed as a small poster using two-colored lithography (Tufte, 2001). It illustrates the story of the Grande Armée's losses on its march to Moscow, represented by a thick brown line and its retreat, represented by a thin black line. As a stand-alone data visualization, it displays multivariate data: the size of the French army, the direction of its movement, spatial and temporal data and the temperature. The title of the map "Carte Figurative" clearly points to image-centricity, with the visual mode dominating. However, text matters as we will see below.

Titles: The main title "Figurative Map of the successive losses in men of the French Army during the Russian campaign 1812–1813" is quite long, but summarizes exactly what the map is about (Figure 5, highlighted with a gray frame). The phrase "successive losses" and years 1812–1813 point to the narrative nature of the visualization: a beginning, an end and a change (successive losses) in between. It is a flow map, also called a thematic statistical graphic, that shows how the size and position of the army changed during the Russian campaign. This combination of space, temporal progression and quantitative data allows Minard's map to function as a narrative visualization. It recounts a historical event—the dramatic losses of the French army over time—represented by a steadily shrinking line. Tufte (2001) referred to it as a "space-time-story graphic". Linked to the march is a second visualization, a temperature graph titled "Graphic table of the temperature in degrees of the Réaumur thermometer below zero." The title specifies the measurement scale used: the Réaumur scale, starting at zero degrees.

Credits: The indented text below the title provides information about the authorship, as well as the date and place of publication (Figure 6): "Drawn up by M. Minard, Inspector General of Bridges and Roads in retirement. Paris, 20 November 1869." It specifies the author's status and expertise, namely, that Monsieur Minard is retired and an expert in civil engineering, which helps the reader place him in a specific professional and temporal context. At the bottom left and right of the map,

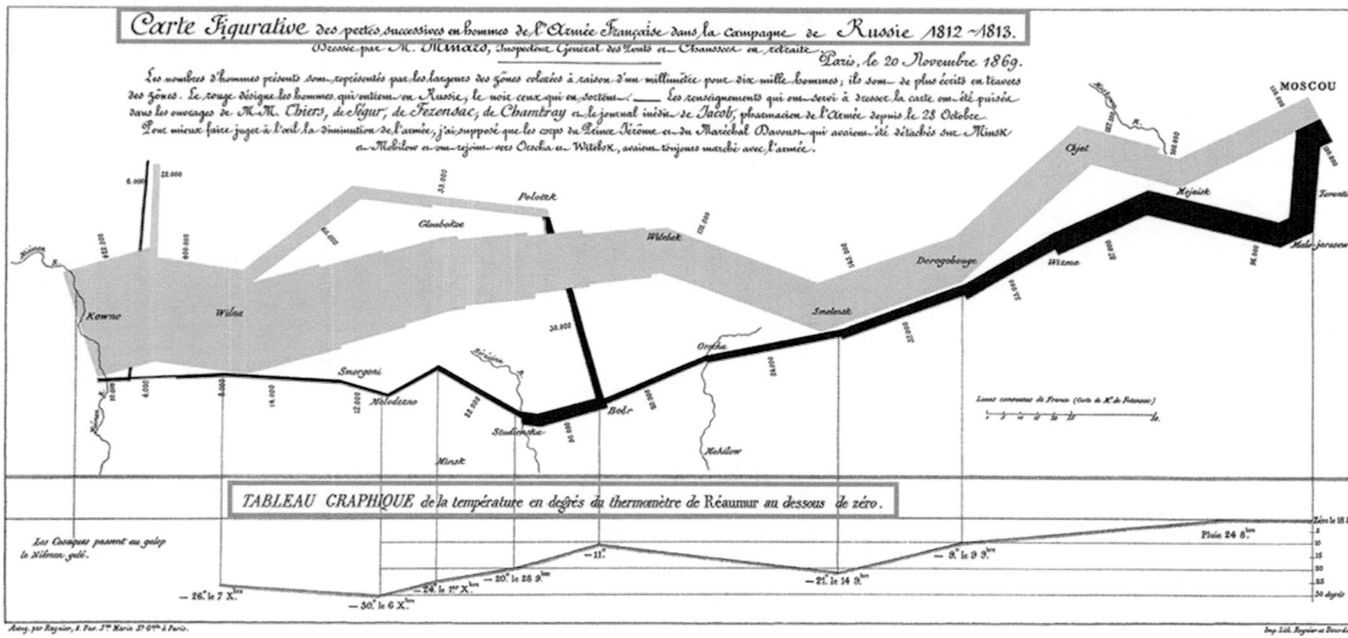

Fig. 5 The original title of the map is "Carte Figurative des pertes successives en hommes de l'Armée Française dans la campagne de Russie 1812–1813". The title of the temperature graph reads: "TABLEAU GRAPHIQUE de la température en degrés du thermomètre de Réaumur au dessous de zero". Both titles are highlighted with gray frames.

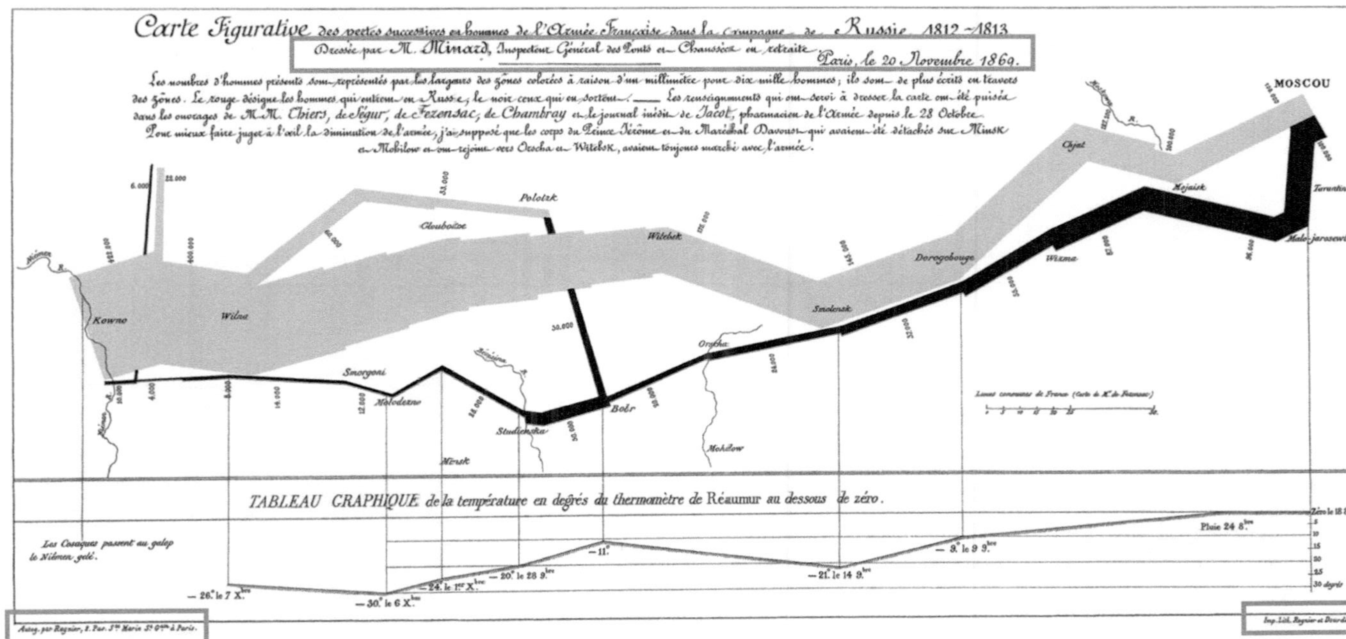

Fig. 6 The French credits, highlighted with gray frames, read: Dressée par M. Minard, Inspecteur Général des Ponts et Chaussées en retraite. Paris, le 20 Novembre 1869. The text in the lower corners indicates the name of the publisher and printer "Regnier et Dourdet" (right) and the address "8 Passage Ste. Marie St. Germain in Paris" (left).

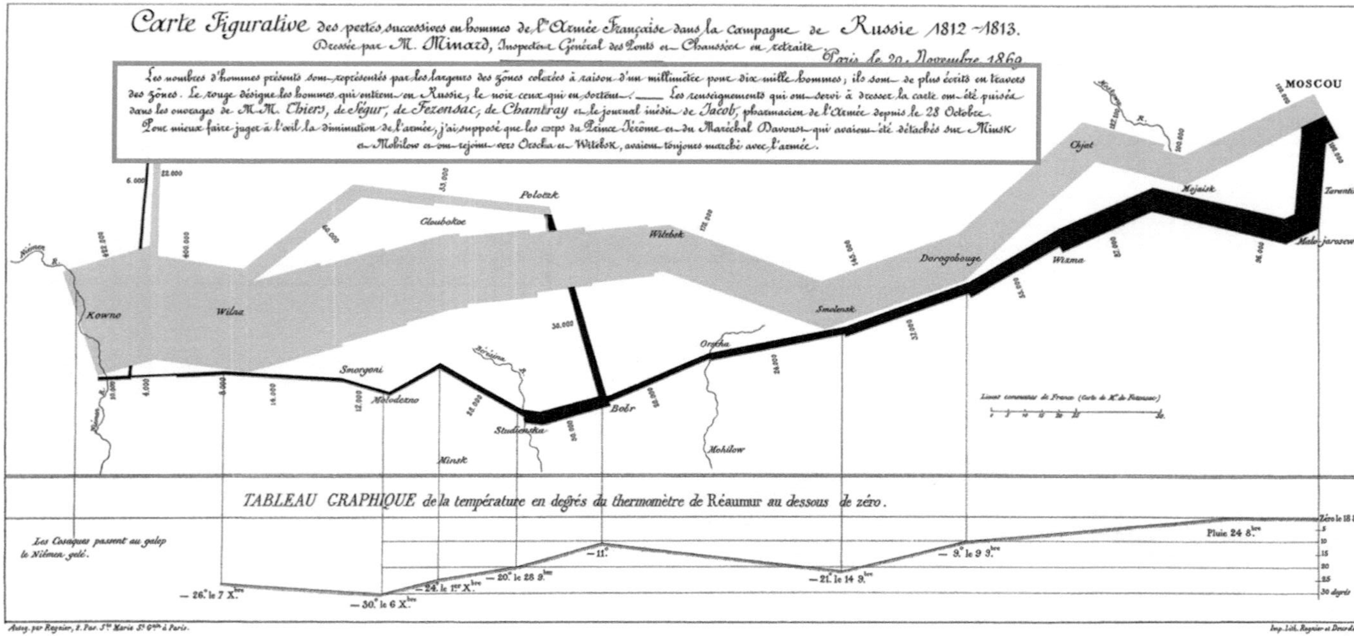

Fig. 7 The original introductory text, highlighted with gray frames, reads: "Les nombres présents sont représentés par les largeurs des zones colorées à raison d'un millimètre pour dix mille hommes; ils sont de plus écrits en travers des zones. Le rouge désigne les hommes qui entrent en Russie, le noir ceux qui en sortent. ___ Les renseignements qui ont servi à dresser la carte ont été puisés dans les ouvrages de MM. Thiers, de Ségur, de Fezensac, de Chambray et le journal inédit de Jacob, pharmacien de l'armée depuis le 28 octobre. Pour mieux faire juger à l'œil la diminution de l'armée, j'ai supposé que les corps du Prince Jérôme et du Maréchal Davout qui avaient été détachés sur Minsk et Mahilyow et ont rejoint vers Orsha et Witebsk, avaient toujours marché avec l'armée."

he also mentioned the publisher and printer in smaller type: "Regnier et Dourdet", a French printing, lithography and publishing firm in Paris. The credits thus serve to establish authority, provide historical context and lend credibility to the data represented in the map.

Introduction: The introductory text (Figure 7) contains different types of text beginning with a kind of legend that reads: "The numbers of men present are represented by the widths of the colored zones at a rate of one millimeter for every ten-thousand men; they are further written across the zones. The red [here rendered as gray W.W.] designates the men who enter into Russia, the black those who leave it." This helps the reader to correctly interpret the colors and widths of the lines. In addition, the text provides guidance on how to read the map by explaining the direction of the army's movement, which is necessary since Minard did not use arrowheads, a typical element in flow maps (Kraak, 2014).

Further, Minard makes his sources transparent: "The information which has served to draw up the map has been extracted from the works of M. M. Thiers, of Ségur, of Fezensac, of Chambray and the unpublished diary of Jacob, pharmacist of the army since October 28th." The next sentence is information about Minard's methodology. Minard comments on his design decision in favor of a simpler, more readable map: "In order to better judge with the eye the diminution of the army, I have assumed that the troops of Prince Jérôme and of Marshal Davout who had been detached at Minsk and Mogilev and have rejoined around Orcha and Vitebsk, had always marched with the army."

Data labels: Looking at the essential properties of the map, we find both the geographic data (places and rivers) and the quantitative data (the size of the army in absolute numbers at selected locations). For example, as highlighted in Figure 8, the Neman River on the left, where the Russian campaign began with 422,000 men; Moscow on the right, with only 100,000 men remaining; the crossing of the Berezina River in the center of the map, which was a disaster for the army (28,000 men); and back to the Neman River with only 10,000 men. The place names along the campaign route help orient the reader geographically. Although the size of the army is represented by the width of the line, the text provides exact numbers to emphasize the devastating losses. A scale bar in the lower right corner of the map indicates that distances are measured using the common French league ("lieues communes de France", which is about 4,4 km). Time becomes visible through selected daily data, as shown in the embedded temperature graph. The graph illustrates the drop in temperature (in Réaumur) on certain days during the army's retreat from Moscow and is linked to the quantitative and geographic data represented by the black line. For example, when the Grande Armée crossed the Berezina on November 28, the temperature was minus 20 degrees Réaumur, which corresponds to minus 25 degrees Celsius (Kraak, 2014). The temporal data provides a sequential structure that helps the reader to understand the chronology and the distribution of losses during the retreat.

Annotations: The map also includes annotations in the left and right corners of the temperature diagram (Figure 9). The left annotations state that the Cossacks

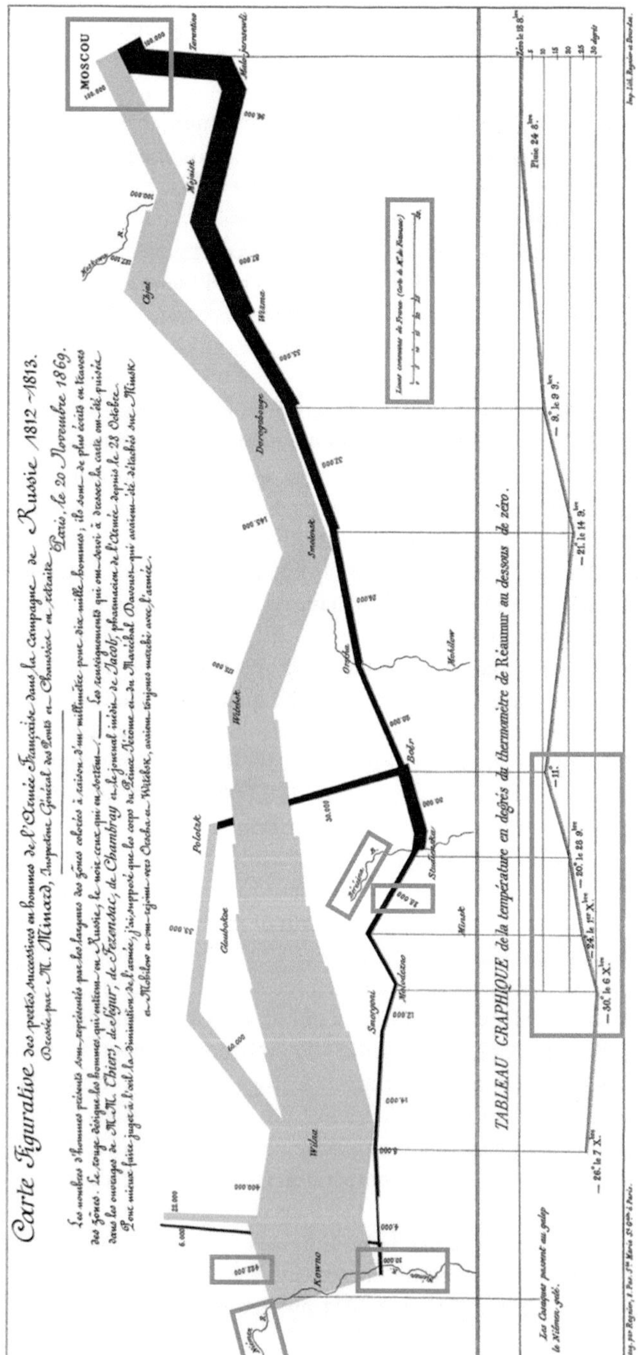

Fig. 8 Examples of the data labels in Minard's map, highlighted with gray frames. The width of the band indicates the shrinking size of the army with exact numbers at specific locations. The temperature graph shows the temperature values at specific days and locations.

pass the frozen Neman River; the right ones inform the reader that it was zero degree (Réaumur) on October 18 and that it rained on October 24. The focus of Minard's map—and thus the key message—is clearly on the shrinking size of the French army during the Russian campaign. However, as Kraak (2014) points out, the map does not say why. The cold during the retreat from Moscow was one reason, as indicated by the declining temperatures. However, the greatest losses suffered by the Grande Armée occurred on the march to Moscow. What caused these losses? Battles, disease, starvation, malnutrition, harsh weather, or desertion? Additional annotations could have answered these questions.

Framing effects: Titles, captions, labels, or annotations can frame the message of a graphic in a specific way. These small elements not only affect the reader's understanding and experience but also influence how the reader interprets the visual elements and reads the graphic as a whole. Framing refers to how information is presented or structured to shape people's perceptions and interpretations of a particular issue. It is defined as the process of "select[ing] some aspects of a perceived reality and mak[ing] them more salient in a communicating text, in such a way as to promote a particular problem definition, causal interpretation, moral evaluation, and/or treatment recommendation for the item described" (Entman, 1993). Framing effects in the media can have a significant impact on social, political and cultural issues. In data visualization, framing can occur through visual design and text. In Minard's map, the title already frames the reader's understanding of the campaign as a data story of space and time. The textual framing reinforces the visual key message by drawing the reader's attention to the "successive losses in men", framing the narrative as a military catastrophe. The catastrophe becomes visible in the dramatic contrast between the broad brown band, representing the initial size of the army at 422,000 as it enters Russia and the thin black line at the end of the campaign, symbolizing just 10,000 men. Minard's design decisions—regarding what to include, how to make some aspects more salient and what to exclude—are also a form of framing. For instance, he included concrete numbers of the army's size at specific locations or events (e.g., battles) and also incorporated temperature data to highlight temperature as an impactful environmental factor in the march from Moscow. However, he excluded any details about the Russian manpower and even excluded the most prominent name of the Russian campaign: Napoleon Bonaparte, the head of the Grande Armée. His focus is solely on the fate of the French soldiers. Thus, the map's framing can be interpreted as a memorial to the soldiers of the Grande Armée. It can also be read as "as an antiwar statement by an engineer who had witnessed the horrors of war in his youth and who, in his final year, was forced to flee his home as the Prussian army approached Paris in September 1870" (Friendly, 2002).

To sum up: Minard's map is an iconic example of an image-centric data visualization. The text that Minard integrated into the map is subordinate to the dominating visual elements: the line graph and the embedded temperature graph. However, without the text the visualization would be difficult to read. The text

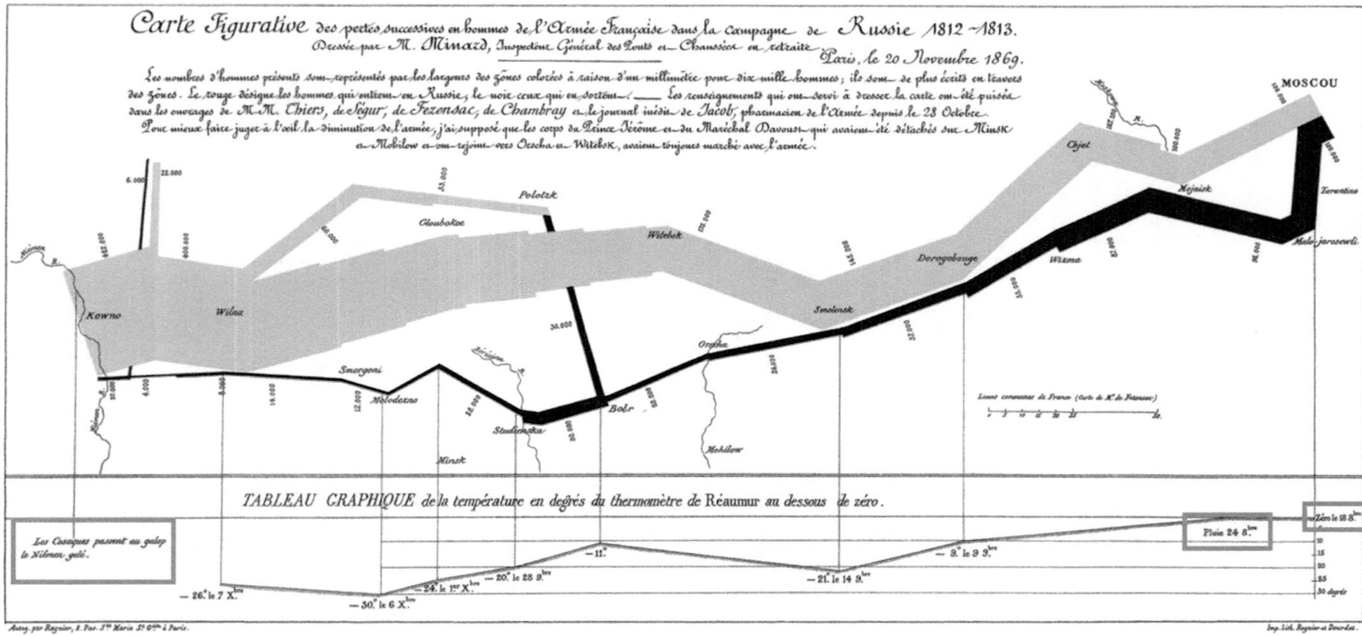

Fig. 9 Annotations in Minard's map, highlighted with gray frames. The French text in the left upper corner reads: "Les Cosaques passent au galop le Niémen gelé". In the right upper corner: "Pluie 24 8.bre and "Zéro le 18 8bre.”

analysis revealed that Minard used different types of texts that can be assigned to all four levels as described in Section 4: from simple but elementary data labels, titles and credits (Level 1) to annotations and how-to-read guidance (Levels 2, 3) to commentary and methodological information (Level 4). Both the text and the design of the map frame it as a memorial to the men of the Grande Armée who lost their lives during the Russian campaign.

Conclusion

Data visualizations are not new, but their social relevance has steadily grown. They can raise awareness of social and political issues, encourage democratic participation and influence decision-making and preference formation. For data visualizations to have an impact on democratic life, i.e., enable citizens to act in an informed and critical manner (Nærland, 2020), they must be available, accessible and comprehensible to the public. In the 19th century, the availability and access to data visualizations such as Minard's work was limited to certain groups – mainly governmental and academic circles. Today, data and data visualizations are publicly available. Open data initiatives, the increasing "datafication" of our social world (Mayer-Schönberger and Cukier, 2013) and new technologies particularly artificial intelligence (AI), have had a transformative impact on data visualization, leading to novel semiotic forms and easier ways to explore and interact with data. Using AI-tools in data visualizations (data viz) can bring efficiency and innovation. Even complex data sets can be visualized in an understandable manner by highlighting relevant patterns. AI enables non-experts, with no prior data visualization skills, to generate graphs and charts and to make data-driven decisions. AI can even recommend the design variants for a given dataset to communicate the data as effectively as possible. With the use of Natural Language Generation (NLG), text can automatically be generated in data visualizations, including elements like titles, labels, captions, annotations and descriptions.

On the other hand, AI-generated text also poses some risks and challenges. AI-generated text may fail to capture the full context of the data which can lead to misinterpretation, deception, or the oversimplification of complex patterns. The text may be too vague, generic, ambiguous, or even biased to clearly communicate the data visualization's message, or it may not meet the needs of the target audience. For example, the same visualization might require a different explanation depending on whether it's intended for domain experts or the general public. Another risk is that producers may rely too heavily on AI and reduce their own critical thinking and engagement with the visualization and accompanying text, missing out on important insights. To mitigate these risks, it is essential to human experts to remain involved in AI-generated data visualizations to ensure accuracy and transparency regarding the methodology and techniques used in the production process. Ethical guidelines are also critical to ensure that the generated text adheres to professional standards and norms.

Whether produced by humans or machines, data visualization can serve as both a resource for informed and critical citizenship, or as "a tool for misinformation and manipulation and consequently contribute to the erosion of informed and critical citizenship" (Nærland, 2020). Text plays a critical role here, as it can add context, provide clarification, guide users on what to focus on in the visualization, promote engagement and interaction, enhance understanding and interpretation and provide transparency. It also has the power to reinforce or counteract visual framing effects.

Drawing meaning from data visualizations involves understanding them as a semiotic interplay of different modes, i.e., as a multimodal, coherent whole. Consequently, data visualization requires data visualization literacy, both on the production side as well as on the reception side. Data visualization literacy is defined as the ability to read and make sense of visual representations of data and to interpret visual patterns, trends and correlations in data visualizations (Börner, Bueckle and Ginda 2019; Boy et al., 2014). Graphicacy is another term that goes along the same lines as data visualization literacy but with a focus on visuo-spatial skills. Balchin and Coleman (1966) define graphicacy as "a skill to be possessed by both those wishing to communicate and those attempting to understand; visual aids, especially maps, photographs, charts and graphs are the media of communication." Both data visualization literacy and graphicacy imply the ability of multimodal reading: to read the visual, the text and the design. Or in the words of Bertin (1983), "We must recognize that graphics is learned, not inherited."

References

Andrienko, N. and Andrienko, G. 2006. *Exploratory Analysis of Spatial and Temporal Data: A Systematic Approach*. Springer, Berlin, Heidelberg.

Balchin, W.G. and Coleman, A.M. 1966. Graphicacy Should Be the Fourth Ace in the Pack. *Cartographica: The International Journal for Geographic Information and Geovisualization, 3*: 23–28.

Barthes, R. 1964/1977. *Image –Music –Text* (S. Heath, Trans.). Fontana, London.

Bateman, J. 2014. *Text and Image: A Critical Introduction to the Visual/Verbal Divide*. Routledge, London. doi: 10.4324/9781315773971.

Bertin, J. 1983. *Semiology of Graphics: Diagrams, Networks, Maps*. University of Wisconsin Press, Madison. [fr. orig.: Semiologie Graphique. Paris: Mouton and Gauthier-Villars 1967].

Bezemer, J. 2023. What modes can and cannot do: *Affordance* in Gunther Kress's theory of sign making. *Text and Talk, 44* (4): 2024, 493–510. doi:10.1515/text-2022-0055.

Boehm, G. 1994. *Was ist ein Bild?* Wilhelm Fink Verlag, München.

Borkin, M.A., Bylinskii, Z., Kim, N.W., Bainbridge, C.M., Yeh, C.S., Borkin, D., Pfister, H. and Oliva, A. 2016. Beyond Memorability: Visualization Recognition and Recall. *IEEE Transactions on Visualization and Computer Graphics, 22*: 519–528.

Börner, K., Bueckle, A. and Ginda, M. 2019. Data visualization literacy: Definitions, conceptual frameworks, exercises and assessments. In: *Proceedings of the National Academy of Sciences of the United States of America, 116*(6): 1857–1864. https://doi.org/10.1073/pnas.1807180116.

Boy, J., Rensink, R.A., Bertini, E. and Fekete, J.-D. 2014. A principled way of assessing visualization literacy. *IEEE Transactions on Visualization and Computer Graphics, 20*(12): 1963–1972.

Bünzli, F. and Eppler, M.J. 2024. How verbal text guides the interpretation of advertisement images: A predictive typology of verbal anchoring. *Communication Theory, 34* (4): 191–204. doi: 10.1093/ct/qtae012.

Caple, H. 2020. Image-Centric Practices on Instagram. Subtle Shifts in "Footing". In: Stöckl, H., Caple, H. and Pflaeging, J. [Eds.], *Shifts towards Image-centricity in Contemporary Multimodal Practices* (pp. 153–176). Routledge, New York.

Caple, H. and Knox, J.S. 2015. A framework for the multimodal analysis of online news galleries: What makes a "good" picture gallery? *Social Semiotics*, *25*(3): 292–321. doi:10.1080/10350330.2014.1002174.

Carpendale, M.S.T. 2003. *Considering Visual Variables as a Basis for Information Visualisation.* University of Calgary, Calgary.

Cheema, G.S., Hakimov, S., Müller-Budack, E., Otto, C., Bateman, J.A. and Ewerth, R. 2023. Understanding image-text relations and news values for multimodal news analysis. *Frontiers in Artificial Intelligence*, *6*: 1125533. doi: 10.3389/frai.2023.1125533.

Engebretsen, M. 2020. Putting the Data Center Stage: Graphs, charts and maps in the news media. *In:* Stöckl, H., Caple, H. and Pflaeging, J. [Eds.], *Shifts towards Image-centricity in Contemporary Multimodal Practices* (pp. 275–296). Routledge, New York. doi: 10.4324/9780429487965-15.

Engebretsen, M. and Weber, W. 2018. Graphic modes. *In:* Cotter, C. and Perrin, D. [Eds.], *Routledge Handbook of Language and Media* (pp. 277–295). Routledge, London.

Entman, R.M. 1993. Framing: Toward Clarification of a Fractured Paradigm. *Journal of Communication*, *43* (4): 51–58. doi:10.1111/j.1460-2466.1993.tb01304.x.

Friendly, M. 2002. Visions and Re-Visions of Charles Joseph Minard. *Journal of Educational and Behavioral Statistics, 27* (1): 31–51.

Gnach, A., Weber, W., Engebretsen, M. and Perrin, D. 2022. *Digital Communication and Media Linguistics*. Cambridge University Press, New York.

Halliday, M.A.K. 1985. *An Introduction to Functional Grammar*. Routledge, London.

Hearst, M. 2023. Show It or Tell It? Text, Visualization and Their Combination. *Communications of the ACM. 66* (10): 68 –75.

Kennedy, H. and Engebretsen, M. 2020. Introduction: The relationships between graphs, charts, maps and meanings, feelings, engagements. In: Engebretsen, M. and Kennedy, H. [Eds.], *Data Visualization in Society* (pp. 19–32). Amsterdam University Press, Amsterdam. doi: 10.5117/9789463722902_ch04.

Kim, D.H., Setlur, V. and Agrawala, M. 2021. Towards understanding how readers integrate charts and captions: A case study with line charts. In: *Proceedings of the 2021 CHI Conf. on Human Factors in Computing Systems* (pp. 1–11). doi:10.1145/3411764.3445443.

Kirk, A. 2016. *Data Visualisation: A Handbook for Data Driven Design.* Sage, London.

Kraak, M.J. 2021. The best map ever? *International Journal of Cartography, 7* (2): 205–210, doi. 10.1080/23729333.2021.1909404.

Kraak, M.J. 2014. *Mapping Time: Illustrated by Minard's Map of Napoleon's Russian Campaign of 1812*. ESRI Press, Redlands California.

Kress, G. 2010. *Multimodality: A Social Semiotic Approach to Contemporary Communication.* Routledge, London.

Kress, G. 2003. *Literacy in the New Media Age*. Routledge, London.

Kress, G. and van Leeuwen, T. 1996. *Reading Images: The Grammar of Visual Design*. Routledge, London.

Lundgard, A. and Satyanarayan, A. 2022. Accessible Visualization via Natural Language Descriptions: A Four-Level Model of Semantic Content. *IEEE Transactions on Visualization and Computer Graphics*, *28* (1): 1073–1083. doi: 10.1109/TVCG.2021.3114770.

MacEachren, A.M. 1995. *How Maps Work: Representation, Visualization and Design*. Guilford, New York.

Martinec, R. and Salway, A. 2005. A system for image–text relations in new (and old) media. *Visual Communication*, *4*(3): 337–371. doi:10.1177/1470357205055928.

Mayer-Schönberger, V. and Cukier, K. 2013. *Big Data: A Revolution that Will Transform How We Live, Work and Think.* Houghton Mifflin Harcourt, Boston, MA.

Mitchell, W.J.T. 1994. *Picture Theory: Essays on Verbal and Visual Representation.* University of Chicago Press, Chicago.

Nærland, T.U. 2020. The political significance of data visualization: Four key perspectives. *In:* Engebretsen, M. and Kennedy, H. [Eds.], *Data Visualization in Society* (pp. 63–73). Amsterdam University Press, Amsterdam. doi: 10.5117/9789463722902_ch04.

Royce, T. 2016. Intersemiotic Complementarity in Print Advertisements. *In:* Klug, N. and Stöckl, H. [Eds.], *Handbuch Sprache im Multimodalen Kontext* (pp. 348–371). De Gruyter, Mouton, Berlin, Boston.

Scarr, S. 2011, Dec 17. Iraq's Bloody Toll. *South China Morning Post.* https://www.scmp.com/infographics/article/1284683/iraqs-bloody-toll.

Scarr, S. n.d. *Information Graphics–Data* Visualisation*. Iraq's Bloody Toll.* https://www.simonscarr.com/iraqs-bloody-toll.

Setlur, V. and Cogley, B. 2022. *Functional Aesthetics for Data Visualization.* Wiley, Hoboken, New Jersey.

Sorapure, M. 2019. Text, Image, Data, Interaction: Understanding Information Visualization. *Computers and Composition, 54* (102519): 1–15. https://doi.org/10.1016/j.compcom.2019.102519.

Stöckl, H. 2020. Image-Centricity: When Visuals Take Center Stage:Analyses and Interpretations of a Current (News) Media Practice. *In*: Stöckl, H., Caple, H. and Pflaeging, J. [Eds.], *Shifts towards Image-centricity in Contemporary Multimodal Practice*s (pp. 19–41). Routledge, New York.

Stöckl, H. 2017. Multimodality in a diachronic light: Tracking changes in text-image relations within the genre profile of the *MIT Technology Review. Discourse, Context and Media: Special Issue 'Media Evolution and Genre Expectations'*, *20*(4): 262–275.

Stöckl, H. and Pflaeging, J. 2022. Multimodal Coherence Revisited: Notes on the Move From Theory to Data in Annotating Print Advertisements. *Frontiers in Communication, 7*: 900994. doi: 10.3389/fcomm.2022.900994.

Stöckl, H., Caple, H. and Pflaeging, J. 2020. Shifts towards Image Centricity in Contemporary Multimodal Practices: An Introduction. *In:* Stöckl, H., Caple, H. and Pflaeging, J. [Eds.], *Shifts towards Image-centricity in Contemporary Multimodal Practices* (pp. 1–16). Routledge, New York.

Stokes, C., Setlur, V., Cogley, B., Satyanarayan, A. and Hearst, M.A. 2023. Striking a Balance: Reader Takeaways and Preferences when Integrating Text and Charts. *IEEE Transactions on Visualization and Computer Graphics, 29*:, 1233–1243.

Tufte, E.R. 2001. *The Visual Display of Quantitative Information.* Graphics Press. Cheshire, Connecticut.

van Leeuwen, T. 2005. *Introducing Social Semiotics*. Routledge, London.

Wainer, H. 1997. *Visual Revelations. Graphical Tales of Fate and Deception from Napoleon Bonaparte to Ross Perot.* Psychology Press, New York.

Weber, W. 2020. Exploring narrativity in data visualization in journalism. In: Engebretsen, M. and Kennedy, H. [Eds.], *Data Visualization in Society* (pp. 295–331). Amsterdam University Press, Amsterdam. doi: 10.5117/9789463722902_ch04.

Weber, W. 2019a. Datenvisualisierungen im Spannungsfeld von Ethik und Ästhetik. In: Schwender, C., Brantner, C., Graubner, C. and von Gottberg, J. [Eds.], *Zeigen –andeuten –verstecken. Bilder zwischen Verantwortung und Provokation* (pp. 145–163). Herbert von Halem Verlag, Köln.

Weber, W. 2019b. Multidisziplinäre Forschungsperspektiven auf Infografiken und Datenvisualisierungen. In: Lobinger, K. [Ed.], *Handbuch Visuelle Kommunikationsforschung* (pp. 335–359). Springer, Wiesbaden.

Weber, W. 2017. Interactive Information Graphics: A Framework for Classifying a Visual Genre. In: Black, A., Luna, P., Lund, O. and Walker, S. [Eds.], *Information Design: Research and Practice (*pp. 243–256). With a foreword by Erik Spiekermann. Routledge, Abingdon.

Weber, W., Burmester, M. and Tille, R. [Eds.]. 2013. *Interaktive Infografiken*. Springer Vieweg, Berlin, Heidelberg.

Weber, W., Engebretsen, M. and Kennedy, H. 2018. Data stories: Rethinking journalistic storytelling in the context of data journalism. *SComS Studies in Communication Sciences*, *18* (1): 191–206.

Zhu, H.H., Cheng, S.S. and Wu, E. 2022. How Do Captions Affect Visualization Reading? *arXiv preprint,* arXiv:2205.01263.

6 Language as Information Design

Peter Gärdenfors[1*]

Introduction

The languages that people use are not products of deliberate design. They have evolved in communities and are still slowly changing over time. Nevertheless, natural languages exhibit several interesting design features.

In the main part of the chapter, these design features will be described. The features will be grouped into pragmatic (rules for how language is used), phonological (constraints on the sounds of a language), semantic (how the meanings of words are organised) and syntactic (how words of a language are combined into sentences). Then, some normative design ideas for language and communication will be discussed.

The first question to ask in a design process is what the product is *made for*. Before answering this question for language, it should be noted that humans are the only animals who have developed languages. So, from an evolutionary perspective, what is the reason humans need a language? Many animal species communicate by various forms of signals. However, their communication deals with the here and now, regarding dangers, food, partners, or offspring. In contrast, human communication largely refers to the non-present: plans, dreams, memories, or gossip.

There are several theories of why only humans have a full-blown language. One account, which will form the basis for the following, is that we need language to improve our *cooperation*, in particular cooperation for future common goals (Gärdenfors, 2004). Animals cooperate, but only about what is present in the environment. In contrast, humans can plan a future hunt, future food procurement, future shelters, etc., that presume cooperation.

Requirements for Communication Systems

The design question for this section is: What properties should a well-designed communication system have? To answer this question, I will present some general

[1] Professor in Cognitive Science, Department of Philosophy and Cognitive Science, Lund University, Box 192, SE-22100 Lund, Sweden.

* Corresponding author: Peter.Gardenfors@lucs.lu.se

DOI: 10.1201/9781032642598-6

constraints that communication systems should fulfil. In order to motivate the constraints, I will start from the evolutionary perspective on language.

- ***Purpose***: The main function of language is to improve cooperation among people.

As will be shown, this constraint has several implications for the structure of language. In particular, it explains why sentences can be classified into *imperatives*, *questions*, and *declaratives*. Imperatives are required to express the commands that may be necessary for successful cooperation to take place. Questions can be seen as a particular form of imperatives: they are requests to obtain information that is needed for the planning and execution of different forms of cooperation. Declaratives, finally, express information that will be shared among the cooperators. It can, for example, be information about where food or water is to be found, or where an enemy or other dangers are lurking. But it can also be information about who can and cannot be trusted in various forms of cooperation. This seems to be one of the main functions of gossip.

- ***Expressiveness***: A language must be able to express all kinds of thoughts.

Animal signalling systems do not fulfil this criterion. The vocabulary and grammar of all human languages, however, are rich enough to express what one wants to say. A language typically has plenty of words which describe what one sees or hears. For example, many nouns and adjectives relate to the shape of things. An exception is how we can describe the face of a person. We are very good at recognising faces, but linguistic descriptions are seldom accurate enough to enable the identification of a person. Sometimes similes or metaphors must be used to catch a particular meaning. For example, languages normally have few words to describe experiences of touch, taste and smell. Here, similes are frequent: "It smells like a rose." "It tastes like a plum." In particular, the vocabulary for describing wines is particularly rich in similes and metaphors.

Fig. 1 What is this?

- ***Efficiency*:** The vocabulary, grammar, and other expressive tools should be designed in an efficient way.

The efficiency of a communication system is a balance between *informativeness* and *brevity*. The expression one uses should convey as much information as is needed for a cooperative task. On the other hand, the expressions should be economical—not being too long or too complicated to be understood. These principles work on several levels of communication as will be shown in the following section.

As an example of the economy of language, look at the picture above and say one word to describe it!

There is no uniquely correct answer, but different levels of concepts that you can choose from in response:

- Animal
- Mammal
- Dog
- Terrier
- Airedale terrier

Most people who are not particularly interested in animals will probably choose the word "dog" to describe the picture, but a dog owner might say "terrier" or even "Airedale terrier". So why do most people choose the middle level? Psychologist Eleanor Rosch investigated this phenomenon and found that for most categories there is a "basic level" that we choose when we want to put into words what we are experiencing or thinking about (Rosch, 1975; Rosch et al., 1976). The word "animal" is far too non-specific to communicate what is in the picture above. The word "dog" contains much more information. The word "terrier" contains even more information, but it assumes that you have learnt to distinguish between different breeds of dogs and know their names, so "terrier" requires more learning and "Airedale terrier" even more. In brief, the choice of the word "dog" is, in most situations, the most efficient way to communicate what Figure 1 shows since it contains comparatively rich information, while at the same time not demanding any special knowledge of the listener.

- ***Learnability*:** A language should be easy to learn.

Linguist Noam Chomsky claims that human minds come with an inborn grammatical structure (Chomsky, 1965). There are, however, strong reasons against this position. In any case, the mapping between words and their meanings must be learned. Neuroscientist Terrence Deacon claims that languages have evolved to be easily learnable by children (Deacon, 1998). Children indeed learn new words miraculously fast. A teenager understands 50,000–60,000 words of her mother tongue by the time she finishes high school. A simple calculation reveals that she has learned an average of 9–10 words per day during her childhood (Carey, 1978). No other form of learning is so efficient. Yet the underlying learning mechanisms are, to a large extent, still unknown.

Language As It Is: Descriptive Features

Linguistics is traditionally divided into *pragmatics* (how language is used in communication, *phonology* (the sounds of language), *semantics* (the meanings of words and expression), and *syntax* (the rules for how words are put together). Here also, *orthography* (how a language is written) will be considered. In this section, these different areas will be analysed from a design perspective. The focus will be on describing how language functions, rather than on how an optimal communication system could be designed. Such normative considerations will be discussed in the following section.

Pragmatics

The design question of this section is how communication can be made efficient. Based on the idea that the main function of language is to support cooperation, the question is what features of language support this function.

The obvious goes without saying. If all partners in a cooperative group perform their tasks as expected by the others, there is no need for communication. It is only when an instruction or a correction is needed that communication plays a role. Language is used to convey nonobvious features of our environment. The basic level of communication is therefore for solving problems of coordinating actions. However, there are situations when the communicators misunderstand each other, because of badly formulated instructions or because they have different mental models of the world. On this level, the goal of communicators is to coordinate their inner worlds. This is also called creating a "common ground" (Clark, 1996). Coordination of inner worlds can also be done as a preparation for future collaboration. In this case, coordination has a value in itself.

There is also another, more severe form of misunderstanding that occurs because the addressee does not understand an expression used by the speaker, or does not understand it in the same way. For example, if you say "I'll talk to the chair" and you mean the chairperson, while *chair* for me just means physical objects, I will not understand your intention. On this final level—the level of semantic coordination—the communicators must negotiate their use of expressions until they find a sufficient agreement.

To sum up, three levels of communication should be distinguished (Winter, 1998; Gärdenfors, 2014):

Level 1: *Instruction*. On this level coordination of action is achieved by instruction (via imperatives).

Level 2: *Coordination of inner worlds* (common ground). On this level, people inform each other in order to reach a richer or better coordination (declarative sentences). It can also be achieved via questions.

Level 3: *Coordination of meanings*. On this highest level, people negotiate the meaning of words (labels) and other communicative elements.

A special case of Level 3 is when somebody also dictates the meaning of a word. Specialists in different fields often have the power to determine the meaning of a particular word and everybody else will simply have to accept that the specialists knows what they are talking about when fixing a meaning.

Phonetics and Orthography

The Sounds of a Language

The most basic form of linguistic communication is speech. When we speak, we use a limited number of speech sounds, so called phonemes. Languages can have close to 40 phonemes (for example, Amharic and Danish) but some have fewer than 20 (in particular, Austronesian languages). The set of phonemes vary quite widely between languages. Swedish contains more vowels than most other languages, Arabic has several different throat sounds, and several African languages have click sounds—Khoisan languages have four different ones. In some so-called "tonal languages", the prosody of a word, that is the speech melody, can be crucial to the meaning of a word. Tonality is most common in Asian languages: Mandarin has four different tones and Vietnamese has six.

There are also sounds that are onomatopoetic, for example, when we imitate the sound of a cat or a cow. Once upon a time, such sounds were probably the origin of vocal communication and we still use them when teaching children about different animals. Having a system of phonemes, makes it possible to create a great number of combinations of sounds, that is, words. The fewer the phonemes there are in a language, the longer the words become. Onomatopoetic constructions are simply not sufficient to generate a sufficient number of different expressions. There are, however, some languages which retain a relatively high proportion of onomatopoetic expressions, for example Japanese and Yoruba.

As infants (and partly already in the womb), we learn to distinguish the phonemes of our mother tongue and we slowly learn to express them. Once we have learnt a phonetic system, it is, for most people, surprisingly difficult to learn a new one. This is why speakers of a second language typically have an accent that reflects the sounds of their first language.

The fact that languages have a limited number of phonemes is an efficient design feature. It helps the listening process, since even if the sounds are pronounced sloppily, our brains assume that the sound sequences can be sorted into the words of the language and thereby the listener can identify the words. This kind of identification process is called "categorical perception" in psychology.

An interesting phenomenon that illustrates this is that when one speaks rapidly, there is sometimes not enough time to pronounce the phonemes correctly, but shortcuts are taken. For example, if the word "sunburn" is pronounced quickly, the 'b' influences the pronunciation of the preceding 'n' so that it becomes 'm'. The word is thus pronounced "sumburn", but the listener does not notice this, since it is not a word in English.

A language also contains constraints for which phonemes may be combined. These constraints help the listening process. In many languages, it is not allowed to have several consonants in a row. In other languages, such as Polish, they are stacked. In German, you can have a consonant after a 'sh' sound, for example "schmerz", but in Swedish this is not possible. In some Slavic languages, 'r' functions as a vowel, for example, 'prst' means "finger" and 'vrt' means "garden" in Croatian. In many African languages, the 'm' as the first consonant in a word may be followed by another consonant: 'Mbwa' (dog) in Swahili has a sound combination that cannot be used in most other languages.

Another design feature concerns how long words are. Efficiency demands that the more frequent a word is, the shorter it should be. In phonetics, the so-called Zipf's law says that a general rule for a language is that the most common word occurs approximately twice as often as the next common one, three times as often as the third, and so on (Zipf, 1945). For example, in English text, the word "the" is the most frequent word, accounting for about 7% of all words. Second is "of", accounting for about 3.5%. In accordance with the efficiency requirement, it can be shown that there is a strong correlation between the frequency of a word in a language and its length.

Written Communication

Written language has developed at least three times in human history. The oldest known is cuneiform, which may be traced back to around 3.400 BCE. Cuneiform was first used for economic transactions reporting on what goods were transported, but was then expanded to a generally expressive language with which stories could be told.

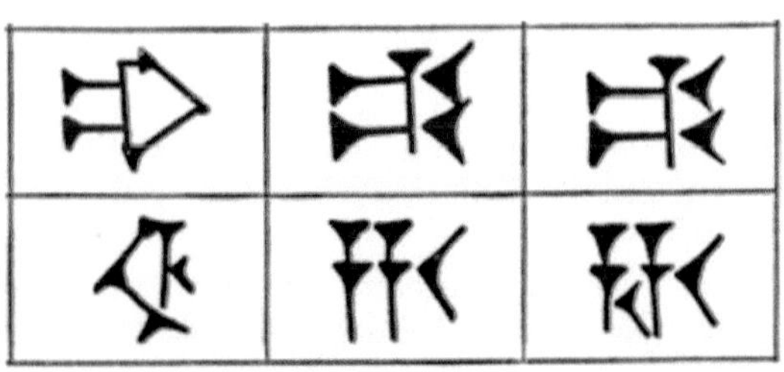

Fig. 2 Cuneiform for 'ox' (upper row) and 'fish' (lower row) during three stages of writing (archaic 2.500 B.C., Assyrian 700 B.C., and late Babylonian 500 B.C.).

A second system is the Chinese characters (ideographs) which were developed 2.500–2.000 BCE. A third is Mayan script where the oldest found examples date to the third century BCE. A fourth possible system is the ideographic Egyptian hieroglyphs which began to be used around 3.300 BCE. It is debated whether cuneiform influenced the hieroglyphs, but the main position is that hieroglyphs developed independently. All four writing systems were originally based on pictures. Cuneiform later developed into a phonemic system with one symbol for each sound (except for vowels). The first one, the Phoenician alphabet, was used most of the first millennium BCE. There are also written systems, for example, modern Korean, which is syllabic, that is, it has one symbol for each syllable of the language.

한	국	어
han	guk	eo

Fig. 3 Three Korean syllables. The word "hangukeo" means "Korean".

The distinction between ideographic, syllabic, and phonemic written languages has consequences for the expressiveness, efficiency, and learnability of a system. The characters of an ideographic system were often *iconic*, that is, they were similar to what they represent. Two examples of this can be seen in the left column of Figure 2, where the signs look like the head of an ox (turned 90°) and a fish. That the characters were iconic entails that it has been easier to understand what an ideographic expression means—even if one does not speak the language that is being represented. Later they developed into more abstract patterns (for example, the second and third columns of Figure 2). Once this happens, all the characters must be *learned*. For a phonemic language, only the symbols for the limited set of phonemes must be learned. There is in general a good correspondence between the letters of words and the way the words are pronounced. This means that once you have learnt a language, it is comparatively easy to learn to *read* an expression of the language.

A phonemic written language need not include all the sounds of the spoken language. For example, many Semitic languages (Arabic, Hebrew) do not represent the vowels of the language, since they normally do not determine the meanings of the words. Furthermore, the stress of a word is often unmarked. For example, in English, the difference between "cóntent" and "contént" is not marked.

Spoken language is tied to a particular context and is expressed by tone and gesture which can be much more informative than the "literal" message. All that is lost when language is pinned to paper. As a consequence, when spoken language was transformed into written, several speech act markers were added. For example, written language uses more words than speech to tell a story. Psychologist David Olson argues that written language must compensate by various means for the part of the linguistic message conveyed by tone, rhythm and emphasis (Olson, 1994). He shows, for example, that linguistic markers for speech acts such as "assert", "declare", and "suggest" first appear in the Middle Ages. Such markers are not needed in an oral tradition in which the speech act is expressed directly by prosody and other means.

Not until we have an alphabetic written language will we be aware of the sound structure of language and its grammatical organisation. It is written language that makes it possible to speculate about the structure of language. The idea that written language has an autonomous meaning independent of the author and the reader is first established in the Middle Ages—this is when the text acquires a literal meaning that does not change when the context changes (Olson, 1994). Grammatical markers become important for making the writing unambiguous.

A language usually has many different dialects, and when a phonological representation is chosen, a choice has to be made as to which dialect is represented. An alphabetic script, a "grapholect", will therefore act as a normative for the spoken language. This is an example of how the design of a new medium can influence the use of an older one. The same is not true of ideographic writing systems. The Chinese dialects are so diverse that it would have been difficult to have a standardised alphabet for writing down the words. But by representing words as ideas rather than sounds, all Chinese people can understand the ideographic writing.

In this way, written languages have come to function as a *norm* for spoken language. As a consequence, dialects have been levelled out in many languages. Developing a standard orthography, that is, a set of rules for writing, is a central step in creating national languages.

Putting a language into a written version means that all words of a language can be collected in dictionaries. Single language users only use a subset of them. As a consequence, users of a written language like English have access to a vocabulary many times larger than that of the language spoken by individuals.

For a written language, the mapping between phonetics and orthography should be transparent. Psychologist Uta Frith with colleagues studied the impact of the transparency of written languages on dyslexia (Landerl et al., 1997; Paulesu et al., 2001). A transparent language is one in which there is a consistent and straightforward relationship between letters and their sounds (for example, Italian or Finnish). In contrast, opaque languages, such as English, have complex and inconsistent spelling rules, making phoneme-to-grapheme relationships less predictable.

Frith's research reveals that the challenges faced by individuals with dyslexia vary significantly based on the transparency of the language they are learning. In transparent languages, individuals with dyslexia encounter fewer difficulties in decoding written language because of the clear phoneme-grapheme correspondence. In contrast, opaque languages present more substantial decoding challenges.

Semantics

The Meaning of Words

A language contains both content words (primarily nouns, verbs, adjectives, adverbs and prepositions) and functional words that determine the syntactic structure. When learning a language, understanding the meaning of the content words is essential. Otherwise, communication would be parrot-like.

There are several theories of how the meaning of words can be described. The classical Aristotelian view says that the meaning of a word can be defined by necessary and sufficient conditions. For example, a person is a grandmother, if, and only if, the person is a mother and at least one of her children also has a child. This kind of characterisation of a concept has often been used in computational applications. For most words, it has, however, turned out to be wellnigh impossible to describe their meaning in the Aristotelian manner.

Instead, the so-called *prototype theory* developed by Eleanor Rosch has come into focus. The main idea of prototype theory is that within a category of objects, certain members are judged to be more representative of the category than others. For example, robins are judged to be more representative of the category "bird" than are ravens, penguins, and emus; and desk chairs are more typical instances of the category "chair" than rocking chairs, deck-chairs, and beanbag chairs. The most representative members of a category are called prototypical members. Prototype theory also claims that concepts show graded membership, determined by how representative the members are. It is well-known that some properties, like "red" and "bald" have no sharp boundaries and for these it is perhaps not surprising that one finds prototypicality effects. However, these effects have been found for most concepts including those with comparatively clear boundaries like "bird" and "chair". A content word typically refers to the prototype of a concept.

Another design feature of language is that the meanings of words can often be analysed in terms of *semantic domains* such as colour, pitch, temperature, force, time, and emotion. I have suggested that basic adjectives all have meanings that belong to a single domain (Gärdenfors, 2000, 2014). A similar analysis can be made of the meaning of verbs and prepositions. Only nouns have meanings that represent several domains. For example, the meaning of the word "dog" contains information about a number of domains, for example, shape, size, sound, smell, colour, and texture.

By combining dimensional domains one can construct "conceptual spaces". The primary role of the dimensions is to represent various properties of objects in different domains. Some dimensions come in bundles, for example, space (dimensions of height, width, and depth); colour (hue, saturation, and brightness); taste (salt, bitter, sweet and sour, and umami); and emotions (positive/negative, arousal). These dimensions are closely connected to what is involved in perceiving and acting. However, there also exist dimensions that are of an abstract, non-sensory character.

Each of the domains is endowed with a geometric structure. As an example, take the dimension of "time". Time is one-dimensional and has the same structure as the line of real numbers. If "now" is seen as the zero point on the line, the future corresponds to the positive real line and the past to the negative line. In contrast, the dimension of "weight" is isomorphic to the positive real line—there exist no negative weights. The differences in dimensional structures show up in the semantics of the adjectives that are used to describe the dimensions.

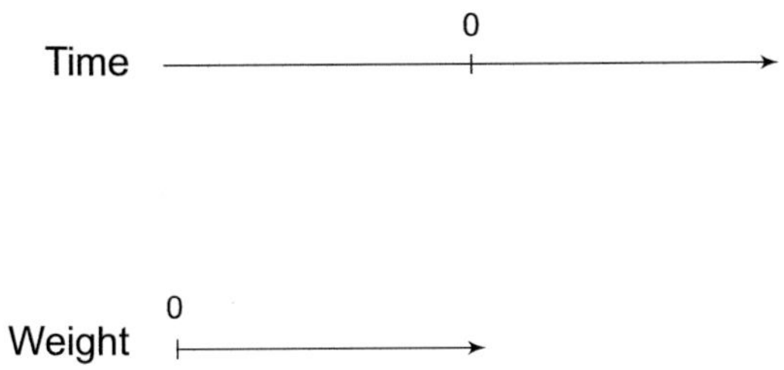

Fig. 4 The structure of the time and weight dimensions.

Another example is the structure of colour space. Human colour perception is composed of three fundamental parameters of colour perception: hue, intensity and brightness (Gärdenfors, 2000; Regier et al., 2007). Any perceived colour can be mapped onto some specific values to these dimensions (Figure 5).

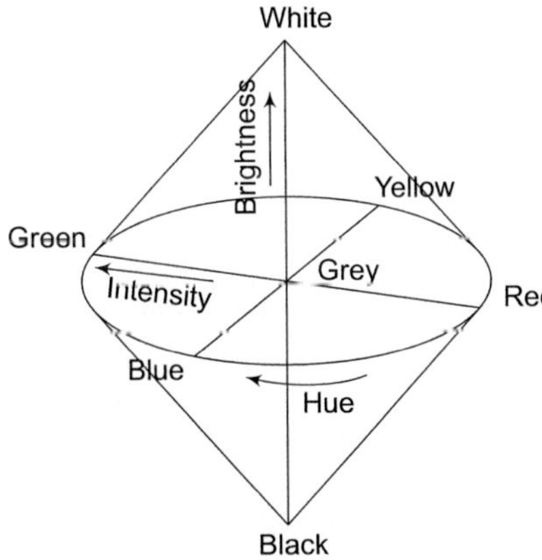

Fig. 5 The colour domain.

Different colour concepts can then be described as *convex* regions of possible values of these three parameters. That a region R is convex means that for any two points x and y in R, all points between x and y are also in R. For example, if one point x in the colour domain is called red and the point y is also called red, then any point between *x* and *y* will also be called red. That concepts correspond to convex regions can be seen as a semantic design principle of languages.

I have argued that the meanings of adjectives can be represented as convex regions of single domains (Gärdenfors, 2000, 2014). Different languages carve up the colour domain in different ways, but all colour regions are convex (Jäger, 2010). In contrast most nouns are represented as a set of regions in a number of domains.

The idea that the meanings of words correspond to convex regions of conceptual spaces can also help explain how children can *learn* so quickly. Learning a concept often proceeds by generalising from a limited number of examples of the concept (Nosofsky, 1988; Langley, 1996). If these examples are spread out in the region for the concept, then the principle of convexity allows the children to generalise a word to new instances. Furthermore, a prototypical instance of the concept can be estimated from the examples (Gärdenfors, 2000).

Combinations of Words

Language is not just made up of single words, but we combine words in shorter or longer sequences. The meanings of word combinations do not just arise from stacking the meanings of individual words. They interact and influence each other to create complex constructions of meaning. There are no simple rules for what happens to the meanings of two words when they are put together. Words are like chemical elements. When hydrogen combines with oxygen, it makes water; when it combines with nitrogen, it makes ammonia, which has completely different properties. In a similar way, the meaning of a word will strongly depend on the other words it is combined with.

We are surprisingly efficient at understanding combinations of words— even completely new and perhaps surprising combinations. A central question is what happens in our heads when we understand so quickly. For an example, let us look at the word "red". For a physicist, this word means light with a wave length of about 740 nanometres. In the colour space presented above, red corresponds to a particular region of the space. There are however, several combinations of words including "red" that do not fit this picture, for example, "red hair", "red face", "red wine", "red drum" (a fish), "redwood", etc. In these combinations, "red" does not denote the typical red colour: red hair is copper coloured, red wine is dark purple, and the red drum is partly reddish brown. We still understand these combinations. The reason is that red hair is red *in contrast* to other hair colours, etc. (Gärdenfors, 2000).

In general, the meaning of an adjective depends on the noun it is combined with. From "The neighbour has a big pug", it cannot be concluded that "The neighbour has a big dog". The meaning of "big" depends on the contrast class that the noun generates. The same applies to all comparative adjectives: "The bath water is cold' does not lead to the conclusion that the water is cold—it can be lukewarm.

A central design feature of language is that words are combined into *sentences*. We all have grammatical intuitions whether a particular combination of words constitutes a grammatical correct sentence that makes sense. I will return to the grammatical aspects in the following section. Most sentences express *events* (or states). (An interesting exception is generics that express relations between concepts.) As mentioned above, one can distinguish between declarative, imperative and interrogative sentences. Here I will focus on the meanings of declarative sentences.

What do declarative sentences have in common and what makes them natural linguistic units? In short, the answer is that they often describe events ("the postman rode his bike over", "the strawberries are boiling in the jam pot", "the wind knocked over the garden furniture") or states of affairs ("the cat is lying on the sofa", "the tower is twenty metres high", "the car is stuck in the mud"). An event (or a state) is the semantic structure that makes a sentence a natural unit of meaning. One reason events are so central for what language means is that one of our memory systems, so-called episodic memories, is structured around events. The event structure also shows up in narratives: When you tell a story, it is organised as a series of events. Thus, the fact the sentences typically express events is a design feature of language that is motivated by the general structure of human cognition.

The principle of expressiveness requires that all possible thoughts can be expressed in a language. Languages contain a limited number of words, but by combining them one can create a large number of new meanings. Nevertheless, there are ideas for which it is difficult to find words. I have already mentioned the use of similes in expressing different smells. Nevertheless, there are new concepts or ideas that are initially difficult to describe with the aid of existing words. To find a way to express a new concept, *metaphors* are frequently exploited. For example, when electricity was a new phenomenon, metaphors relating to the flow of water were introduced. Thus, one spoke of electric "currents" and "resistance" against the flow of electricity. Another example is the metaphor of "computer virus", which is a figurative description of programs that are self-replicating on a computer and that spread over large parts of computer memory. New terms were then coined using "virus" as a starting point. For example, there are "vaccination programs" against computer viruses and "malignant" and "benign" viruses. Social media posts can spread like a virus – they can go "viral".

Syntax

The ease with which a child learns a language is partly due to the fact that the structure of language has been adapted to make it easy to learn. Some linguists, notably Noam Chomsky, have argued that the grammar of language is so complicated that it cannot be learned (Chomsky, 1965). He argues that there are so many possible systems of rules that children cannot possibly figure out which system their native language follows just from the limited variety of linguistic constructions they hear from parents, siblings, and others in the environment. Chomsky and his followers therefore argue that certain basic grammatical structures must be innate, that is, prepared in the genetic structure of humans, and that these genetic components are not present in other animals.

This is, however, an untenable theory. First, Chomsky assumes that a single mutation at some point in human evolution gave rise to the innate grammatical structures. But language is too complex to be controlled by a single gene. Even if more genes are involved, it is difficult to explain the selection mechanisms that created the grammar genes.

Second, Chomsky is only concerned with the grammar of the language and is not interested in how you learn the meanings of words or how you choose to put words together to express a particular thought. It seems obvious that if you don't know the semantics of a language, there is no point in knowing its grammar. If you cannot communicate with words and understand their meanings, then there will be no point in having grammar. Third, linguistic skill is not an independent "module" of thinking, as Chomsky has argued, but is a consequence of the general human talents for pattern recognition, learning, and mind-reading.

Children do not just learn language by listening to others and (unconsciously) figuring out what the correct grammatical rules are. Children learn language by using it in everyday contexts which gives cues to what is talked about. Chomsky is only interested in the child's grammatical competence, but the focus should be on their ability, or rather abilities, to communicate.

One might ask how a language with grammar is better as a means of communication than a language without grammar. The big advantage is that syntactic markers make the message less ambiguous. If a two-year-old coming from kindergarten tells you "not eat fish", you don't know whether it is the child who has not eaten the fish (or didn't want to eat it), or whether it is the aquarium fish that has not eaten the candy the child tried to give it. Grammar also makes language more independent of the present because, for example, it is easier to mark whether one is talking about the present, the past, or the future.

In brief, the two main design features of grammar are that it can be used to eliminate ambiguities in communication and that it extends expressivity, for example, by marking when something happened. In the section on semantics, it was argued that sentences express events. As a matter of fact, how we structure events mentally, has a strong influence on how language is structured. The main components of events are actors, action, patients (what is acted on), and results. Expressed in language, they correspond to subject, object, verb, and prepositional phrases. Despite this influence of event structure, there are many options for how a syntax is built up.

One design factor for a grammar concerns how the elements of the language are put together. In *agglutinative* languages (for example, Turkish) words contain multiple morphemes concatenated together, while in *fusion* languages (for example, English) words are generally kept shorter but added on to each other. An example from an agglutinative language is the Persian phrase "mashinhashunra". The word can be broken down as follows: *mashin* (car) + *ha* (plural suffix) + *shun* (possessive suffix) + *ra* (suffix marking location "at"), thus meaning "at their cars". A slightly bizarre example is the German word "Hottentottenpotentatenattentaten", which can be broken down into *Hottentott* (noun) + *en* (plural suffix) + *Potentat* (noun) + *en* (plural suffix) + *Attentat* (noun) + *en* (plural suffix), thus meaning "attacks on potentates of the Khoekhoe people".

Another design choice is between having a syntax of *cases* such as nominative, accusative, dative and genitive, versus having a syntax based on prepositional phrases. For example, Latin is rich in cases: Apart from the four already mentioned, there is the ablative case, which in English is often translated with the aid of the preposition "from". "Vulnere mortuus est" means "He died from a wound" where "vulnere" is the ablative form of "vulnus" which means "wound". Finnish is even more extreme having 15 cases that do the semantic work that is done by prepositions in English.

A third design choice is how many *tenses* exist in a language, that is, which aspects of time that can be marked grammatically. For one extreme, French has about 20 different ways of expressing time by verbs. For the other extreme, Mandarin has no tenses, at least in the sense that the verbs are not marked for time. There are, however, other ways to express time-frames in Mandarin.

A fourth design feature is to what extent nouns (and other words) are marked by *gender*. Many languages, such as English, only mark gender for humans and some animal species. German has three genders—masculine, feminine, and neuter—and all nouns have a gender. Many African and Australian languages have several genders.

There are just some examples of design choices that can be made when constructing a language. There exist many other grammatical design features that vary between languages. These variations make it possible to express a particular thought in different ways. When translating from one language to another, a particular linguistic form must often be reformulated in the translation. A general rule is therefore that a translated text is normally longer than the original.

Language As It Should Be: Normative Design Considerations

The Economy of Language

There is a cost to communicating. Speaking in itself is not that exhausting (some people find it too easy), but we have to choose what to say and how to say it (some people do not care about this either) and it requires some thought. Above all, we must *learn* a language. Anyone who has tried to learn a foreign language knows the efforts involved. Even the mother tongue has to be learnt. For the young child, it is easy and happens without the child noticing the learning. But it takes time.

One aspect of the economy of language is the number of words used. An average language speaker understands around 50,000 words, but the active vocabulary rarely exceeds 5,000 words. With those words one has to express everything one wants to say about the world, including one's memories, fantasies, and dreams. Using just 5,000 words to do this may seem like a very small amount.

The problem with the large vocabulary of a language is that memory is limited. When one hears or reads one of the 50,000 words, one retrieves its meaning from memory. And when one speaks, one has to activate a word whose meaning captures what one wants to say. We have all experienced situations where we cannot find a suitable word. (This is most noticeable in a non-native language.) So, when using the language, one has to find a balance between the words one understands and the cost of activating them from memory. This balance is a central part of the economy of language.

Of course, if one understood more words, one could find richer and more precise ways to express oneself. How can as few as 5,000 words be enough for everything one wants to say? The key is that words are not used in isolation, they are combined – and combined in ever new ways. For example, it is very probable that the sequence of words in this particular sentence has never been expressed before. The expressive richness of a language therefore comes from the way words are put together.

Why is language not precise? Would our communication work not better if each word had a precise meaning? These questions lead to a third aspect of the economy of language. In brief, the reason is that if the meanings of words are flexible, they can be adjusted in relation to the other words with which they are combined, thus creating more opportunities for new and richer meanings. An analogy is that a word is like a spanner that can be adjusted to the desired width to fit a new nut. The alternative is to have a set of block spanners (Figure 6), each of which has a precise width that can be effective in certain situations, but such a set is more expensive and requires a larger toolbox. If the meanings of words were precise like block keys, our learning and memory management would be cognitively more costly. If each word had a single meaning, our communication would be very limited.

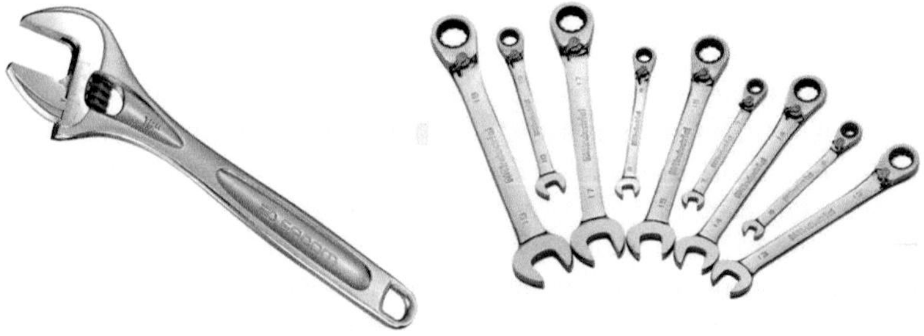

Fig. 6 Words as spanners or block keys.

For example, take the many uses of 'red' that were mentioned earlier. Instead of saying "red hair", "red face", and "red wine", one could say "copper-coloured hair", "pinkish beige face", and "dark purple wine". However, this would presume that both speaker and listener master an extensive colour terminology and communication would therefore be cognitively more demanding. Using 'red' in all cases is like using an adjustable spanner. The word "red" is adjusted to its intended meaning when it is combined with the different nouns.

Philosopher Gottfried Leibniz dreamt of an exact language that would allow everything to be expressed precisely. This dream has been one of the reasons why analytical philosophers have tried to provide necessary and sufficient conditions for concepts. However, the meanings of words do not work like that. They have central meanings that are then expanded or changed depending on the context—similar to the function of an adjustable spanner.

The Dynamics of Language

Languages change over time. New words are added – often borrowed from other languages – and some words become obsolete. Who remembers the meanings of "snoutfair" and "crapulous"? Slang expressions become a common-day language. Those who write dictionaries or grammars of a language must constantly update the contents. Words that have fallen out of use are removed and new ones added. The grammar of a language changes slowly. The changes are often seen as simplifications, but because of the principle of expressiveness, disappearing grammatical features are often replaced by other constructions. An eternal principle is that each generation complains that the following generation speaks and writes ungrammatically and sloppily.

From a design perspective, a question is to what extent language should be normed. Many nations have institutions or organisations that give recommendations concerning what is "correct" in a language. However, they do not have legal status (it is very difficult to command people how they should speak) so changes in the language will still occur. For some languages, for example Icelandic and Modern Hebrew, authorities have tried to keep the language pure, that is, free from "foreign" influences. Because of increasing globalisation, however, such attempts have in general failed.

To a large extent, natural spoken languages are as they should be in the sense that they have developed to satisfy the communicative principles presented in Section 2. Exceptions are that most languages contain vestiges of earlier structures, for example, irregular verbs or traces of case systems.

From a design perspective, such vestiges should be counteracted. In fact, the principle of learnability has as a consequence that such vestiges will, in the long run, disappear. For example, when children are learning their mother tongue, they first tend to generalise verb forms so that all verbs are treated as they are regular. Only later do they learn the irregular forms. Over time, it will become more accepted to replace the irregular form by the regular ones.

Designing Languages

If one were to design a language from scratch, what would it look like? To be sure, there have historically been several attempts to create "constructed" languages. The most well-known examples are Esperanto and Ido. The goals of such a language are to optimize the design criteria of expressiveness, efficiency and learnability. These languages have not been used very extensively (although a fair number of people speak or understand Esperanto). One reason is that their vocabularies are often based on European language and are, therefore, difficult to learn by people from other language groups. Another reason is that these languages have very few native speakers.

An interesting example is the language Klingon that is spoken in the Star Trek universe. This language has deliberately been crafted by linguist Marc Okrand to *violate* natural design principles as much as possible. For example, Klingon uses a very rare Object-Verb-Subject word order, which is uncommon in human languages. It also contains grammatical patterns that feel unintuitive to speakers of common languages. Phonetically, Klingon includes sounds that are not found in most languages and are difficult to pronounce for many people. It also combines sounds in ways that feel unnatural. Despite these quirks, there are people who have learnt to speak the language, including one boy who was raised with Klingon as his first language.

Many sign languages have been created in modern times and they are therefore interesting to study from a design perspective. Here the use of a visual signing space instead of an auditory channel makes different design features available. A large portion of the signs have an iconic root, that is, they are similar to the object or action they represent. For example, in American Sign Language, when signing "ball" one puts the fingertips from each hand together with the palms separated but facing each other. It looks like one has a grapefruit-sized ball in one's hands. This means that sign languages resemble more ideographic languages than alphabetic languages. However, there are also signs for the letters of the alphabet, so any spoken word can also be expressed in a sign language.

In spoken languages, it is common to refer to an already mentioned person by a pronoun (which makes language more efficient). In sign languages, this is often achieved by assigning a person a place in the space in front of the signer. Then the person can be referred to by pointing to that location.

Different forms of technology also allow new methods of communication. Here normative considerations are more applicable. The design principles of written languages have already been discussed. A more modern example is Morse coding which is a way of transforming the letters of an alphabetic language into sequences of short and long beeps. When Morse chose how to represent the letter of the English alphabet, he followed a version of Zipf's law so that the more frequent letters are expressed by short sequences of beeps, while the uncommon letters are expressed by longer sequences. For example, a single short beep expresses the letter E, while the sequence for Q is long, long, short, long. This choice of coding makes the Morse alphabet more efficient.

A relatively new type of linguistic constructions are *emojis*. Originally, they were sequences of characters that could be typed on a standard keyboard. The combinations ;-), :-) and :-(were the first to be used in text messages and emails. They then became a system of specialised picture-like symbols that have now developed into a rich flora. Emojis are a kind of ideographs, similar to Chinese characters. They can be seen as a substitute for the expressions of emotion found in spoken language. In the past, because text messages had to be short, there was not enough room to describe emotions in ordinary words.

Emojis have quickly developed their own meanings. The Unicode Consortium that determines which emoji exist does not decide what different emojis mean, but this is a more or less democratic process. Many emojis have moved from concrete meanings to more abstract ones. A typical example is the skull, which in 2012 was associated with words like "zombie", "corpse" and "murder", but from 2013 is more commonly used for "laughing till I die". Other examples, reminiscent of how slang words arise, are aubergine and peach, which have come to be used as signs for the male and female genitals respectively.

Conclusion: Should Language be Designed?

One explanation for the constant change in language may simply be that young people do not want to talk like their parents. A group is united by having its own words and its own phrases. If you use expressions from the wrong generation, you will inevitably reveal yourself as antiquated. The slang you learnt as a teenager just seems daft to the younger generation.

There has always been, and always will be, language police who insist on telling us what is correct in a language. There is a constant battle between generations about how to speak or write. Older people often think that the younger generation speaks carelessly and they are horrified by incorrect and lax uses of the language.

Should one impose norms? The answer is: It depends. It would be futile to try to stop the slow dynamic changes of a language. There are, however, some situations where norms can be helpful. For example, when changes of the pronunciation of words in a language make them differ too much from the existing spelling system, a spelling reform is motivated. As was explained earlier, a good correspondence between phonology and orthography (sound and writing) makes it easier to learn to read for children and for dyslectic persons.

A particular case of how much norms should be imposed concerns how language should be taught in schools. Teachers will have to find a middle road between the norms of spelling and grammar they once were taught and the language currently used by their students. There is no universal solution to this tension, but if teachers (and other writers and speakers) can coordinate their standards, communication will become more uniform.

In summary, this chapter has presented descriptive and normative aspects of the design of languages. Languages are like living organisms, so it is rather hopeless to control its changes. On the other hand, some norms for how a language is pronounced and written within a community will improve the success of communication.

References

Carey, S. 1978. The child as word learner. In: M. Halle, J. Bresanan and G.A. Miller (Eds.), *Linguistic Theory and Psychological Reality* (pp. 264–293). Cambridge, MA: MIT Press.

Chomsky, N. 1965. *Aspects of the Theory of Syntax* (No. 11). MIT Press.

Clark, H.H. 1996. *Using Language*. Cambridge University Press.

Deacon, T.W. 1998. *The Symbolic Species: The Co-evolution of Language and the Brain*. WW Norton & Company.

Gärdenfors, P. 2000. *Conceptual Spaces: The Geometry of Thought*. Cambridge, MA: MIT Press.

Gärdenfors, P. 2004. Cooperation and the evolution of symbolic communication. In: K. Oller and U. Griebel (Eds.), *The Evolution of Communication Systems* (pp. 237–256). MIT Press.

Gärdenfors, P. 2014. *Geometry of Meaning: Semantics Based on Conceptual Spaces*. Cambridge, MA: MIT Press.

Jäger, G. 2010. Natural colour categories are convex sets. In: *Amsterdam Colloquium 2009* (pp. 11–20). *Lecture Notes in Artificial Intelligence, 6042*.

Landerl, K., Wimmer, H. and Frith, U. 1997. The impact of orthographic consistency on dyslexia: A German-English comparison. *Cognition, 63*(3): 315–334.

Langley, P. 1996. *Elements of Machine Learning*. San Francisco, CA: Morgan Kaufmann.

Nosofsky, R.M. 1988. Similarity, frequency, and category representations. *Journal of Experimental Psychology: Learning, Memory, and Cognition, 14*(1): 54–65.

Olson, D.R. 1996. *The World on Paper: The Conceptual and Cognitive Implications of Writing and Reading*. Cambridge University Press.

Paulesu, E., Démonet, J.F., Fazio, F., McCrory, E., Chanoine, V., Brunswick, N., ... & Frith, U. 2001. Dyslexia: Cultural diversity and biological unity. *Science, 291*(5511): 2165–2167.

Regier, T., Kay, P. and Khetarpal, N. 2007. Color naming reflects optimal partitions of color space. In: *Proceedings of the National Academy of Sciences, 104*(4): 1436–1441.

Rosch, E. 1975. Cognitive representations of semantic categories. *Journal of Experimental Psychology: General, 104*(3): 192–233.

Rosch, E., Simpson, C., and Miller, S. 1976. Structural bases of typicality effects. *Journal of Experimental Psychology: Human Perception and Performance, 2*(3): 491–502.

Winter, S. 1998. *Expectations and Linguistic Meaning*. Lund University Cognitive Studies 71, Lund.

Zipf, G.K. 1945. The meaning-frequency relationship of words. *The Journal of General Psychology, 33*(2): 251–256.

7 The Role of Prior Knowledge in Understanding and Applying Technical Communication

Jonatan Lundin[1*]

Introduction

In both professional and everyday contexts, individuals who use tools—referred to here as users—sometimes turn to manuals when they need support in operating those tools. Yet they occasionally find that they do not fully understand what they read, which hinders their ability to act effectively. Understanding why this occurs is a concern within technical communication, as both practitioners and researchers seek insight into what makes manuals understandable, and how they can be designed to effectively support tool use.

This concern is particularly important in contemporary society, where users' professional and everyday activities increasingly depend on the effective use of tools such as industrial hardware and digital applications, including AI (Mishrif and Khan, 2023; Sindakis and Showkat, 2024)[#]. The ability to use—and continuously learn to use—these tools effectively and safely is essential not only for participation in society, but also for ensuring organisational competitiveness, strengthening national resilience in the face of demographic change, and responding to global crises and the pressures of globalisation. As tools become more specialised, they demand domain-specific knowledge and, in many cases, familiarity with the associated terminology. In many contexts, however, users lack this knowledge, which either causes them to avoid using the tool altogether or limits their ability to use it correctly and efficiently without assistance (Ceaparu et al., 2004; Mendoza and Novick, 2005)[†].

Such assistance is provided by an information designer, such as a technical communicator. The task of the technical communicator is to design technical communication—for example, instructions and descriptions—with the main purpose of helping users to quickly find relevant information, understand it, and apply that understanding to use the tool effectively, efficiently, and safely, while

[#] See also, Følstad et al. (2017); Janssen et al. (2019).

[†] See also, Novick et al. (2007), Welty (2011).

[1] Ph.D Information Architect, Excosoft, Sweden. and affiliated to Department of Health Science, Innovation and Design, Department of Eskilstuna and Västerås, Mälardalen University, Box 325, 631 05 Eskilstuna, Sweden.

* Corresponding author: Jonatan.lundin@mdu.se

DOI: 10.1201/9781032642598-7

minimising negative environmental impact. Nevertheless, there are reports of users who, despite being motivated to use technical communication, experience difficulties in finding the information they need, understanding what they read, and applying that information into correct actions. This prevents them from using the tool as intended (Martin et al., 2005; Novick and Ward, 2006a)[#], causing frustration at an individual level, as well as inefficiency, anxiety, and social exclusion on a broader scale.

For the purposes of this chapter, technical in technical communication refers to the tools used by various professionals and non-professionals—such as everyday users in domestic settings, service technicians, industrial operators, or farmers—to carry out their work tasks. These tools may consist of hardware, software, or a combination of both, for example a welding machine or software for measuring electric current. Communication in technical communication refers to information that functions as a mediating agent which, under the right conditions, enables users to construct mental representations and actions they would not be able to form independently. Designed information may be delivered as a multimodal media product in which text (written or spoken) and images (such as illustrations, photographs, or animations) interact—for example, instructions and descriptions in a digital user manual for an industrial product.

Technical communication takes many forms, each tailored to a specific audience and situation. At one end, local technical communication addresses immediate, situated needs—for example, an instruction written by co-workers and taped to the office kitchen refrigerator to explain how to use the old coffee machine. At the other end, global technical communication supports users of standardised, mass-produced, configured-to-order, or customised industrial and consumer tools intended for worldwide markets. Global tools often have long lifecycles, during which diverse users perform tasks such as installation, commissioning, operation, and maintenance. Some tools are technologically simple and require little prior knowledge, while others are advanced and demand specialised expertise. A single tool model may involve both simple and complex tasks, and because tool models are distributed globally, they are used in a wide variety of workplaces. In some contexts, installers or operators may have limited knowledge of the tool domain and its terminology, while in others most users may be highly skilled professionals. Workplaces also constitute socio cultural contexts, where norms evolve regarding how tasks are carried out, what things are called, and who is considered knowledgeable. Consequently, technical communicators designing global technical communication for a specific tool model must anticipate use by both novice and expert users across diverse cultural backgrounds, workplace contexts, knowledge levels, work practices, beliefs, motivations, and experiences.

Technical communication must be useful to support the user—here understood as an information design that enables users to efficiently find what they need (such as an instruction), understand it, and apply that understanding effectively

---

[#] See also, Novick and Ward (2006b); Wieringa et al. (1998).

and safely in their situation to accomplish a goal. However, many factors must be considered when designing information to increase its usefulness. Previous research emphasises that one such factor playing a crucial role is the user's prior knowledge (Caillies et al., 2002; Cain et al., 2001)[#]. For information to be useful, it must connect to and activate the user's relevant prior knowledge. This, in turn, shapes the extent to which users can understand and apply the information design. It is therefore essential for technical communicators to understand why and how a user's prior knowledge shapes their cognitive capacity to understand and apply information. This chapter explores how users' prior knowledge influences their cognitive capacity to understand and apply technical communication, and discusses the implications for the design of such technical communication to increase its usefulness.

Prior Knowledge Shaping the Cognitive Capacity to Use a Tool and Technical Communication

The concepts of knowledge, understanding, and learning, and the processes involved, are multifaceted and complex subjects discussed within the fields of instructional design and technical communication (Gagné, 1985; Mayer, 2002)[†]. In this chapter, the concept of knowledge is limited to, and treated as, cognitive structures stored in long-term memory. Furthermore, the process of learning new knowledge to apply information when using a tool is viewed from a cognitive constructivist perspective, which is common in learning and instructional design theories (Smith and Ragan, 2005). From this perspective, users construct their knowledge through active learning processes that involve using different tools in various contexts. Such learning situations often involve processes of perceiving, seeking, understanding, and applying designed information. Within the scope of this chapter, the knowledge a user possesses in long-term memory before beginning to use a tool and reading and applying information is referred to as prior knowledge. It is worth noting that users do not have a single, unified prior knowledge. Each individual user holds knowledge of many things—concepts, events, procedures, and aspects of life and the world—which are organised in long-term memory. There are also different conceptual models of how knowledge is organised in long-term memory, such as procedural, semantic, and episodic memory, as well as schemas, mental models, and visual images (Johnson-Laird, 1983; Kosslyn, 1994)[§]. Moreover, no two users have exactly the same knowledge, nor is it feasible to objectively capture or measure their knowledge in precise numerical terms.

[#] See also, Elbro and Buch-Iversen (2013); Garnham and Oakhill (1996); Johnson-Laird (1983); Kintsch (1988); McNamara et al. (1991); McNamara and Kintsch (1996); McNamara et al. (1996); Rawson and Kintsch (2004); Demir et al. (2024).

[†] See also, Carroll (1990).

[§] See also, Bartlett (1932); Paivio (1986); Rumelhart (1986); Tulving (1972).

Using a tool effectively requires a certain level of prior knowledge, which in turn depends on the complexity of the tool. Because much of what users do in both professional and everyday contexts depends on their ability to use tools efficiently and accurately, their prior knowledge plays a crucial role in shaping their performance. The alignment—or mismatch—between the user's prior knowledge and the knowledge required by the tool influences what they can accomplish independently to achieve a desired outcome efficiently, effectively, and with satisfaction (ISO 9241-11, 2018), without relying on external support such as technical communication. Here, the distinction and relationship between different types of knowledge is relevant when examining how prior knowledge shapes a user's ability to use tools. There are many conceptual models for categorising knowledge as learning outcomes. One of the early and pioneering classifications was developed by Benjamin Bloom, who developed a taxonomy of learning outcomes (Darwazeh and Branch, 2015).

In what follows in this text, Robert M. Gagné's (1985) classification of learning outcomes is used as a foundation, as it is useful for examining how prior knowledge shapes a user's ability to use tools, understand and apply technical communication. Gagné is also an educational psychologist, who is known for his influential work on the conditions of learning and instructional design theory. Drawing from both behaviourist and cognitive psychology, Gagné's work bridged the two traditions and helped establish modern instructional design as a scientific discipline. He developed a comprehensive model for understanding how people learn and how instruction can be structured to facilitate learning effectively. Gagné proposed that different kinds of learning—such as declarative knowledge, intellectual skills, cognitive strategies, motor skills, and attitudes—require distinct internal and external conditions to occur. Drawing on his categories of learning outcomes, the remainder of this chapter focuses on two of these categories: declarative knowledge (knowing that)—what a learner can explicitly recall and verbalise—and intellectual skills (knowing how)—what a learner can do. Declarative knowledge relates to the process of semiosis and to knowledge about language, such as knowing what words signify and what things are called. Intellectual skills involve conceptual, principles, and procedural knowledge, which relate to the ability to apply knowledge in new situations not previously encountered during learning.

Instructional design literature emphasises that learners can learn to recall and verbalise the steps involved in a procedure and perform well on assessments that test this knowledge (Smith and Ragan, 2005). In such cases, they have acquired declarative knowledge—that is, they know what to do. When a learner has developed conceptual and principles knowledge, they can, outside the training situation, recall and explain why a step is performed. However, such knowledge does not necessarily mean they can identify the relevant procedure and apply it in a specific situation. Successful application requires procedural knowledge, which involves knowing which procedure is relevant to apply in a given situation, selecting the appropriate steps, and performing the task through coordinated cognitive and motor processes. To learn how to use a tool effectively, it appears necessary that the underlying concepts of the procedure are already within the learner's possession (Smith and Ragan, 2005). However, it does not appear necessary for declarative knowledge to be fully acquired before procedural learning can occur.

A user's prior knowledge can be understood as comprising varying degrees of declarative knowledge and intellectual skills, both of which influence their ability to use a tool. To deepen this understanding, it is relevant to relate declarative knowledge and intellectual skills to the concepts of tacit and explicit knowledge when examining how prior knowledge shapes a user's ability to use tools. Tacit and explicit knowledge were described by Michael Polanyi, who argued that not all knowledge can be explicitly stated or codified in language or rules (Collins, 2010). Much of what humans know is tacit—embodied in skills, experiences, and personal understanding—summarised in Polanyi's phrase, "we know more than we can tell". Nevertheless, to use explicit knowledge meaningfully, a user must rely on tacit processes such as interpretation, judgement, and perception. Polanyi maintained that all explicit knowledge ultimately rests on a tacit foundation, meaning that explicit knowledge does not stand alone ; it is always rooted in tacit knowledge (Collins, 2010). Although Gagné and Polanyi approached knowledge from different disciplinary perspectives—Gagné from instructional psychology and Polanyi from epistemology—their frameworks can be meaningfully compared when discussing instructional design. Gagné's distinction between declarative knowledge (knowing that) and intellectual skills (knowing how) resonates with Polanyi's differentiation between explicit and tacit knowledge. Both describe a transition from symbolic, articulable understanding toward embodied, context-dependent competence.

From this perspective, a user can be a highly skilled practitioner, demonstrating excellence in the craft of using a tool. Such a user possesses substantial tacit knowledge in the form of intellectual skills within their domain. However, having extensive tacit knowledge does not necessarily mean possessing extensive explicit—or declarative—knowledge about the same domain. A lack of explicit knowledge implies limited declarative understanding, which restricts the ability to verbalise tacit knowledge to others. Conversely, a practitioner who can verbalise and explain what they do with a tool, and how they do it, in a way that others can understand, possesses explicit knowledge. Yet a practitioner may also have extensive explicit knowledge about a tool but limited tacit knowledge, which restricts their ability to use the tool effectively in specific work situations. This reasoning suggests that one user may possess substantial intellectual skills—knowledge demonstrated through action—but limited declarative knowledge, while another user may possess substantial declarative knowledge—knowledge demonstrated through recognition, recall, elaboration or explanation—but limited intellectual skills. Here, it is relevant to emphasise that declarative knowledge becomes observable in different ways. Instructional design literature emphasises a difference between surface recall and deep conceptual integration, and between knowing that and being able to explain what one knows. An individual can perform well on written assessments that test recognition or factual recall, yet struggle to verbalise or explain the same information coherently.

Through learning processes, users construct knowledge that enables them to perform specific tasks independently with a tool. Drawing on Gagné's (1985) classification of learning outcomes, the following list illustrates abilities related to the use of tools.

- **Declarative knowledge**
  - Understand the meaning and implications of central terms in the tool domain, and be able to explain them in one's own words.
  - Know the manufacturer's intended purpose of the tool and its main areas of application.
  - Describe, at a general level, how the tool functions and is designed.
  - Explain the risks associated with using the tool and the necessary safety precautions.
  - Identify and describe how the user interface is structured—naming icons, buttons, and menus, and locating specific interface elements.
  - Recall and verbalise the procedural steps required to accomplish a task goal using the tool, including the expected outcome.
- **Conceptual intellectual knowledge**
  - Distinguish and describe differences between similar tool models.
  - Understand and explain how various system outputs—such as messages, error codes, and feedback—relate to each other.
  - Explain the abstract and concrete design concepts on which the tool is based, and identify which user needs they address in specific use situations.
  - Evaluate and select an appropriate tool for a specific user situation.
  - Interpret, analyse, and justify when and why a specific tool can or cannot be used.
  - Assess whether the tool supports a particular task goal, and explain what can and cannot be achieved with it in a given situation.
  - Interpret and explain the reasons behind specific tool feedback (for example, system messages, error codes) and their implications in a user situation.
  - Explain to others the procedural steps required to accomplish a task goal in a particular user context, and why they are required.
- **Principles intellectual knowledge**
  - Predict outcomes and consequences of specific actions or inputs when interacting with the tool.
- **Procedural intellectual knowledge**
  - Evaluate and select appropriate task goals for the tool in a given situation.
  - Analyse and justify why a particular task goal is relevant and appropriate in the current user context.

- Plan, articulate, and perform a sequence of actions to achieve a task goal using the tool (for example, installation, operation, repair) in a given situation.
- Identify the steps necessary to avoid hazards and potential damage during tool use.
- Read and comprehend text and image-based instructions, and apply that understanding to complete a task in a given situation.
- Explain and guide another person in accomplishing a task goal using the tool in a given situation.

If users have not yet learned all the procedural knowledge required to use a tool in a particular situation, they may avoid using the tool altogether or use it incorrectly—either knowingly or unknowingly (Ceaparu et al., 2004; Mendoza and Novick, 2005)[#]. A user may have learned procedural knowledge but forgotten some aspects of it, which can lead to uncertainty, prompting the need for verification or reminders. Users in such situations become dependent on external support, which they may seek either before starting to use the tool or when they encounter difficulties while using it. Such support is typically provided by information designers—such as technical communicators—who design instructions and descriptions to help users, whether professional or non-professional, quickly find relevant information, understand it, and apply that understanding to use the tool effectively, efficiently, safely, and with minimal environmental impact. How individuals perceive, understand, and apply such information depends on factors such as socio economic and cultural background, gender, age, education level, cognitive and perceptual abilities, and motivation to engage with the information (Eriksson and Carlsson, 2022). In the same way that users require prior procedural knowledge to use the tool itself, they also need relevant prior knowledge to understand and apply the information provided in technical communication. Drawing on Gagné's (1985) classification of learning outcomes, prior declarative, conceptual, and principles knowledge enables a user to understand an instruction and restate it in their own words within the reading situation. Depending on the quality of this prior knowledge, it may also enable the user to apply the procedure in a specific context while reading and understanding the instruction. This involves the ability to translate textual or visual information into effective and context-appropriate action.

Importantly, the ability to understand and apply a procedure from reading an instruction does not always indicate that learning has occurred in the sense of constructing new cognitive structures in long-term memory (Smith and Ragan, 2005). In some cases, users may simply aim to complete a one-off task—such as assembling a piece of furniture—without being motivated to retain the knowledge for future use. A user may therefore complete a task without deeper reflection or conscious awareness of what they are doing (Mayer and Fiorella, 2021). To learn from instruction, educational and learning theories emphasise persistence

[#] See also Novick et al. (2007); Welty (2011).

and change over time. Gagné (1985) defined learning as "a change in human disposition or capability that persists over a period of time and is not simply ascribable to processes of growth". Similarly, Mayer (1982) described learning as "the relatively permanent change in a person's knowledge or behaviour due to experience". Mayer's definition highlights three essential components of change for learning to occur: 1) the duration of the change is long-term, rather than short-term; (2) the locus of the change is the content and structure of knowledge in memory or the behaviour of the learner; (3) the cause of the change is the learner's experience in the environment rather than fatigue, motivation, drugs, physical condition, or physiological intervention.

How Lack of Relevant Declarative Knowledge Shapes the Capacity to Understand

A central factor shaping a user's cognitive capacity to understand and apply technical communication is the relationship between their prior knowledge and the prior knowledge implicitly assumed by the information design. This section focuses specifically on why users who possess relevant intellectual skills, but lack declarative knowledge of the tool domain may struggle to understand and apply technical communication that contains words they do not recognise. In this examination, the user's processes of perceiving, seeking, understanding, and applying information when using a tool are examined from a cognitive constructivist perspective, which is commonly adopted in learning and instructional design theories (Smith and Ragan, 2005).

A user who possesses substantial prior intellectual skills related to a tool, but lacks prior declarative knowledge of its language, is likely to be unfamiliar with certain texts and images used in the information design. This unfamiliarity arises because central concepts the user already knows may be expressed through specific domain terminology that they do not recognise. An information design may also include terminology that is overly abstract, rendering it vague because it can refer to several different things. This ambiguity can leave the user uncertain about what a term actually denotes. As a result, the user may only partially understand, misunderstand, or fail to understand the information altogether. In contrast, users who possess the assumed prior declarative knowledge of the specific terminology will find the information partly or fully understandable. Therefore, technical communicators are advised to select texts and images that users are likely to be familiar with, avoid using overly abstract terminology, and provide supportive aids—such as lists of definitions or explanations—to help users interpret words and images that may not be familiar to them.

Viewed from another perspective, every designed information comprising text and images either explicitly states or implicitly assumes that the user possesses certain prior declarative knowledge of the domain's language. The words and nominal phrases written in the definite form signal that the designer assumes the user knows what these terms signify in order to achieve partial or full understanding.

This occurs because reading is an active cognitive process in which the user engages with elements of the design that trigger associated prior knowledge. The process of reading text and images in technical communication displayed on a screen, and applying what is understood when using a tool, is illustrated in Figure 1 and further described as follows.

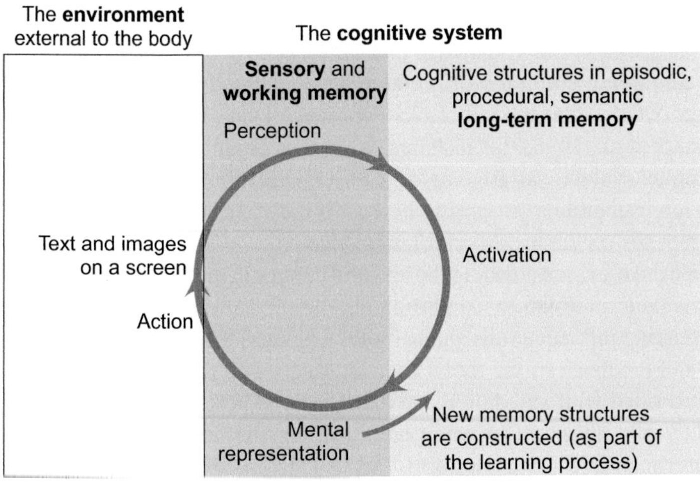

Fig. 1 A user's active process of reading text and images on a screen results in mental representations in working memory by activating prior knowledge stored in long-term memory. These mental representations, in turn, influence the user's motor actions when using a tool. The white area represents the user's external environment—for example, a maintenance workshop—while the grey areas represent the user's cognitive system. The precise shapes of the boxes and the defined endpoints of the arrows may suggest a highly structured and predictable cognitive process, although this is not always the case.

Starting from the text and images on a screen in the user's environment, shown to the left in Figure 1, the user's attention is guided by factors such as beliefs, motivation, intentions, and situational context, which shape what they notice in the designed information (Kintsch, 1998; van Dijk and Kintsch, 1983). The perceived textual and visual elements that enter sensory memory activate associated cognitive knowledge structures in long-term memory, leading to the construction of mental representations in working memory (Mayer and Fiorella, 2021). This activation of prior knowledge is most likely a process of which the user is seldom aware or able to control. The resulting mental representations in working memory are dynamic and evolve as reading progresses. As the user continues to read, the text and images activate additional related knowledge structures in long-term memory, which are integrated into the evolving mental representation to create a more coherent and organised whole—a process often referred to as assimilation and accommodation (Hanfstingl et al., 2021). Such mental representations are often new constructions that the user has not previously formed. The activation of prior knowledge is closely connected to the semiotic process, in which text and images function as signs that signify objects, concepts, or abstract ideas (Kress and Van Leeuwen, 2020). A user's prior learning, experiences, attitudes, and interests shape which knowledge structures are activated during reading, as these factors influence how the user has come to associate particular signs with specific objects, concepts, or ideas. To be familiar with and know words is therefore understood to mean that the user possesses prior knowledge that can be activated. However, an information design containing familiar words and images may still be difficult to understand. This can occur when the text lacks coherence (McNamara and Kintsch, 1996; McNamara et al., 1996), for instance, when it is organised into long sentences or has complex syntactic arrangements. Consequently, the relevance of the resulting mental representations—meaning how well they correspond to the design of the tool—depends on the user's prior knowledge.

Once the user stops reading the text and images in the technical communication and turns their attention to using the tool, the mental representations constructed during reading influence subsequent motor actions. These actions, in turn, generate events and feedback in the tool's user interface. The mental representations that the user formed while reading the text and images before using the tool influence what they pay attention to when later reading the text and images within the tool's user interface (Johnson-Laird, 1983). This process is illustrated in Figure 1, where the grey arrow extending from the mental representation process reaches the environmental area and connects with the grey arrow representing perception. As the user reads the text and images in the tool's user interface, they construct new mental representations, which may then influence what they notice if they subsequently return to the technical communication. Thus, reading, understanding, and applying information can be viewed as a cyclical process of interpretation.

The grey arrow in Figure 1, extending to the right from the mental representation area to the long-term memory area, represents a learning process. If the conditions for learning are met, the mental representations formed in working memory may be internalised into long-term memory as new knowledge. Importantly, previous research emphasises that what is constructed is not the exact wording of the text and images in the information design, but rather the mental representations in working memory that are created during reading (Garnham and Oakhill, 1996).

From this perspective, understanding and applying information have both subjective and functional dimensions. Subjectively, a user experiences understanding when a mental representation feels coherent and makes sense, even if it does not accurately correspond to the design of the tool. This may lead to misunderstandings that they may not recognise. Functionally, applying what is understood is demonstrated through action—the ability to apply information correctly (or incorrectly)—which can be externally assessed and measured (Morrison et al., 2011).

The following two text examples demonstrate an information design that is assumed to be difficult to understand for a user who possesses the relevant intellectual skills but lacks declarative knowledge of the central words it contains.

Each example consists of two versions: an original version that is difficult to understand, and a revised version in which key terms have been replaced to make the text more understandable. In the first example, the initial version represents a text from a fictitious piece of technical communication aimed at a specific audience. It describes a common task in a software tool used to create particular types of documents. This version is difficult to understand for a user who possesses the necessary intellectual skills but is unfamiliar with the specific jargon and slang.

To insert a gizmo in the brief, click the flux operator in the second ribbon lane from the left, then select the gizmo asset in the panel that appears.

This version is assumed to be understandable for an audience familiar with the words gizmo, brief, flux operator, ribbon lane, gizmo asset, and panel. This may be the case in local technical communication written for a small group of users within a workplace who have agreed upon the meaning of these terms.

The second, rewritten version below is likely to be easier to understand for a user who also possesses intellectual skills related to handling image files on a hard drive, editing documents in word-processing software, and inserting visualisations into a document by selecting an image file. However, such a user may not be familiar with the specific jargon and slang used in version one above. The revised version therefore replaces the central words—gizmo, brief, flux operator, ribbon lane, gizmo asset, and panel—with terms the user is more likely to recognise, such as image, document, insert command, menu, image file, and window.

To insert an image into the document, click the insert command in the second menu from the left, then select the image file in the window that appears.

The first version of the second example below describes a common procedure that many users are familiar with. It illustrates the difficulty of understanding that arises when central words are deliberately replaced with overly abstract terminology capable of signifying multiple things.

The procedure is as follows:

1. Group the items according to their general characteristics, such as colour, appearance, texture, and overall condition. One group may be sufficient depending on the number of items.
2. Place one group of items inside the device.
3. Turn on the device.
4. Choose an appropriate setting to treat the items. The suitable setting depends on the items' colour, appearance, texture, and overall condition.
5. Run the chosen setting to complete the treatment.
6. Remove the items from the device, when the treatment is finished.
7. Apply additional treatment to the items so that they are ready for use again.
8. Put the items away where they belong for later use.

The version above does not effectively help the user determine which prior knowledge to activate for the central words items, device, setting, and treatment. These words belong to common vocabulary and do not represent terminology specific to any particular domain. The difficulty lies in their abstraction—they can signify many different things. As a result, the user may struggle to know which prior knowledge to activate when constructing a mental representation, or may form one that does not correspond to the information designer's intention.

In the second version below, a title and a visualisation have been added to help the user identify which prior knowledge to activate (assuming they possess intellectual skills related to washing clothes). It is likely that this second version is easier to understand, as it guides the user towards activating the appropriate intellectual skills and relevant prior knowledge.

How to wash clothes

The procedure is as follows:

1. Group the items according to their general characteristics, such as colour, appearance, texture, and overall condition. One group may be sufficient depending on the number of items.
2. Place one group of items inside the device.
3. Turn on the device.
4. Choose an appropriate setting to treat the items. The suitable setting depends on the items' colour, appearance, texture, and overall condition.
5. Run the chosen setting to complete the treatment.
6. Remove the items from the device, when the treatment is finished.
7. Apply additional treatment to the items so that they are ready for use again.
8. Put the items away where they belong for later use.

The second version of the second example contains exactly the same words as the first version—no replacements have been made—since the aim is to demonstrate how titles and images can assist the user in determining which prior knowledge to activate and in contextualising overly abstract words such as items, device, setting, and treatment. This assumes that the user possesses intellectual skills related to washing clothes, which most adults As this excludes many knowledgeable individuals are likely to have. However, although this second version is easier to understand, technical communicators are advised to avoid using overly abstract terminology.

How Lack of Relevant Intellectual Skills Shapes the Capacity to Understand

The previous section focused on why users who possess relevant intellectual skills but lack declarative knowledge of the tool domain may struggle to understand and apply technical communication containing words they do not recognise. This section specifically focuses on the opposite case: users who lack relevant intellectual skills but possess declarative knowledge of the language used in the technical communication. Such users may also struggle to understand and apply the information when the technical communication assumes certain intellectual skills but does not contain unfamiliar terms.

To begin with, some tools are simple gadgets with low task complexity, requiring only a basic level of intellectual skills and imposing a low intrinsic cognitive load. Other tools, by contrast, are advanced and specialised industrial systems with high task complexity, demanding a high level of intellectual skills and imposing a high intrinsic cognitive load. From this perspective, the amount of intellectual skill required to use a tool independently depends both on the task's complexity and its goal. For example, troubleshooting a problem in a complex industrial system requires different intellectual skills from those needed to locate a menu in the user interface of a simple gadget. Furthermore, experts may occasionally fail to use a simple gadget as intended, and novices may sometimes succeed in using an advanced tool—perhaps without realising it. Nevertheless, users' prior intellectual skills related to a specific tool vary widely: some are beginners with very limited skills, while others are experts with extensive experience of the domain.

Viewed from another perspective, every designed information—comprising text and images—either explicitly states or implicitly assumes that the user possesses certain intellectual skills. When some words and nominal phrases are written in the definite form, they signal that the designer assumes the user knows what these terms signify in order to achieve full, or at least partial, understanding. If the user does not possess the assumed intellectual skills, they will be hindered to construct a relevant mental representation from the text, even when it contains familiar words. A user with extensive prior intellectual skills will activate more relevant knowledge in long-term memory from text and visuals than a user with limited intellectual skills. This enables the knowledgeable user to connect sentences and create a coherent whole while reading. It also means that the information designer can omit certain details related to prior knowledge that the user is assumed to possess.

The following example, a fictitious warning from technical communication for industrial machinery, illustrates this point. The first version assumes certain intellectual skills in order to be understandable. For a user who lacks these skills, constructing a mental representation to achieve understanding becomes difficult.

This large industrial machine is extremely heavy. During winter, do not attempt to move it across the desert.

The first version above, although it does not contain any difficult words, is not understandable to a user who lacks the intellectual skills it assumes. Such a user will not know why the industrial machine mentioned in the text should not be moved across the desert. For a user lacking these intellectual skills, the words winter and desert activate incompatible prior knowledge structures. They create cognitive dissonance: winter evokes a cold environment with snow or ice, whereas desert typically signifies heat and dryness. These conflicting cues prevent the user from constructing a coherent mental representation of the text. The lack of semantic and causal coherence disrupts understanding, as the user cannot reconcile why a frozen condition would exist in a desert. However, the first version is likely to be understandable for users who possess expert intellectual skills related to deserts. Although a user may typically associate deserts with heat and sand, the scientific definition of a desert is an area that receives very little precipitation, generally less than 250 mm per year. By that definition, both the Arctic and Antarctica are cold deserts (also called polar deserts). In such regions, lakes can exist that are often frozen. When reading the warning in a situation involving the transport of a machine over lakes in an Arctic region, the first version above is likely to be understandable for users with such expert intellectual skills. In that context, they would correctly infer that the warning means the machine should not be moved across frozen lakes, as it might break the ice and fall into the water.

In the rewritten second version below, one word has been replaced (desert has been changed to lake):

This large industrial machine is extremely heavy. During winter, do not attempt to move it across the lake.

The second version above is likely to be understandable for users who possess intellectual skills in the concepts and principles about how lakes freeze during winter, how gravity causes heavy objects to break thin ice, and that if objects fall into a lake, they are lost—something an owner of expensive industrial machinery naturally wants to avoid. Most users with such intellectual skills will, therefore, understand why the industrial machine mentioned in the text should not be moved across the lake. However, none of this is stated explicitly in the text; the user is, in effect, reading between the lines. Thus, a given information design may be understandable to some users who possess the assumed prior intellectual skills but incomprehensible to others who lack them.

A user's intellectual skills may also come into conflict with what is conveyed in the information. When there is a contradiction between what is stated and the user's prior knowledge, the user may, in many cases, adhere to what they already know. This can lead them to ignore, reject, or misinterpret the information they read. Such behaviours are related to the cognitive phenomena of misconceptual reading (Kendeou and van den Broek, 2005, 2007; van den Broek and Kendeou, 2008)[#] and confirmation bias (Nickerson, 1998; Klayman, 1995)[†]. According to research on confirmation bias, users tend to engage in selective exposure, biased interpretation, and biased memory recall. The following example of a user interacting with tools and their technical communication illustrates these cognitive phenomena. The mental representations constructed before and during tool use will influence what the user later understands when reading text and images in the technical communication.

In this example, the user is located in an industrial workplace and intends to operate an automated robotic software system used to assemble and weld large steel frames for car production. They are using a new version of the software that they have not used before. The user possesses excellent intellectual skills regarding the functioning and terminology of the previous version of the software. In the new version, some functions have been slightly modified, the layout has changed, and several terms in the user interface have been replaced by others. Although the user is aware that they are using a new version, they do not know what has changed—for example, they are unfamiliar with the new terminology.

The user's goal is to operate the software to initiate an operational sequence in the robot, perform a series of production tasks, and then restart the system. As part of their preparation, the user gathers the necessary tools—such as screwdrivers—and places them in a toolbox, a sturdy container used for storing and carrying small hand tools. The user then puts on protective gloves and a helmet. Drawing upon Johnson-Laird (1983) and Norman (2013), as the user interacts with the workplace environment before starting to use the software, they construct mental representations based on their prior knowledge. These mental representations may influence what the user later pays attention to when reading the software interface. As the user perceives elements of the interface—icons, buttons, labels, and other visuals—mental representations are formed that guide how the interface is understood. After some time, the interface displays a message that reads:

> The production sequence has finished. Dismount the robotic arm and place it in the toolbox to load the configuration data before starting the next sequence.

The user's mental representation when reading this message does not include the instruction "place the robotic arm in the toolbox", as this statement contradicts the user's already activated prior knowledge. A robotic arm is a large piece of equipment that clearly cannot fit into the small toolbox the user has placed nearby. In this case, the user's reading is influenced by confirmation bias: they have

---

[#] See also, Chinn and Brewer (1993).

[†] See also, Lord et al. (1979); Kahneman (2011).

rejected, ignored, or misinterpreted what they read because it conflicts with their existing understanding.

Later, the user seeks to end the robotic sequence by shutting down and restarting the software system. The new software version introduces minor modifications to the user interface layout, and the icons the user recognises from the previous version no longer represent the same functions—something the user is not entirely sure of. This uncertainty prompts them to consult the technical communication for clarification. To locate the relevant instruction, the user must search, skim, and judge the relevance of what they find—a process that also relies on prior knowledge (Lundin, 2020). Once a potentially relevant instruction is identified, the user reads it to understand. The instruction includes the text:

Activating this function will automatically repart the system after shut down.

Before reading, the user had already activated prior knowledge and constructed a mental representation related to restarting the system. They attempt to fit what they read in the instruction into that pre-existing representation, reinterpreting the word repart as restart. The mental representation constructed from reading the technical communication thus leads the user to believe that activating the function will automatically restart the system after shutdown. However, in this context, the word repart refers to a new feature in the updated software version—repartitioning the hard drive to change how the storage space is divided into partitions. When the user clicks the icon to activate the repart function, the software does not shut down and restart as expected. The behaviour of the software contradicts the user's mental model, leaving them uncertain about how to proceed and possibly experiencing confusion, frustration, or anxiety.

Implications on the Design of Technical Communication

The knowledge gap between a user's prior knowledge and the knowledge required to use a tool effectively influences the information design. The amount of information to include, and how to organise it, aims to bridging this gap. An information design intended for users with limited prior declarative knowledge and intellectual skills who are using a highly complex tool will differ substantially from one aimed at users who possess extensive prior knowledge and skills when using a less complex consumer gadget.

When the gap between what the user already knows and what they need to know is large, the user must engage in learning new knowledge. In this context, learning is viewed as an active cognitive process of constructing new knowledge, which occurs when the conditions for learning are met. Following Gagné (1985), these conditions involve three key components: a change in cognitive structures in long-term memory that can be activated later in new situations, stability or durability of this change over time, and the occurrence of the change as a response to active and motivated behaviours directed toward learning. Instructional design theory further emphasises that learning requires certain supportive conditions, such as motivation, appropriate materials, and conducive contexts (Gagné, 1985). A user with limited prior knowledge cannot bridge the gap by processing and internalising large amounts of new information at once, because working memory is limited (Sweller, 1988; Sweller et al., 2011). Instead, learning must occur in segments aligned with specific learning goals, where later segments build on earlier ones. As users construct new knowledge toward successive learning goals, they gradually accumulate understanding and develop the capacity to benefit from more advanced instruction. This capacity—representing the distance between what they can achieve independently and what they can achieve with guidance—is referred to as their zone of proximal development (Cole et al., 1978). According to instructional design models, a typical learning curriculum consists of multiple learning goals that can be organised as concept-related or prerequisite-related. Learners typically first acquire declarative knowledge, such as the meanings of terms. Once this knowledge has been acquired, it becomes relevant to continue developing intellectual skills, including concepts, principles, and procedures, and later to progress towards higher-level problem-solving abilities (Gagné, 1985). Effective instructional designs encourage users to think, repeat, elaborate, discuss, and express concepts in their own words. Research on generative learning shows that learners understand and recall more effectively when they create their own associations rather than receiving them passively. Such activities require time, effort, and cognitive investment, as deep learning involves actively relating new information to prior knowledge. This effect is explained by the depth of processing framework (Craik and Lockhart, 1972): the more learners integrate new information with their existing cognitive structures through elaboration, the deeper the processing and the more durable the learning.

From an information design perspective, the interaction between text and images can, under certain conditions, enable users to learn more deeply than from text or images alone. This idea builds on the view that the human cognitive system functions as a dual-channel, limited-capacity system, consisting of verbal and visual channels that process information in parallel (Mayer and Fiorella, 2021; Paivio, 1986)[#]. Verbal information—such as written or spoken text—is processed primarily through the verbal channel, whereas visual information—such as photographs, illustrations, or animations—is processed through the visual channel. Users tend to learn more deeply when engaging with information designs in which text and images are meaningfully integrated, as this interaction requires cognitive effort to build coherent connections between verbal and visual representations. Because depth of processing is crucial for learning, such integrated processing generally promotes greater understanding than processing text or images in isolation. However, for users with extensive prior knowledge, this type of design can occasionally have the opposite effect. Redundant information may impose unnecessary cognitive load—a phenomenon known as the expertise reversal effect

[#] See also, Baddeley (1992); Sweller (2011).

(Kalyuga et al., 2003; Sweller, 2011). This highlights the importance of adapting information design to users' prior declarative knowledge and intellectual skills.

The resolution of the action steps in an instruction—that is, the degree of specificity in how many steps the instruction is divided into—has implications for both initial performance and subsequent learning and transfer (Eiriksdottir and Catrambone, 2011). High-resolution instructions tend to improve initial performance, but learning and transfer often decline. In contrast, low-resolution instructions may impede initial performance but typically result in improved learning and transfer. High-resolution instructions appear to encourage users to rely heavily on the instructions, whereas low-resolution instructions require users to engage more actively with the task, since they must infer the specific actions needed.

Building on the examination in the previous sections, several suggestions can be offered for how information design in technical communication can be made useful for users in different situations—whether the gap between their prior knowledge and the knowledge required to use the tool is small or large. The following descriptions serve as personas, each accompanied by suggestions for how technical communication can be designed to meet their needs. Figure 2

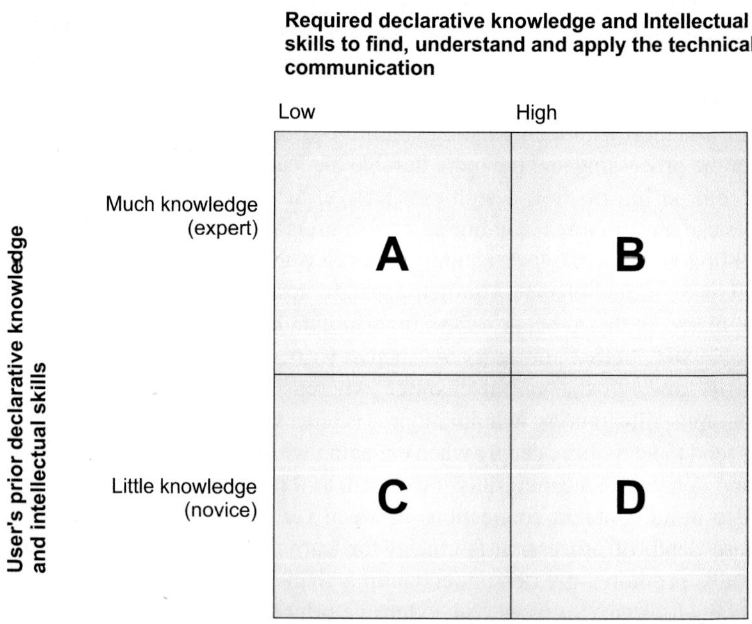

Fig. 3 The gap between the user's existing prior knowledge in the usage situation and the knowledge required for technical communication to be useful—as a support for completing a tool-related task—increases from situation A to D. The precise lines and contours of the boxes may give the impression of a clearly defined division of a user's knowledge, which is seldom the case.

visualises a simplified categorisation of four situations illustrating the relationship between a user's prior knowledge (the vertical axis) and the knowledge required for technical communication to be useful as a support for completing a tool-related task (horizontal axis).

Situation A is characterised by an expert user who possesses extensive prior declarative knowledge and intellectual skills related to the tool domain. In this case, the task to be performed using the tool has relatively low complexity—such as, recording work hours in a basic office software application—which requires little additional knowledge. This type of user generally does not need support from technical communication, as they have used the tool many times and are familiar with almost all procedures. When they encounter a task they have not previously performed, they can usually figure it out based on their expertise. However, there may be occasions when they have forgotten a specific detail of a step learnt some time ago—such as the name of a menu command—and wish to be reminded. Alternatively, they might feel slightly uncertain and seek confirmation that their knowledge and procedures are correct. A typical user behaviour in such cases is to begin the work task, realise they have forgotten a detail, and then pause tool use to search for and read supporting information in the technical communication. For users in situation A, suitable information designs include descriptive resources such as reference material—for example, tables of menu commands. Instructions may have high specificity, meaning they contain few overall steps. The information design can employ specific domain terminology and omit lengthy explanations of concepts or principles, as the user is assumed to be capable of bridging sentences and filling in contextual gaps. Because these users possess substantial prior knowledge, they are unlikely to experience difficulty locating the relevant instruction or description, even in extensive technical communication containing numerous instructions.

Situation B is characterised by an expert user who possesses extensive prior declarative knowledge and intellectual skills within the tool domain. In this case, the task to be performed using the tool has relatively high complexity—such as configuring an advanced industrial product—which requires substantial prior knowledge. This type of user may have used the tool extensively and therefore knows which tasks can be performed and most of the associated procedures. Similar to users in situation A, a user in situation B may occasionally forget a detail in a step learnt some time ago—such as the name of a system parameter—and wish to be reminded. Alternatively, they might feel uncertain and seek confirmation that their knowledge and intended procedure are correct. A typical user behaviour when encountering a task they have not previously performed is to begin using the tool and only later realise that they do not know how to complete it. In such situations, they may initially believe they can work it out based on their expertise, but when they try, they fail and become stuck. Another behaviour pattern is that the user recognises beforehand that they do not know how to proceed and therefore searches for and reads supporting technical communication before starting the task. For users in situation B, suitable technical communication designs include instructions with low step specificity, in which the user is expected to infer the intermediate steps. The design can employ domain-specific terminology and omit detailed explanations of concepts and principles, as the user is assumed to bridge sentences and fill in contextual gaps. Because these users possess substantial prior knowledge, they are unlikely to have significant difficulty locating the relevant instruction or description, even within extensive technical communication containing numerous instructions. A user in this situation may also appreciate descriptive information explaining how a specific component of the tool functions, which allows them to draw their own conclusions about how to apply that understanding in practice.

Situation C is characterised by a beginner user who possesses little prior declarative knowledge or intellectual skills within the tool domain. In this case, the task to be performed using the tool has relatively low complexity—such as recording work hours in a basic office software application—which requires minimal prior knowledge. This type of user may not know any of the procedures, as they have never used the specific tool before. They might be aware of the kinds of tasks that are typically possible but not of how to perform them. Because they have never used the tool, they must search for and read supporting information when attempting to complete a task. Such users may also assume that a certain procedure is possible in the tool when, in fact, it is not supported. If the tool is one they use only occasionally—or perhaps just once—they may have little motivation to learn. A typical user behaviour in this situation is to attempt the task through trial and error, especially when no risk of damage is involved. As they lack relevant intellectual skills, they may believe they can work out how to proceed based on prior experience with similar tools. However, the mental representations they form may be misconceptual, leading them to attempt actions that the tool does not support. When such attempts fail, the user becomes stuck. Another pattern is that the user recognises beforehand that they do not know how to complete the task and therefore searches for and reads the technical communication before starting work. For users in situation C, suitable technical communication designs include instructions with high step specificity, since these users are not expected to infer intermediate steps. The design should avoid highly specific domain terminology; if such terminology is necessary, it should be accompanied by explanations that provide declarative knowledge. The information design may include descriptions of concepts and principles, but these descriptions must not omit essential information or refer to domain knowledge the user does not possess. Because users in this category have limited prior knowledge—and because locating a relevant instruction itself requires some prior knowledge (Lundin, 2000)—they are likely to encounter difficulties finding the information they need in large collections of instructions. The design should, therefore, avoid creating a "chicken-or-the-egg" problem, where the user requires prior knowledge to find an instruction, yet the purpose of reading the instruction is to acquire that very knowledge. In such cases, a minimalist design approach may be appropriate, as it encourages exploration and discovery, supporting the user's gradual learning.

Situation D is characterised by a beginner user who possesses little prior declarative knowledge or intellectual skills within the tool domain. They may be new to a job that involves the use of specific tools. In this situation, the task to be performed using the tool has high complexity—such as operating or configuring an advanced industrial product—which requires substantial prior knowledge. This type of user may never have used the specific tool before and might not know what the tool is for, which tasks are relevant to perform, or how to carry them out. If the tool is one they will use frequently, such as daily in their work, they may be motivated to learn. A possible learning strategy could be to explore and discover the tool independently through trial and error. However, such an approach is unlikely to be effective and may instead increase the user's confusion and frustration. For this situation, providing a comprehensive set of technical communication containing all instructions for every possible task—organised in a hierarchical structure—is not appropriate. Such a user will not know what information they need or what to search for. The technical communication design for users in situation D therefore cannot rely on typical instructions consisting of a list of steps for an advanced task. Instead, these users require an instructional design that supports learning before doing: first acquiring declarative knowledge, followed by the development of intellectual skills such as concepts, principles, and procedures. In global technical communication, the information design can explicitly state the level of prior declarative knowledge and intellectual skills assumed of the user. It can also provide references to information that enables the user to acquire the required knowledge. Such guidance can be presented in an advanced organiser or pre training section placed before the procedural instructions. In digital environments, domain-specific terms can be highlighted so that definitions appear when hovering over them (for example, as tooltips). This supports learning in context and helps bridge the gap between the user's existing knowledge and what is required to apply the information effectively.

Discussion and Conclusions

This chapter has examined the cognitive, linguistic, and design-related conditions that determine whether users can understand and apply technical communication to use tools effectively, efficiently, and safely. Central to this examination has been the relationship between users' prior knowledge—both declarative knowledge and intellectual skills—and the prior knowledge implicitly assumed by the information design. Understanding is not simply a matter of decoding linguistic or visual representations; it involves constructing meaning through mental representations, where new information is integrated with existing knowledge from long-term memory. The success of this integration determines whether a user can perform a task independently and with confidence.

From a cognitive perspective, users' prior knowledge functions both as an enabler and a constraint on understanding. When information design aligns with users' prior knowledge, understanding proceeds smoothly because users can draw upon their long-term memory to construct coherent mental representations in working memory (Kintsch, 1998). When the design presupposes knowledge the user lacks, cognitive overload may occur and understanding may fail, leading to partial or incorrect comprehension. The result is often hesitation, error, or reliance on external assistance. The gap between what designers assume users know and what users actually know represents a persistent challenge in technical communication.

A user's likelihood of understanding and applying technical communication depends both on their prior knowledge and on how the information is designed for its intended audience. A user lacking the prior knowledge assumed by technical communication written for specialists will struggle to understand it fully and may only grasp fragments of the information. Conversely, experts reading information intended for novices may experience cognitive interference and boredom—a phenomenon known as the expertise reversal effect (Kalyuga et al., 2003).

This raises an important question about the purpose of technical communication: should it aim to support users when there is a large gap between what they know and what they need to know to use the tool, or should it primarily facilitate effective task completion when that gap is relatively small? It can be argued that technical communication is less relevant for users with very limited prior knowledge who must operate advanced tools requiring substantial expertise. For such users, an instructional design approach is more appropriate—one that supports deep learning through the construction of new knowledge. This type of information design engages users in learning for future performance rather than immediate task execution. Minimalist instructional design (Carroll, 1990) is relevant in these contexts, as it encourages exploration, error recovery, and learning by doing. However, such learning demands time, cognitive effort, and supportive pedagogical environments—conditions typically unavailable to users performing work tasks under time pressure.

Users in situations where the knowledge gap between what they know and what they need to know to use the tool is relatively small can operate within the flow of work. They may seek information to remind themselves of procedures they have already learned or to confirm assumptions about a task. In these contexts, technical communication is most effective when it provides concise, contextually relevant support that enables immediate action, rather than promoting extensive conceptual learning.

Even when technical communication is not explicitly designed for learning, users may still treat it as a learning resource to deepen understanding or reflect on their actions. Thus, the boundaries between learning and use are fluid, shaped by users' goals, contexts, and cognitive readiness. Technical communication may also aim to influence user behaviour. Furthermore, even when users fully understand the information, they may still disregard it if it conflicts with their prior knowledge, particularly when that knowledge is embedded in socio cultural practices in their workplace. This illustrates that understanding alone does not guarantee compliance or effective action—the socio cultural context mediates how information is interpreted and enacted.

Designing technical communication that effectively supports users requires an understanding of their prior knowledge and the conditions under which they interact with the information. Technical communicators must assess not only what users know, but also how they think, behave, and collaborate within their socio cultural environments. Methods such as design thinking, participatory design, and research through design enable communicators to investigate users' prior knowledge, needs, behaviours, and contexts through empirical and iterative inquiry. This involves testing and refining designs based on user feedback until they align with real expectations and experiences. Such methods often assume that technical communication follows a linear design process in which the designer creates and refines the information before it is distributed to users. However, this linearity can be challenged by situated and collaborative design processes that occur directly within the environment of use. In such settings, roles merge: users become co-designers, and designers learn from users' lived practices. This perspective aligns with a dialogic, participatory view of design, where information evolves through interaction rather than being finalised and delivered.

Identifying users' prior knowledge is particularly challenging when the audience is global and heterogeneous, as is often the case in global technical communication. The task for the technical communicator is then to design information that remains effective across variations in prior knowledge, language, and context. This becomes especially demanding when global technical communication is developed in parallel with tool development, aiming to be ready for market release. Technical communicators must therefore design information without knowing exactly when, where, or by whom it will be used. This temporal and geographical distance between design and use makes it extremely difficult to assess the prior knowledge users will bring to the task. The result is often information based on assumed user models that do not accurately represent the diversity of actual users.

In summary, the usefulness of technical communication depends on how effectively it aligns with users' prior knowledge, cognitive capacities, and situational contexts. Understanding and applying information require the activation of relevant prior knowledge and the construction of coherent mental representations. Technical communication is most valuable when it helps users recall and apply what they already know, or when it confirms their assumptions in the course of task execution. For deep learning, however, instructional design is more appropriate. For technical communicators, this underscores the importance of a cognitively informed, context-sensitive approach that integrates insights from cognitive psychology, instructional design, and participatory design. It also highlights their ethical responsibility to create information that is accurate, cognitively accessible, and meaningful to users working under diverse conditions. Future research should continue to explore why and how a user's prior knowledge shapes their cognitive capacity to understand and apply information in an increasingly complex, tool-dependent world.

References

Baddeley, A.D. 1992. Working memory. *Science, 255*(5044): 556–559.

Bartlett, F.C. 1932. *Remembering: A Study in Experimental and Social Psychology*. Cambridge University Press.

Caillies, S., Denhière, G. and Kintsch, W. 2002. The effect of prior knowledge on understanding from text: Evidence from primed recognition. *European Journal of Cognitive Psychology, 14*(2): 267–286.

Cain, K., Oakhill, J.V., Barnes, M.A. and Bryant, P.E. 2001. Comprehension skill, inference-making ability and their relation to knowledge. *Memory and Cognition, 29*(6): 850–859.

Carroll, J.M. 1990. *The Nurnberg Funnel: Designing Minimalist Instruction for Practical Computer Skill*. MIT Press.

Ceaparu, I., Lazar, J., Bessiere, K., Robinson, J. and Shneiderman, B. 2004. Determining causes and severity of end-user frustration. *International Journal of Human-Computer Interaction, 17*(3): 333–356.

Chinn, C.A. and Brewer, W.F. 1993. The role of anomalous data in knowledge acquisition: A theoretical framework and implications for science instruction. *Review of Educational Research, 63*(1): 1–49.

Cole, M., John-Steiner, V., Scribner, S. and Souberman, E. 1978. *Mind in Society: The Development of Higher Psychological Processes*. Harvard University Press.

Collins, H. 2010. *Tacit and Explicit Knowledge*. University of Chicago Press.

Craik, F.I. M. and Lockhart, R.S. 1972. Levels of processing: A framework for memory research. *Journal of Verbal Learning and Verbal Behavior, 11*(6): 671–684.

Darwazeh, A.N. and Branch, R.M. 2015. A revision to the revised Bloom's taxonomy. In: *2015 Annual Proceedings–Indianapolis*, (Vol. 2, pp. 220–225).

Demir, B., Haverkamp, Y.E., Braasch, J.L. and Bråten, I. (2024). Investigating the role of prior knowledge in comprehending intratextual and intertextual relationships when reading multiple texts. *Learning and Individual Differences, 111*: 102442.

Eiriksdottir, E. and Catrambone, R. 2011. Procedural instructions, principles and examples: How to structure instructions for procedural tasks to enhance performance, learning and transfer. *Human Factors, 53*(6): 749–770.

Elbro, C. and Buch-Iversen, I. 2013. Activation of background knowledge for inference making: Effects on reading comprehension. *Scientific Studies of Reading, 17*(6): 435–452.

Eriksson, Y. and Carlsson, A.L. 2022. *The Challenge of Designing Meaningful Information: In Different Perspectives in Design Thinking* (pp. 67–87). CRC Press.

Følstad, A., Engen, V., Haugstveit, I.M. and Pickering, B. 2017. Automation in human-machine networks: How increasing machine agency affects human agency. In: *International Conference on Man-Machine Interactions on Advances in Intelligent Systems and Computing*.

Gagné, R.M. 1985. *The Conditions of Learning and Theory of Instruction* (4th Edn.). Holt, Rinehart and Winston.

Garnham, A. and Oakhill, J. 1996. The mental models theory of language comprehension. In: A.C. Graesser and B.K. Britton (Eds.), *Models of Understanding Text* (pp. 313–339). Erlbaum.

Hanfstingl, B., Arzenšek, A., Apschner, J. and Gölly, K. I. 2021. Assimilation and accommodation. *European Psychologist*.

International Organization for Standardization. 2018. ISO 9241-11:2018 – *Ergonomics of Human-System Interaction – Part 11: Usability: Definitions and Concepts*. ISO.

Janssen, C.P., Donker, S.F., Brumby, D.P. and Kun, A.L. 2019. History and future of human-automation interaction. *International Journal of Human-Computer Studies, 131*: 99–107.

Johnson-Laird, P. 1983. *Mental Models: Towards a Cognitive Science of Language, Inference and Consciousness*. Harvard University Press.

Kahneman, D. 2011. *Thinking, Fast and Slow*. Farrar, Straus and Giroux.

Kalyuga, S., Ayres, P., Chandler, P. and Sweller, J. 2003. The expertise reversal effect. *Educational Psychologist*, *38*(1): 23–31.

Kendeou, P. and van den Broek, P. 2005. The effects of readers' misconceptions on comprehension of scientific text. *Journal of Educational Psychology*, *97*(2): 235–245.

Kendeou, P. and van den Broek, P. 2007. The effects of prior knowledge and text structure on comprehension processes during reading of scientific texts. *Memory and Cognition*, *35*(7): 1567–1577.

Kintsch, W. 1988. The role of knowledge in discourse comprehension: A construction–integration model. *Psychological Review*, *95*(2): 163–182.

Kintsch, W. 1998. *Comprehension: A Paradigm for Cognition*. Cambridge University Press.

Klayman, J. 1995. Varieties of confirmation bias. *Psychology of Learning and Motivation*, *32*: 385–418.

Kosslyn, S.M. 1994. *Image and Brain: The Resolution of the Imagery Debate*. MIT Press.

Kress, G. and Van Leeuwen, T. 2020. *Reading Images: The Grammar of Visual Design*. Routledge.

Lord, C.G., Ross, L. and Lepper, M.R. 1979. Biased assimilation and attitude polarization: The effects of prior theories on subsequently considered evidence. *Journal of Personality and Social Psychology*, *37*(11): 2098–2109.

Lundin, J. 2020. *Shaping Thought through Action: A Study of the Use and Design of Technical Information* [Doctoral dissertation, Mälardalen University]. Mälardalen University Press Dissertations.

Martin, A.P., Ivory, M.Y., Megraw, R. and Slabosky, B. 2005. *Exploring the Persistent Problem of User Assistance* (Technical Report IS-TR-2005-08-01). University of Washington.

Mayer, R.E. 1982. Learning. In: H.E. Mitzel (Ed.), *Encyclopedia of Educational Research* (5th Edn., Vol. 3, pp. 1040–1058). Free Press.

Mayer, R.E. 2002. Multimedia learning. In: *Psychology of Learning and Motivation* (Vol. 41, pp. 85–139). Academic Press.

Mayer, R.E. and Fiorella, L. (Eds.). 2021. *The Cambridge Handbook of Multimedia Learning* (3rd Edn.). Cambridge University Press.

McNamara, D.S. and Kintsch, W. 1996. Learning from texts: Effects of prior knowledge and text coherence. *Discourse Processes*, *22*(3): 247–288.

McNamara, D.S., Kintsch, E., Songer, N.B. and Kintsch, W. 1996. Are good texts always better? Interactions of text coherence, background knowledge and levels of understanding in learning from text. *Cognition and Instruction*, *14*(1): 1–43.

McNamara, T.P., Miller, D.L. and Bransford, J.D. 1991. Mental models and reading comprehension. In: R. Barr, M. L. Kamil, P. B. Mosenthal and P.D. Pearson (Eds.), *Handbook of Reading Research* (Vol. 2, pp. 490–511). Erlbaum.

Mendoza, V. and Novick, D. 2005. Usability over time. In: *Proceedings of the 23rd Annual international Conference on Design of Communication: Documenting and Designing for Pervasive Information* (pp. 151–158). ACM.

Mishrif, A. and Khan, A. 2023. Technology adoption as a survival strategy for small and medium enterprises during COVID-19. *Journal of Innovation and Entrepreneurship*, *12*(1): 1–23.

Morrison, G.R., Ross, S.M., Kalman, H.K. and Kemp, J.E. 2011. *Designing Effective Instruction* (6th Edn.). John Wiley and Sons.

Nickerson, R.S. 1998. Confirmation bias: A ubiquitous phenomenon in many guises. *Review of General Psychology*, *2*(2): 175–220.

Norman, D. 2013. *The Design of Everyday Things: Revised and Expanded*. Basic Books.

Novick, D., Elizalde, E. and Bean, N. 2007. Toward a more accurate view of when and how people seek help with computer applications. In: *Proceedings of the 25th Annual ACM International Conference on Design of Communication* (pp. 95–102). ACM.

Novick, D. and Ward, K. 2006a. Why don't people read the manual? In: *Proceedings of the 24th Annual ACM International Conference on Design of Communication* (pp. 11–18). ACM.

Novick, D. and Ward, K. 2006b. What users say they want in documentation. In: *Proceedings of the 24th Annual ACM International Conference on Design of Communication* (pp. 84–91). ACM.

Paivio, A. 1986. *Mental Representations: A Dual Coding Approach*. Oxford University Press.

Rawson, K.A. and Kintsch, W. 2004. Exploring encoding and retrieval effects of background information on text memory. *Discourse Processes*, *38*(3): 323–344.

Rumelhart, D.E. and McClelland, J.L. (Eds.). 1986. *Parallel Distributed Processing: Explorations in the Microstructure of Cognition* (Vol. 1: Foundations). MIT Press.

Sindakis, S. and Showkat, G. 2024. The digital revolution in India: Bridging the gap in rural technology adoption. *Journal of Innovation and Entrepreneurship*, *13*(1): 29.

Smith, P.L. and Ragan, T.J. 2005. *Instructional Design* (3rd Edn.). John Wiley and Sons.

Sweller, J. (1988). Cognitive load during problem solving: Effects on learning. *Cognitive Science*, *12*(2): 257–285.

Sweller, J. 2011. Cognitive load theory. In: J.P. Mestre and B.H. Ross (Eds.), *The Psychology of Learning and Motivation* (Vol. 55, pp. 37–76). Academic Press.

Sweller, J., Ayres, P. and Kalyuga, S. 2011. *Cognitive Load Theory*. Springer.

Tulving, E. 1972. Episodic and semantic memory. In: E. Tulving and W. Donaldson (Eds.), *Organization of Memory* (pp. 381–403). Academic Press.

van den Broek, P. and Kendeou, P. 2008. Cognitive processes in comprehension of science texts: The role of co-activation in confronting misconceptions. *Applied Cognitive Psychology*, *22*(3): 335–351.

van Dijk, T.A. and Kintsch, W. 1983. *Strategies of Discourse Comprehension*. Academic Press.

Welty, C.J. 2011. Usage of and satisfaction with online help vs. search engines for aid in software use. In: *Proceedings of the 29th ACM international Conference on Design of Communication* (pp. 203–210). ACM.

Wieringa, D., Moore, C. and Barnes, V.E. 1998. *Procedure Writing: Principles and Practices*. Battelle Press.

8 Communication Spaces
A Framework for Information Design

Jennie Schaeffer[1*]

Introduction

In this chapter, we explore the concept of *information design*. By examining different forms of information and their relationships within the communication spaces that we experience, the chapter offers a theoretical framing that expands the understanding of information design. The chapter presents the *Communication Space Framework* and proposes a new tentative definition of information design.

For students of information design and design practitioners, this chapter offers alternative perspectives on the role of information design and aims to provide new entry points to the creation of wayfinding systems, signage, text-based environments, illustrations, exhibitions, and other communicative elements embedded in physical or virtual settings. It emphasises how we read, interpret, speak, and act within space, how signage and text do not simply deliver information, but participate in shaping our meaning-making.

The chapter contains examples that explore how the *Communication Space Framework* might be translated into action. It does not offer fixed answers, but rather suggestions and starting points for how one could begin to experiment with putting the thoughts into practice in different contexts. The chapter opens ways of thinking about the relationship between communication, environment, and embodied experience.

Background: Information Design

Just what is information design, anyway? said Rob Waller, a founder and editor of *IDJ Information Design Journal*, in a publication on the origins of the Information Design Association in 1996. He describes information design as cross-disciplinary and integrative in its approach (Waller, 1996).

The term *information design* has several descriptions and emerged as a multidisciplinary area of study in the 1970s. The term has its roots in graphic design professionals' view of the corporate and surficial view of graphic design at the time (Waller, 1996). *The Information Design Journal* started in 1979 with the

---

[1] Ph.D. Associate Professor at Departement of Health Science, Innovation and Design, Museum director, Västmanlands Läns Museum, Karlsgatan 2, 722 14 Västerås, Sweden.

* Corresponding author: jennie.andersson.schaeffer@regionvastmanland.se

DOI: 10.1201/9781032642598-8

intention to use "processes of design (that is, planning) to the communication of information (its content and language as well as its form)" (Waller, 1996). The related International Institute for Information Design (IIID) started 1987 and Information Design Association (IDA) was established in 1991. In 1982, Edward Tufte's book on information design, *The Visual Display of Quantitative Information* was published, a classic book on statistical graphics, charts, and tables. In 2002, the professor in Information Design at Mälardalen University, Sweden, Rune Pettersson, wrote the book, *Information Design : An Introduction*, explaining verbo-visual communication, information, and message design principles, design processes, and design tools. Based on theories for verbo-visual communication, his book offered practical guidelines for information design including the use of text, symbols, visuals, typography, and layout in information and learning materials.

Historically, information design has often focused on graphical form, visual forms of information, signs, text, and images in relation to each other. But, the definition of information design has also included the environment, as in Simlinger (2007: 8):

> *Information design is the defining, planning, and shaping of the contents of a message and the environments in which it is presented, with the intention of satisfying the information needs of the intended recipients.*

With the perspective of information design as a cross-disciplinary area of study and the fact that the produced information design artefacts in our everyday lives are experienced and integrated in a multimodal physical space, this chapter aims to spread light on the relations in this multimodal communication space.

The impact that the design of spaces has on our experiences and meaning making is considerable. To say that message is *presented* in an environment to *satisfy information needs*, as in the citation above, is just a half truth. From a phenomenological perspective, the experience of space is not merely about physical dimensions or measurable distances, but is deeply rooted in the lived, embodied perception of the individual. Phenomenologists such as the French philosopher Maurice Merleau-Ponty emphasize that space is experienced through the body as a situated and intentional agent, where perception is shaped by movement, orientation, and interaction within the environment (Merleau-Ponty, 1962). Rather than being a static container, space emerges as a dynamic field of meaning, co-constituted by the subject and the surrounding world (Casey, 1997). This means that a space is not perceived neutrally or abstractly, but through the atmosphere it evokes, the affordances it presents, and the existential sense it produces (Böhme, 2013). For example, the same room can be felt as welcoming or oppressive depending on lighting, spatial layout, and/or prior experiences, which underline the subjective and affective nature of spatial perception. Merleau-Ponty's ideas are particularly useful within the field of information design, because he offers a way of understanding how users perceive and interpret information within spatial and embodied contexts.

Situated Meaning-making

In the line of reasoning from the theories of Merleau-Ponty, information design cannot be reduced to choices of color, typography, or icon placement, it concerns how a person experiences being in relation to information. How the body moves through a room, how the gaze is guided by visual elements, the evoked emotions, or how a digital interface feels intuitive are all part of this experience.

A common area of focus for research within the field of information design, which already addresses information design as situated meaning making in various studies, is wayshowing and wayfinding (for further reading on this topic, see also Uggla's chapter in this book). Many studies within the field concentrate on wayshowing and/or wayfinding, as these processes play a central role in how users navigate, interact with, and make sense of complex environments. These concepts go beyond the simple transmission of directional information, they involve the design of cues, spaces, and narratives that support users in constructing meaning based on their immediate context. In this sense, wayshowing and wayfinding serves as a practical example of how information design is inherently situated, shaped by both physical surroundings and user experience in the moment. This makes it a rich site for exploring how meaning is co-created between designed artefacts, space, and experience.

For example, the associate professor in Wayfinding/Inclusive Design at Birmingham City University, Colette Miller, and designer David Lewis (2000) identify several environmental factors that influence individuals' directional decision-making in healthcare settings. They show how poorly designed physical settings can lead to feelings of confusion, stress, and even impact the quality of care. Through their analysis of 19 healthcare and 8 non-healthcare facilities, they propose guidelines for designing effective wayfinding systems, emphasizing clear signage, consistent design, and logical spatial organisation to reduce stress and confusion among visitors. Coming back to the thinking of Merleau-Ponty, all understanding is situated. Information design is, therefore, a part of the environment and context and the user experience. It is therefore not enough to create a general logic; meaningful information is locally and bodily anchored. To take another example from research that emphasizes what people do, i. e., how they act in relation to wayfinding systems in a hospital in a state of stress, urgency, and the need for clear direction, design researcher Kelli Smythe (2018) gives voice to the individual's experience in a complex hospital setting, where wayfinding is not only a technical matter but also a cognitive and emotional process. In her study, we follow the person in the space: searching for signs, asking for help, interpreting spatial cues. These moments highlight how meaning is made in action and in context.

Wayshowing and wayfinding, in this light, becomes more than the presentation of information, it is about enabling orientation and understanding in real-time. It illustrates how information design operates as *situated meaning-making*, shaped by physical surroundings, individual needs, and emotional states. As such, it serves as a powerful example of how design both supports and is interpreted through human experience.

Consequently, we do not experience any single object in isolation, no printed material, no small illustration, no set of instructions. Rather, everything is perceived within a context. It is the entire environment that surrounds us and shapes our perception, where elements, for example, light, sound, scale, materials, and objects come together as a unified whole. This context influences how we feel and how we make meaning in the situation, forming the room not just a collection of things, but a living atmosphere that guides our experience.

Extending Merleau-Ponty's thinking, particularly through the work of the German philosopher Gernot Böhme, design can be understood as communicating through atmosphere: a feeling or mood that emerges within a space or interface. For example, the experience of a visitor in a museum exhibition space is affected by the whole environment, since museum spaces are inseparable from their contents (Simonsson, 2014). A text panel on the wall in a museum, for instance, is not just read as text, but also in relation to its position, lighting, and spatial integration. The room "speaks" through atmosphere, which, as Böhme (2013) suggests, is not only an aesthetic quality but a felt presence that envelops and shapes the visitor's experience. Thus, content, text, and space merge into a continuous experiential field rather than separate elements (Figure 1).

For example: Imagine a child entering a dimly lit room at a museum designed to immerse them in eerie tales of sickness, curses, and tangled threads that guide

Fig. 1 Space emerges as a dynamic field of meaning, co-constituted by the subject and the surrounding world. (Photograph: Lasse Fredriksson)

the search for the right remedy. Soft, flickering lights cast long shadows across the walls, evoking an unsettling yet mesmerizing ambiance. A slow, haunting melody hums quietly in the background, setting a tense mood. Along the walls, symbols that hint at medicine are lit from behind. Towering medicinal flowers, larger than the child, loom mysteriously from shadowed corners, their oversized petals catching the dim light like strange, otherworldly guardians.

In this setting, the atmosphere is not merely decoration, it is an active agent in shaping the visitor's perception and engagement. Following Gernot Böhme's (2013) theory, the atmosphere arises as a felt space between subject and environment. It envelopes the visitor and creates an emotional and sensory context in which meaning is formed. The exhibition does not simply present facts about illnesses in history; it invites the visitor to feel the cultural and historical presence through their bodily experience.

From a Merleau-Pontian perspective, this experience is grounded in embodied perception. The visitor does not observe the exhibition from a detached viewpoint, they move, hear, feel, and see. The understanding of the historical treatments that is the content of the exhibition, is shaped not only by the texts on the wall but also by how their body resonates with the rhythms, textures, voices, and spatial layout. This makes the information in the exhibition situated and lived, rather than abstract or neutral.

Thus, the atmosphere in an exhibition is not a secondary layer, it is central to how the information is understood. It creates a perceptual space where historical facts, cultural narratives, and emotional resonance intertwine in the visitor's experience.

So far, we have discussed a few studies that focus on more obvious spatial contexts, such as hospitals and museums, settings where the relationship between space, movement, texts, illustrations, and meaning- making is both visible and immediate. However, this chapter aims to expand the perspective by introducing a framework that also encompasses the production and planning of various forms of verbo-visual communication. This includes illustrations, texts, and graphic design elements that may not so obviously be tied to physical space in the same way. By including these dimensions, the aim is to offer a broader understanding of situated meaning-making within information design, one that bridges both spatial and traditional representational information design practices.

The Communication Space Framework (CSF)

The phenomenological perspective, where body, space, perception, and meaning are deeply interwoven, offers a rich understanding of how information and design are experienced. However, this holistic view can at times feel overwhelming within a practical design process. When everything is interconnected, it becomes difficult to navigate the different elements that influence one another. To support the design process, a framework is, therefore, introduced here to help structure the thinking and discussion throughout the design journey. Although the lived experience itself is holistic and characterised by a close interlacing of spatial, visual, and embodied dimensions, it can be helpful to temporarily mentally disentangle the whole with several possible perspectives. This is not to reduce the richness of the phenomenological approach, but rather to enable designers to engage with its complexity in a more manageable and focused way. The framework is thus not a replacement for holistic thinking, but a supporting structure that allows for clarity without losing depth.

Therefore, in this chapter, a possible theoretical framing for the role of space in information design through different forms of information will be described. The forms of information, interlaced with each other, form *a communication space* (Figure 2). The model is useful for students and practitioners in Information design, since it gives guidance to analyse different forms of information in a lived and experienced real-life setting. According to Marcia J. Bates (2006), professor emerita at Department of Information Studies at the University of California, a source of information can be anything that human beings interact with, or observe. This broad definition challenges narrow notions of information as merely text-based or digitally encoded and instead opens for a more embodied and spatial understanding of communication. In her foundational paper, *Fundamental Forms of Information*, Bates (ibid.) identifies several distinct but interrelated categories of information, including *Experienced form of information*, *Enacted form of*

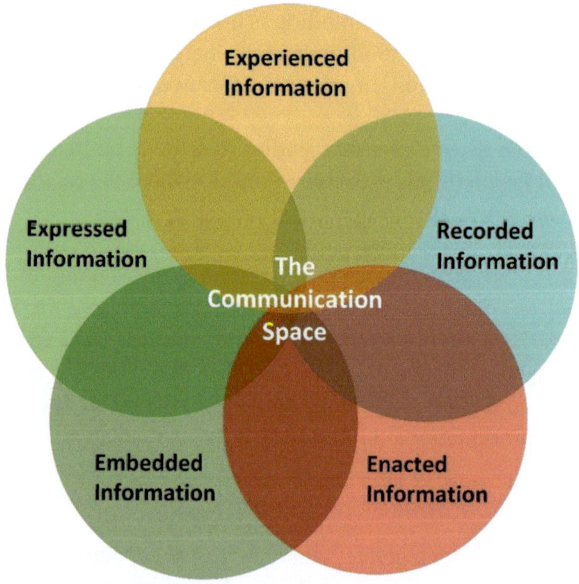

Fig. 2 The Communication Space. In the intersection of the different categories of information, a communication space emerges, a layered, dynamic environment where experience, social behaviour, architecture, objects, spoken worlds, text, and visuals intertwine.

information, *Expressed form of information*, *Embedded form of information*, and *Recorded form of information*.

When applied to the context of information design, this typology becomes especially powerful. For example, *experienced information* relates to the sensory and bodily perception of the environment, which resonates with Merleau-Ponty's (1962) phenomenology of perception. As mentioned in the introduction, information is never encountered in abstraction, but always through our bodily engagement with the world. We see, hear, touch, and move, through which meaning emerges. Thus, experiencing a room, a visual display, or a flow of signage is already a form of meaning-making.

Enacted information refers to actions and practices performed by individuals within a space, what people do, how they behave, how they navigate or interact based on their values. In an airport setting, for instance, this could include how passengers move through security checkpoints, interpret signage to find their gate, or respond to real-time announcements and digital displays. These interactions highlight how information is not just consumed but actively used and interpreted in context. This echoes Gernot Böhme's (2013) idea of atmosphere as a relational quality that arises between the subject and their environment, shaping how actions are perceived and carried out. In participatory design, enacted information also emerges through collaborative practices where users and stakeholders actively engage in shaping the design process, contributing their experiences, interpretations, and actions as integral parts of how information and environments are co-created.

Expressed information involves intentional communication, such as spoken words or body language. In contrast, *embedded information* refers to the meanings carried in objects, architecture, or materials, meanings that are not always explicitly stated but are perceived and interpreted in context. Finally, *recorded information* represents more conventional forms such as text, drawings, photographs, film, or audio.

In the intersection of the different categories of information, a communication space emerges, a layered, dynamic environment where experience, social behaviour, spatial form, and signs and symbols intertwine. This aligns closely with the phenomenological idea of perception as an active, situated process. Rather than isolating discrete channels of communication, the design challenge becomes to orchestrate these forms of information so that they reinforce and enhance one another.

In design practice, this requires a careful balance: recognizing the richness and complexity of lived experience, while also structuring it in ways that support communication.

The term communication space, as presented and described in this section, is thus used to verbalise different aspects of an information design context. For example, the values (enacted information) that take form in the expressed information (how people talk about a message or a phenomena) and affect how the built space and photos, texts, and signs (the embedded and recorded information) relates to each other in the process of designing, evaluating, and experiencing.

Communication Space Framework in Practice

The *Communication Space Framework* is built on the interplay between different forms of information. Together, these forms constitute a *communication space*. The framework provides a useful lens for analyzing and structuring a current situation or communication environment. Rather than providing fixed answers, it can help draw attention to the different ways information is being expressed and interpreted, through language, visuals, spatial arrangements, physical elements, and bodily actions. Using the framework in this way might reveal patterns or tensions that are not immediately visible and open up questions about what is working well and where there may be gaps or inconsistencies. It invites a closer look at how meaning is created in practice and could potentially support more grounded reflections on how to move forward or make informed changes.

Figure 3 is created to illustrate an example case, based on findings from case studies in the manufacturing industry (Schaeffer, 2011). A meeting in an improvement place is taken as an example of how the different forms of information can relate to each other in an everyday situation in a place designed for communication regarding improvements in manufacturing industry.

When the different forms of information are not aligned with each other, the improvement processes can be disturbed and become difficult. This was typically what happened in the cases studied.

In Figure 3, a production leader, operators, and a company's direction are involved in the communication (the company's direction is present, since the values of the company's direction influence the enacted information). The improvement processes are considered important, but the place for improvement (embedded information) is placed in an environment where it is difficult to concentrate on communication over improvement. The graphical material on the board (recorded information) is a cluster of information considered important (enacted information) by the information provider, for example, the manager. However, it is hard to read from the users' distance and surrounding the space are also disturbances such as heat, sound, or movement (experienced information). In the texts (recorded information), nothing is visually structured in order to be emphasized as more important than anything else. The operators are told (expressed information) to write on the board, but they do not write on the board; the lines stay empty (enacted information). There are markings on the floor (embedded information) with connection to the lean tool 5S, but the operators have weak connections between the markings on the floor and what 5S means (enacted information). Here, communication space may be less supportive for its purpose than it could have been. The intention was to gather the personnel, to be united around the processes of improvements in the production, to expose relevant data, and to create an awareness of the production system. In visual management, the immediate awareness is seen as important (for further reading on this topic, see also Söderlunds's chapter in this book). When the improvement places were not used as planned, they became symbols of something else; not of a living improvement work, but of stagnation or of less importance. A

Fig. 3 Example of the Communication Space in the manufacturing industrial context. The figure illustrates how different categories of information overlap in a real-world production environment. It highlights the complexity of communication in industrial settings, where instructions, visual cues, layout, tools, and embodied actions all contribute to the meaning-making. (Illustration: Jan Brandt. Image previously published in Schaeffer 2011.)

contributing reason for why the improvement places were not working to their full potential in the three studied cases can be found in the values and beliefs that were acted out in the communication space, maybe on a sub conscious level. The low level of use can be connected to disturbances such as heat, sound, or movement. These conditions did not support communication between the users. In one case, despite the intention of creating an improvement culture and process by designing a place visible from a distance, the dedicated space became a storage area rather than a place for meetings and learning. It was considered as a better use of space at the time, space for boxes instead of humans. In the studied companies, there was an attempt to create places to meet and improve the production, and the worker was seen as an analyst, as in Liker (2004). To make the places supporting improvement work, a suggestion is to make the values of the company conscious and to analyse how the values dynamically interact with the other forms of information.

From the overarching example above, we now turn to explore different entry points into *the Communication Space* in greater detail. By breaking down and examining the components more closely, it becomes possible to deepen our understanding of how the various forms of information interact in practice. The aim is not to isolate each form, but rather to see how they emerge in different situations and how they work together to support meaning-making in specific contexts.

With Experienced Information as an Entry Point

As a man steps into the open landscape, his posture might subtly shift, shoulders relax, gaze lifts, and his pace slows. The surrounding environment, with its vast horizon and textured terrain, seems to invite a moment of reflection. This encounter with the landscape is not merely visual; it is affective. Figure 4 is an attempt to illustrate this kind of emotionally charged moment, where the landscape is not just seen but felt. It captures how emotional experience can become visible through enacted gestures and spatial orientation, offering insight into how environments can evoke deeply personal reactions shaped by memory, mood, and sensory engagement.

Fig. 4 Experienced information includes memory, the body, and the senses, which help us understand the world. (Illustration generated with AI [Copilot], based on the author's prompt.)

Experienced information reflects an individual's subjective, emotional, and embodied experience within an environment (Bates, 2006). This concept, as outlined by Bates (2006), might be aligned with Merleau-Ponty's (1962) phenomenology of perception, which emphasises the centrality of bodily experience and sensory engagement in shaping our understanding of the world. Rather than viewing information as a purely cognitive or symbolic entity, experienced information recognizes that perception is intertwined with feelings, memories, and physical sensations. This holistic approach is essential for information design, as it foregrounds how humans not only interpret but also emotionally inhabit a space or interface, thus influencing comprehension, engagement, and behaviour.

Experienced information allows for the inclusion of sensory aspects such as scent in the analysis, as demonstrated, for example, in the research of the late professor and design theorist Robert Spence (2004) in the article, "Sensitivity, residue and scent: Concepts to inform interaction design for the support of information space navigation". Spence introduces concepts such as *sensitivity*, *residue*, and *scent* to describe how users navigate and interpret information spaces, emphasizing how interaction is shaped by subtle cues and accumulated traces in the environment. These concepts highlight the importance of the spatial and temporal dimension in information design, not only what information is presented, but how it is sensed and recalled in space.

To thoroughly analyze the aspect of experienced information, researchers and designers use a range of qualitative and participatory methods to learn about humans sensory and emotional responses, for example:

- ***Emotional mapping*** invites participants to identify and communicate where they feel particular emotions within a space, such as calm, anxiety, comfort, or stimulation. This spatially grounded method highlights how design elements evoke emotional patterns, providing valuable insights for refining environments to better support user wellbeing (Thibaud 2013; Pánek, 2018).
- ***Sensory ethnography*** invites researchers to explore how multisensory experiences, such as light, sound, temperature, texture, and smell, shape the way a communication space is perceived and felt. Drawing on the work of anthropologist Sarah Pink (2009), this emphasises embodied and affective dimensions of experience.
- ***Experience sampling methods*** (*ESM*) or ecological momentary assessments (EMA) involve participants at random or scheduled intervals to report their immediate feelings and sensory impressions within a space or while using an interface, capturing real-time data on experienced information (Csikszentmihalyi and Larson, 1987).
- ***Psychophysiological measures***, such as heart rate variability, skin conductance, or eye-tracking, can provide data on participants' emotional arousal and attentional focus in response to environmental stimuli, complementing subjective reports and enriching understanding of embodied experience (Bolls et al., 2025).

Together, these methods offer multifaceted possibilities for exploring how experienced information shapes, and is shaped by the physical and sensory qualities of an environment.

With Enacted Information as an Entry Point

Enacted information emerges through human actions, both conscious and habitual, that express cultural values, social norms, or behavioural expectations (Bates, 2006). Enacted information refers to the meanings and messages that emerge through human behaviours and interactions within a space. These actions, whether deliberate or habitual, serve to express underlying cultural values, social norms, and behavioural expectations embedded in the environment (Bates, 2006). To engage with the entry point of enacted information, it is essential to observe how people physically and socially engage with their surroundings.

Fig. 5 An image from the exhibition Collaboration and Resistance at the Norrköping City Museum, which highlighted local instances of collaboration, resistance, and contemporary connections to far-right extremism. (Photograph by Anders Ristenstrand, used with permission from the Norrköping City Museum.)

For example, when a single visual artifact, in this case, a newspaper photo, enters a charged context like a museum exhibition on Nazism, with local examples of collaboration and resistance, it does not remain neutral (Figure 5). This is one example of how different cultural and social actors publicly enact their values and convictions. In this case, The Sweden Democrats filed a police complaint against the City Museum in Norrköping, Sweden after a photo of their local party board appeared in an exhibition about Nazism. The image, part of a 2018 newspaper article, was displayed alongside other clippings to illustrate contemporary connections to far-right ideology. It showed a board member later expelled for ties to Nazi environments. The image has since been removed, and the police case dropped. However, author and journalist Johan Perwe, whose research underpins the exhibition, criticises both the political party and the museum. He calls it regrettable that politicians interfere with cultural expression and that the museum didn't defend the inclusion. Perwe argues the article highlighted relevant, documented history and says linking the 2018 image to the entire exhibition is a misinterpretation. The Sweden Democrats declined to comment (SVT, 2020).

This process with different points of view illustrates how a visual artifact can trigger strong reactions and bring about a form of *enactment* shaped by conflicting value systems. Different actors, the museum, the political party, and the author and journalist, interpret the meaning and implications of the image through their respective perspectives on history, accountability, and cultural freedom. The visual presence of the photo does not only communicate content but also mobilises ethical, political, and historical narratives in tension with one another. In the example, an artefact such as an exhibition, a form of embedded information, is then not complete once the exhibition opens. Negotiations, a form of enacted information, about its content, interpretations, and meanings may continue even after the inauguration. The exhibition, thus, becomes an ongoing process rather than a finished product.

An example of how people *physically* enact information can be found in the renowned graphic designer and educator Roman Duszek's article, "The Warsaw Subway Signage System" (2004). The article examines how the design and spatial arrangement of signage in the Warsaw subway influence users' ability to navigate and process information. It highlights the relationship between experienced, embedded, and enacted information, where signage is not only visually perceived but also physically interacted with, as users move through space, make decisions, and adjust their behavior based on their own beliefs and how the recorded information is presented within the environment. This demonstrates how bodily movement and spatial context play a crucial role in how information is enacted in real-world settings.

To analyze the aspect of enacted information, researchers and designers can employ a range of qualitative and participatory methods to gain insight into how information is performed, embodied, and co-created through human action. This perspective is understood as something done, emerging through practices, interactions, and situated activities. By observing and engaging with how people enact information in real contexts, through gestures, routines, collaboration, and material engagement, researchers in information design and related fields can explore how meaning is dynamically produced and negotiated. Methods such as ethnographic interviews and participant observation can be used to understand more about the performative nature of information:

- ***Ethnographic interviews*** offer a qualitative approach to understanding how users themselves interpret their behaviours and experiences in the space, capturing personal narratives that reveal the subjective meaning behind enacted actions. Provides a nuanced understanding of how design shapes, and is shaped by, the behaviors and social agreements enacted by its users (Spradley, 2016).

- ***Participant observation*** is an ethnographic method that involves carefully documenting how individuals move through, interact with, and adapt to a space in real-time, providing rich insights into lived experience and social dynamics (for example, Musante, 2015).
- ***Behavioural mapping*** complements the methods above by visually tracing patterns of user movement, pauses, and interactions, allowing researchers to identify common flows and hotspots within an environment (Gehl and Svarre, 2013; Petrtýlová and Jaššo, 2022).

With Expressed Information as An Entry Point

Expressed information encompasses all forms of human communication, both verbal and non-verbal, that facilitate interaction and the co-construction of meaning within a space (Bates, 2006). This includes spoken language, gestures, posture, facial expressions, and other bodily movements that mediate social exchanges and help individuals convey intentions, emotions, and ideas. In the context of information design, understanding expressed information is crucial because it highlights how the physical and social environment supports or constrains communicative behaviors and shared understanding.

Fig. 6 A woman gestures toward an architectural diagram during a presentation. The image, published in Örebro-Kuriren on May 6, 1967, exemplifies expressed information through posture, gesture, and spatial interaction. (Örebro läns museum OLM-2012-8-10655 [(Public Domain].)

In the image in Figure 6, we likely see agency director Britten Melchert, captured during a presentation or review of architectural plans. The photograph was published in the newspaper, *Örebro-Kuriren*, on May 6, 1967, in connection with the article "Möblerna tränger ut barnen" [eng. Furniture Crowds Out the Children].

The image offers an example of expressed information, as defined by Bates (2006), since, expressed information includes all forms of human communication, verbal and non-verbal, that facilitate interaction and the co-construction of meaning within a space. In this case, Melchert is seen pointing to a diagram on a large board while holding a paper in the other hand. This gesture of reference, combined with her posture and orientation toward the board, suggests an act of explanation or instruction, likely in a professional context.

The physical setup, the person gestures, the board, and the architectural drawings, forms a communication space where meaning is enacted through bodily movement and spatial arrangement. This moment illustrates how expressed information operates in real-world settings, where gestures and physical presence are integral to the transmission and negotiation of knowledge. It also highlights how the social and physical environment, in this case, public presentation, supports communicative behavior and shared understanding.

To analyse expressed information, a variety of qualitative and observational methods can be explored:

- ***Interaction analysis*** focuses on detailed observation and transcription of face-to-face or mediated communication events, capturing the sequence, timing, and content of exchanges. This approach helps uncover patterns in how people coordinate conversations, manage turn-taking, and use verbal and non-verbal cues to negotiate meaning, but the method also has constraints (for example, Heritage 1990; Jordan and Henderson, 1995).
- ***Video ethnography*** utilizes audiovisual recordings to document both verbal and non-verbal communication within naturalistic settings. This method allows researchers to analyze body language, eye contact, gestures, and group dynamics in rich detail, often revealing subtle communicative behaviors that are missed in purely textual analysis (for example, Mitsuhara and Hauck, 2021).
- ***Conversational analysis*** is another rigorous method that systematically studies the structure and organization of talk-in-interaction, helping to understand how social actions are performed through language and how meaning is collaboratively constructed (for example, Liddicoat, 2021).
- ***Whole-body mapping*** is a qualitative and participatory method that combines creative expression with embodied reflection to explore lived experience. The process involves tracing a person's body to create a life-sized outline, which participants fill with colours, symbols, and text to represent emotional, physical, and social aspects of their lives. Used across research, therapeutic, and educational contexts, body mapping allows participants to express complex experiences visually, fostering inclusive knowledge production and social awareness, though further research is needed to clarify its applications (for example, Jager et al., 2016.).

- ***Discourse analysis*** broadens the scope by exploring how language and non-verbal communication reflect and reproduce social norms, power relations, and cultural values within communicative contexts (for example, Gee, 2025) for an extensive guide.

Together, these methods offer possibilities to investigate how expressed information unfolds in social settings, and how the context can facilitate or hinder communication.

With Embedded Information as an Entrance Point

Embedded information refers to meaning physically encoded in the material and spatial features of environments, such as tools, architecture, and designed objects (Bates, 2006). In this section, we examine how designed objects can embody and communicate knowledge beyond written or spoken language. Through material, spatial arrangement, and form, objects can carry traces of use, intention, and context, what Bates (2006) refers to as *embedded information.* The following example illustrates how a historical architectural model encodes meaning through material, form, and context.

Fig. 7 The image shows a preserved model of a sports cabin probably from the 1930s, an example of embedded information, meaning physically encoded in its material, form, and construction logic. It reflects both the historical context of its origin and the design knowledge embedded in its modular structure (Photo: Västmanlands läns museum. Vlm-30421 [CC BY-SA].)

The sport cabin, shown as a model in Figure 7 was produced by Knut Lewenhaupt at Geddeholm, Sweden in the wake of the devastating 1928 storm, which left large amounts of fallen timber with limited commercial use. As a creatively gifted landowner, Lewenhaupt sought to give new value to this material by designing practical and aesthetically refined cabins, and even coffins, using the storm-felled wood.

The cabins were prefabricated at Geddeholm, disassembled, numbered, and transported for reassembly at the buyer's chosen location. They were sold through outlets such as the PUB department store in Stockholm during the 1930s and came in four different models. This preserved model, the only one known today, measures 84 x 111 cm. The primary material of the model is wood, which is used for the facade, roof, and details such as the steps and the small porch above the door. The wooden panels run horizontally, the door is painted a bright blue with white trim, while the window frames are also white and feature a decorative grid pattern that mimics traditional mullioned windows. The chimney, painted white, appears to be made from a different material, possibly a painted wooden block or cardboard. The roof is covered with a flat wooden surface, likely thin plywood or veneer, with no visible roofing material.

In terms of style, the house model has a rustic character. The wooden siding, simple gable roof, and modest porch with structural supports evoke the feel of rural architecture, reminiscent of Swedish country cottages or early 20th-century American cabins. The design is functional and unadorned, focusing on structure rather than elaborate decoration. However, the decorative windowpanes suggest a certain aesthetic consideration, aiming to imitate older architectural styles.

The model also embeds a story of resourcefulness, and early prefab innovation. The materials themselves speak of the storm and recovery, if you know the story behind it. The form reflects both artistic vision and practical function, embedding layers of historical, economic, and personal meaning into the object. In its construction, layout, and materiality, the model communicates design intentions, spatial thinking, and cultural values. It is a form of embedded information, a physical object that stores and expresses the story of its origin, purpose, and context.

When this detailed model of the sports cabin was displayed at PUB department store in Stockholm during the 1930s, visitors could look inside, explore the layout, and get a sense of the space. The model was designed to allow customers to imagine how their own furniture and belongings might fit, helping them to visualise their dreams and lifestyle within the cabin.

With its partially fixed interior elements, such as built-in benches, a stove, and cupboards, the model demonstrated practical living solutions (Figure 8). Visitors could move around the rooms, consider how to arrange their furniture, and see how their life in their free time (if they had some) might unfold in this compact designed space.

Fig. 8 Interior view of the model house, revealing a simple and unfinished wooden structure (Photo: Västmanlands läns museum. Vlm-30421 [CC BY-SA].)

The model does not just represent a building; it embodies a way of responding to material abundance. It tells of a time when craftsmanship, practicality, and aesthetics worked together in direct response to environmental and economic conditions. Themes of resilience, modular thinking, and vernacular innovation emerge, as the sport cabin becomes more than a shelter, it becomes an idea. In this way, the model becomes a carrier of cultural memory, and an example of how narratives can be stored and understood through material form.

Unlike many prototypes today, which are often used as a way to explore and test new ideas, this model was not only created for conceptual development or experimentation. In contemporary design practice, prototypes are frequently employed as part of an iterative and exploratory process, where designers and architects test spatial ideas, user interaction, and material solutions before arriving at a final form (Buchenau and Suri, 2000). The model of the sports cabin functioned purely as a marketing product, intended to visually present and promote a finished design to potential clients or the public. Its purpose was to showcase the possible charm of the building, rather than serve as a tool for architectural iteration or innovation.

When analysing embedded information within communication spaces, it is important to employ methods that capture both the tangible and intangible ways in which for example tools, architecture, or furniture convey meaning. Understanding how objects, architectural features, and spatial arrangements shape human behavior and social interaction requires a combination of observational and interpretive techniques. Observing physical traces allows designers to systematically map the organization of space and identify implicit cues that guide user movement and interaction (Zeisel, 2006). Complementing this, design heuristics evaluation draws on the concept of affordances, how design features suggest possible uses or behaviors, to assess whether elements such as seating, pathways, or barriers promote desired social or functional outcomes (Norman, 2013). Additionally, photo elicitation provides a powerful visual tool, enabling reflection on how physical features carry symbolic or practical significance over time and across contexts (Rose, 2016). Architectural analysis can uncover the underlying organizational principles of a building to reveal the designer's intentions. It involves dissecting elements such as geometry, spatial volume, circulation, key axes, and structural systems, while also considering materials, environmental strategies, and technical services to understand the project's overall design logic (Baker, 2006). Archival research methods involve the systematic investigation of documents and texts produced by or about organisations. Traditionally focused on historical records that offer insight into past events, people, and institutions, these methods are also used in contemporary studies to analyze current materials such as architectural drawings, models, reports, emails, databases, and web pages (Ventresca and Mohr, 2017). Together, these methods offer possibilities to dissecting the layers of meaning embedded within a communication space and understanding their impact on user experience:

- ***Observing physical traces***: Mapping how spatial layouts or object placements suggest intended behaviors or hierarchies (Zeisel, 2006).
- ***Design heuristics evaluation***: Assess how affordances (Norman, 2013) are embedded in spatial arrangements.
- ***Photo elicitation***: Use photography to document and reflect on how physical elements carry meaning over time and in human experience (Rose, 2016).
- ***Architectural analysis***: Architectural analysis can be applied as a method to uncover the underlying organizational principles of a building and to interpret the intentions guiding its design. Following Baker's (2006) approach, it becomes possible to reveal the conceptual logic and design reasoning embedded in architectural form. Similar analytical perspectives can be found in the works of Ching (2025), *Architecture: Form, Space, and Order*, and Unwin (2020), *Analysing Architecture*, both of which provide frameworks for understanding spatial organization, structure, and meaning in design.
- ***Archival research*** can stand alone or complement other methods, providing valuable contextual and documentary evidence for understanding the embedded information in different practices over time (Ventresca and Mohr, 2017).

With Recorded Information as a Starting Point

Recorded information plays a vital role in shaping how people access, interpret, and act upon information within an environment. This category encompasses all forms of symbolic media that preserve knowledge and meaning over time, ranging from written texts and images to audio and video recordings (Bates, 2006). The image (Figure 9) presented as the first example in this section exemplifies how informal or handwritten notes can serve as vessels of cultural memory and social documentation. Captured on what appears to be a personal or community record, the text reflects a moment in time, August 14th (year unknown).

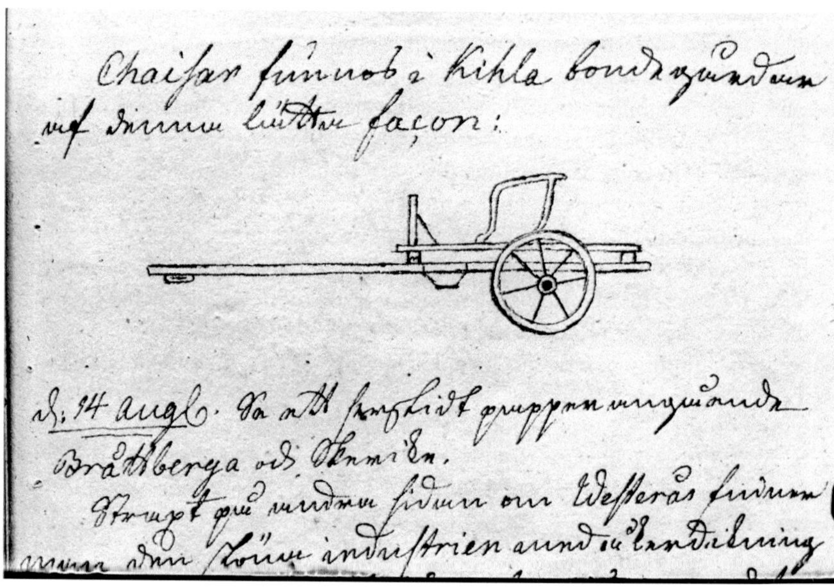

Fig. 9 This handwritten with a drawing of a "schäs" note serves as an example of recorded information in everyday contexts (Vlm-E 8822 [Public Domain].)

The drawing in Figure 9 depicts a traditional "schäs", a type of light, horse-drawn carriage common in rural Sweden during the 18th and 19th centuries. The drawing with the handwritten text offering insights into local transportation practices and design traditions in the Swedish countryside. The text is written in Swedish, though it includes some archaic spellings and possibly dialectal variations. As a piece of recorded information, it contributes to our understanding of rural material culture and informal knowledge transmission in pre-industrial Sweden.

Beyond textual documentation, the field of visual communication offers critical insights into the interpretive power of recorded information. The late professor in semiotics, Gunther Kress, and the professor and social semiotician, Theo van Leeuwen's (2020), seminal work on the grammar of visual design elucidates how images operate as a language, capable of encoding and conveying complex meanings. Their analysis supports the understanding of illustrations not merely as decorative elements, but as integral recorded information that shapes perception and cognition.

Fig. 10 Promotional poster for PUB's sports cabin exhibition, featuring a stylised illustration of a log cabin and descriptive text. The poster exemplifies recorded information by documenting the event and conveying cultural ideals of leisure, nature, and modern rural living through both visual and verbal design (Photo: Västmanlands läns museum. Vlm-30421 [CC BY-SA].)

Returning to the model of the sports cabin previously discussed in the section on embedded information, we now shift our focus to the accompanying materials that exemplify recorded information. These materials not only document the cabin's features and context but also communicate cultural values and aspirations through visual and verbal design. The promotional poster for PUB's sports cabin exhibition (Figure 10) serves as an example of recorded information in a commercial and cultural context. Drawing on Kress and van Leeuwen's (2020) framework for visual communication, the image can be analysed through its representational, interactive, and compositional meanings. Representationally, the illustration of a log cabin surrounded by trees evokes a sense of tranquillity and escape, aligning with the accompanying text's message about finding refuge from urban stress. The image functions as a symbolic representation of the ideal lifestyle being promoted. From an interactive perspective, the absence of human figures and direct gaze positions the viewer as an observer, invited to consider the offer rather than respond to a demand. Compositionally, the layout can be understood within an "ideal-real structure": the upper part of the page presents the aspirational image (the ideal), while the lower section provides factual details about the exhibition (the real). On the left side of the spread, a magenta background frames two stylized black silhouettes, a man smoking a pipe and a woman with possibly a blanket, walking toward a cabin and a sportscar, visually reinforcing the theme of rustic life and gendered roles in leisure. This structured interplay between image and text exemplifies how recorded information, here in the form of a printed advertisement, communicates the underlying values about lifestyle, aspiration, and consumer culture through the design.

Fig. 11 Illustration of a smartphone interface as an example of recorded information. (Illustration generated with AI [Copilot], based on the author's prompt.)

In archival studies, the late professor in the Archival Studies Program at the University of Manitoba, Terry Cook (2013), articulates a paradigm shift from perceiving recorded information merely as static documents to understanding them as dynamic evidentiary materials that shape collective memory and identity. He emphasizes that texts and visual artifacts must be interpreted contextually, acknowledging their embeddedness within social, historical, and technological frameworks. This contextualization is particularly pertinent in the digital era, where the proliferation of media forms challenges traditional archival practices. The interface of a modern smartphone (Figure 11) exemplifies recorded information in a dynamic form. Every element on the screen, from app icons to notification banners, represents recorded form of information (Bates, 2006). The home screen functions as a curated gateway to various types of recorded content, including text messages, emails, images, and especially video.

One of the most prominent features is the seamless access to moving images. Applications such as YouTube, TikTok, and streaming platforms are often placed within immediate reach, reflecting their central role in everyday information consumption. These apps not only store vast libraries of recorded video content but also allow users to create, edit, and share their own recordings.

The interface design supports this fluid interaction with recorded media through intuitive icons, touch gestures, and personalized content feeds. Visual cues such as thumbnails, play buttons, and progress bars guide the user's navigation and interpretation of video content. In this way, the smartphone interface becomes conduit for recorded information, optimised for immediacy, mobility, and multimodal engagement.

From an organisational standpoint, professor of Business Administration at the University of Michigan, James Walsh and his co-writer Geraldo Ungson (1991) identify recorded information, such as documents, reports, diagrams, and illustrations, as critical elements of organizational memory. Their framework illustrates how recorded knowledge supports decision-making, learning, and continuity within institutions, distinguishing it from tacit knowledge that resides primarily in individuals. Visual information facilitates comprehension and dissemination of complex organizational processes. This distinction between tacit and explicit knowledge is further elaborated by Ikujiro Nonaka (he was a Japanese organizational theorist and Professor at the Graduate School of International Corporate Strategy of the Hitotsubashi University) and Hirotaka Takeuchi, professor of management practice at Harvard Business School (1995), who argue that recorded information serves as a tangible representation of explicit knowledge. Through the systematic documentation of insights and processes in texts and visuals, organizations convert tacit knowledge into forms that can be shared and leveraged for innovation. Visual models and illustrations thus play a pivotal role in making abstract knowledge accessible and actionable.

Collectively, these examples illuminate the multifaceted nature of recorded information. Whether situated within archives, digital systems, organizations, or educational environments, texts and illustrations are not passive repositories but active agents in knowledge creation, interpretation, and communication. We can see that the relationship between recorded information and the other forms of information is interwoven. Although there are many methods available, I recommended a multimodal analysis approach as it can capture multiple aspects. Analysing recorded information requires attention not only to the content itself but also its multimodal contexts.

- ***Multimodal analysis*** offers a comprehensive account of the grammar of visual design, explaining how images communicate meaning across diverse media, from children's drawings and photojournalism to digital interfaces and games. It serves as a key resource for understanding and analysing images in contemporary multimodal contexts (Kress and van Leeuwen, 2020).

So far, we have examined the various forms of information as an entry point into the concept of communication space. As a conclusion, we will now discuss the implications that this reasoning has for the definitions and understanding of information design. By reflecting on how different types of information, whether experienced, embedded, expressed, enacted, or recorded shape communication, we can better grasp the evolving role of information design. This discussion will help clarify how information design must adapt to accommodate multiple layers of meaning, context, and user engagement within communicative spaces.

Discussion: Toward a Spatial Understanding of Information Design

To approach *Communication Spaces* as a framework for Information Design opens up new ways of understanding how information operates, not merely as visual content or structured data, but as something deeply entangled with space, embodiment, and experience. This spatial lens encourages us to shift from a narrow focus on clarity and efficiency, toward a broader view of information design as a practice that shapes how people move, orient, act, and make meaning in situated environments.

Information does not only exist in static forms. It is experienced, enacted, and embedded. People engage with information as they move through environments, touch surfaces, interpret signs, follow paths, and participate in social interaction. Texts, illustrations, signage, and spatial arrangements are not only visual carriers, but they are also active participants in how meaning is negotiated. As such, the power lies not just in the forms of information themselves as an analytical tool, but in their intersections, where visual, spatial, and embodied elements overlap and meaning is dynamically produced.

While this interwoven nature reflects the richness of human experience, breaking it down into analytically distinct categories can help information designers gain insight and have the framework as support to verbalise their processes.

To approach *Communication Spaces* as a framework for Information Design is to shift our understanding of what information is, how it is experienced, and how it operates in the world. Traditionally, information design has been defined as the clear and efficient communication of information, typically focusing on legibility, usability, and function. But when we consider information as something that is always situated, anchored in space, shaped by context, and navigated through physical and social environments, a broader, more dynamic picture emerges.

Information is never neutral or static. It is embedded in environments, layered with cues, and shaped by the way people believe, how they move, speak, look, and interact. Wayfinding systems, signage, texts, public displays, and even illustrations placed within a book or on a screen are not only carriers of content, but they are also actors in a larger choreography of meaning-making. This means the power of information design lies not just in the forms of information themselves as an analytical tool, but in their *intersections*, where visual, spatial, social, and linguistic elements converge and meaning is actively produced.

To take this perspective seriously, we may need to reconsider how we define information design. Rather than seeing it solely as the structuring of data or visual content, we can begin to define it as a *relational practice*, a design mode that organizes interactions between people, information, and spatial contexts. This suggests that information design is also a form of spatial literacy: a way of shaping how people orient themselves, navigate meaning, and make sense of their surroundings. Here are a few points of departure for rethinking the definition of information design:

From Form to Relation

Information design is not only about the design of forms, but also to take in consideration the relationships between form, content, environment, human values, and actions.

From Transmission to Situated Meaning

Rather than seeing information as something transmitted from sender to receiver, we can understand it as something produced in context through movement, attention, gesture, experience, and positioning in space.

From Static Structure to Dynamic Choreography

Information unfolds in time and space, experienced through the body, shaped by architecture, signage, and social interaction. Design becomes a way to choreograph these encounters.

The Definition of Information Design Reconsidered

In exploring the concept of information design, it becomes clear that this field is multifaceted and continuously evolving. It is productive to approach information design as an open, probing practice, one that considers the rich and interconnected layers of communication. This perspective invites us to reconsider and expand the very definition of information design. Rather than seeing it as a fixed or narrow concept, we can view information design as a dynamic and multifaceted practice that embraces the complexity of different information forms within communication spaces. This opens up possibilities for designing environments and systems that not only convey data but also engage users in meaningful ways, adapting to diverse contexts and evolving needs. Such a definition encourages ongoing reflection and innovation, challenging us to explore new ways of shaping information experiences.

> *Information design is the practice of developing communication situations that acknowledge and integrate the complex, interwoven aspects of communication spaces, encompassing visual messaging, material and spatial contexts, embodied experience, enacted behaviours, and spoken language.*

The definition proposed here invites us to see information design not merely as the creation of isolated messages or visuals, but as a holistic and situated practice. It embraces the full spectrum of communication elements within a given space and experience, guiding us toward a deeper understanding of how design can actively shape the ways we perceive, interpret, and interact with information.

Conclusion

Viewing information design as a spatial situated practice challenges us to think beyond the page or screen, and to consider how text, image, sound, and structure shape not only understanding, but movement, behavior, and social meaning. For students and practitioners alike, this approach offers both a critical lens and a practical toolkit for designing more responsive, meaningful, and context-aware information environments.

This chapter introduces and presents the Communication Space Framework, a comprehensive model for analyzing how information is conveyed, experienced, and interacted with in both physical and digital environments. The framework is structured around five interrelated fundamental forms of information: *Embedded, Recorded, Enacted, Experienced*, and *Expressed information* (Bates, 2006). Each form offers a unique lens on how communication spaces function and how users engage with information.

The framework offers a methodological entry for designers and researchers to map, analyse, and improve communication spaces by considering the dynamic interplay between environment, behavior, perception, and social interaction. This holistic perspective enables the creation of more intuitive, engaging, and context-sensitive information design situations.

References

Baker, G.H. 2006. *Design Strategies in Architecture: An Approach to the Analysis of Form*. Routledge, Taylor & Francis Group, Oxon.

Bates, M.J. 2006. Fundamental Forms of Information. *Journal of the American Society for Information Science and Technology*, *57* (8): 1033–1045. Retrieved from https://doi.org/10.1002/asi.20369.

Buchenau, M. and Suri. J.F. 2000. Experience prototyping. In: *Proceedings of the 3rd Conference on Designing Interactive Systems: Processes, Practices, Methods and Techniques* (pp. 424–433). ACM.

Bolls, P.D., Lee, Y.I. and Lee. Y.J. 2025. Biometric Measures as Tools for Investigating the Embodied Mind "on" Advertising: The Importance of Grounding Research in the Psychophysiology Paradigm. *Journal of Advertising*, 1–19.

Böhme, G. 2013. Atmosphere as the fundamental concept of a new aesthetics. *Thesis Eleven*, *113*(1): 5–21. https://doi.org/10.1177/0725513613512540.

Casey, E.S. 1997. *The Fate of Place: A Philosophical History*. University of California Press, Berkeley.

Ching, F.D. 2023. *Architecture: Form, Space and Order*. John Wiley & Sons, Hoboken.

Csikszentmihalyi, M. and R. Larson. 1987. Validity and reliability of the experience-sampling method. *The Journal of Nervous and Mental Disease*, *175*(9): 526–536.

Cook, T. 2013. Evidence, memory, identity and community: Four shifting archival paradigms. *Archival Science*, *13*(2–3): 95–120. Retrieved from: https://doi.org/10.1007/s10502-013-9195-0.

Duszek, R. 2004. The Warsaw subway signage system. *Information Design Journal*, *12*(3): 181–186.

Gee, J.P. 2025. *How to Do Discourse Analysis: A Toolkit* (3rd Edn.). Routledge, Oxon. Retrieved from: https://doi.org/10.4324/9781003537991.

Gehl, J. and Svarre. B. 2013. *How to Study Public Life* (Vol. 2), Island Press, Washington DC.

Heritage, J. 1990. Intention, meaning and strategy: Observations on constraints on interaction analysis. *Research on Language and Social Interaction*, *24* (1–4): 311–332.

Kress, G. and Van Leeuwen. T. 2020. *Reading Images: The Grammar of Visual Design* (3rd Edn.). Routledge, Taylor & Francis Group, Oxon.

Liddicoat, A.J. 2021. *An Introduction to Conversation Analysis*. Bloomsbury Publishing, London.

Liker, J.K. 2004. *The Toyota Way: 14 Management Principles from the World's Greatest Manufacturer*. McGraw-Hill, New York.

Jager, A.D., Tewson, A., Ludlow, B. and K. Boydell. 2016. Embodied ways of storying the self: A systematic review of body-mapping. *Forum Qualitative Sozialforschung/forum: Qualitative Social Research*, *17*(2).

Jordan, B. and Henderson. A. 1995. Interaction analysis: Foundations and practice. *Journal of the Learning Sciences*, *4*(1): 39–103.

Merleau-Ponty, M. 1962. *Phenomenology of Perception*. Routledge & Kegan Paul, London.

Miller, C. and Lewis. D. 2000. Wayfinding in complex healthcare environments. *Information Design Journal*, *9*(2–3): 129–160. Retrieved from https://doi.org/10.1075/idj.9.2-3.04mil.

Mitsuhara, T.V. and Hauck. J.D. 2021. Video ethnography: A guide. In: Perrino, S. and S. Pritzker (Eds.), *Research Methods in Linguistic Anthropology* (pp. 223–259). Bloomsbury, London.

Musante, K. 2014. Participant observation. In: H. Russell Bernard and C. Gravlee (Eds.), 2014. *Handbook of Methods in Cultural Anthropology* (pp. 251–292). Rowman & Littlefield, London, UK.

Nonaka, I. and Takeuchi. H. 1995. *The Knowledge-creating Company: How Japanese Companies Create the Dynamics of Innovation*. Oxford University Press, Oxford.

Norman, D.A. 2013. *The Design of Everyday Things*. MIT Press, Cambridge.

Pettersson, R. 2002. *Information Design: An Introduction*. John Benjamins Publishing: Amsterdam.

Petrtýlová, R. and M. Jaššo. 2022. Behavioural mapping and online data as tools for socio-spatial analysis of public spaces–Bratislava, Slovakia waterfront case study. *GeoScape*, *16*(1): 39–54.

Pánek, J. 2018. Emotional maps: Participatory crowdsourcing of citizens' perceptions of their urban environment. *Cartographic Perspectives*, *91*: 17–29.

Pink, S. 2009. *Doing Sensory Ethnography*. Sage, London.

Rose, G. 2016. *Visual Methodologies* (4th Edn.). Sage, London.

Schaeffer, J.A. 2011. *Communication Space: Spatial Design in Manufacturing Industry*. (Licentiate thesis), School of Innovation, Design and Engineering, Mälardalen University, Eskilstuna.

Simlinger, P. 2007. idX development of international core competencies and student and faculty exchange in information design within the EU/ US cooperation program in higher education and vocational education and training. International Institute for Information Design, Wien.

Simonsson, M. 2014. *Displaying Spaces: Spatial Design, Experience and Authenticity in Museums*. (Dissertation), Umeå universitet, Umeå.

Smythe, Silva K.C.A. 2018. How users find their way in hospitals. *Information Design Journal*, *24*(3): 236–253. https://doi.org/10.1075/idj.24.3.03sil.

Spence, R. 2004. Sensitivity, residue and scent: Concepts to inform interaction design for the support of information space navigation. *Information Design Journal and Document Design*, *12*(3): 163–180.

Spradley, J.P. 2016. *The Ethnographic Interview*. Waveland Press, Long Grove.

SVT. 2020. Författare kritisk mot Norrköpings stadsmuseum: "Beklagligt att man inte står på sig". Retrieved from https://www.svt.se/kultur/forfattaren-bakom-norrkopingsutstallningen-beklagligt-att-museet-inte-star-pa-sig. Published 13 February 2020.

Thibaud, J.-P. 2013. The backstage of urban ambiances: When atmospheres pervade everyday experience. *Emotion, Space and Society*, *15*: 39–46.

Unwin, S. 2020. *Analysing Architecture: The Universal Language of Place-making*. Routledge, London.

Waller, R. 1996. Annual Report of the IDA, 1996. In: *The Origins of the Information Design Association* (PDF). University of Reading. 2008.

Walsh, J.P. and Ungson. G.R. 1991. Organizational memory. *Academy of Management Review*, *16*(1): 57–91. https://doi.org/10.5465/amr.1991.4278992.

Ventresca, M.J. and Mohr. J.W. 2017. Archival research methods. In: J. A. C. Baum (Ed.), *The Blackwell Companion to Organizations* (pp. 805–828). Blackwell Publishers Ltd., Oxford.

Zeisel, J. 2006. *Inquiry by Design: Environment/ Behaviour/ Neuroscience in Architecture, UInteriors, Landscape and Planning*. W.W. Norton, New York.

9 Wayfinding in Information Design
The Promise of Smooth Navigation

Karolina Uggla[1*]

Introduction

Navigating in unknown terrain and in new locations is an essential part of our lives. Our mobility and infrastructure offer a diversity of navigational challenges that we are faced with every day in urban space, both indoors and outdoors. Information design encompasses the theory and practice of designing information with the aim of making people's lives easier. Well-designed information helps us navigate in built environments when using services such as public transport, healthcare, and education. This is, of course, a rather simplified model of how information design as a knowledge domain and as a set of design practices functions. To dig deeper into the problem when it comes to the theory and practice of designing navigation, we need to know more about the spatial context, the situation, the user needs and what we mean by the concept of information. Under the umbrella term of wayfinding a set of design principles is provided, mainly aimed at making navigation easier in built environments, indoors and outdoors. Wayfinding incorporates a systemic view on navigation, and overall planning of functions in buildings and complexes, like hospitals, schools, universities, airports, commuter traffic, and large offices, for instance. The design of locational, spatial, and geographical information—that is, designing visual overviews and signage systems—has, since the outset of the knowledge domain, been a design activity for the information designer.

The more complex the structures, the more difficult they are to navigate, and built structures are not static. They are continuously being altered and rebuilt, and functions and institutions within a building complex change and move. Thus, the need for navigational aids like maps and wayfinding systems is increasing in densely populated and complex urban areas. This need for on-site navigation assistance persists despite digital aids, apps, and maps, and continues to be a knowledge area for the information designer. However, this area of knowledge and practice is vast and borders on other major areas of knowledge, just as an interdisciplinary field should do (Thompson-Klein, 2023). Involved fields may be architecture, urban design, graphic design, cognitive science, geography, visual studies, and many more. At the same time, it can now be of interest to ask, as of

[1] Ph.D. Senior Lecturer in Information Design, Department of Health Science, Innovation and Design, Mälardalen University, Box 325, 631 05 Eskilstuna, Sweden.

* Corresponding author: karolina.uggla@mdu.se

DOI: 10.1201/9781032642598-9

today, how the academic discipline of information design relates to the different fields of knowledge and how these relationships have developed. By examining concepts and their historical meanings, this can lead to a fruitful and constructive critical discussion within the subject of Information design.

Compared to the large and diversified field of geography, information design is a relatively young discipline with a strong practical tradition characterised by the use and development of guidelines and principles, as well as a dynamic discourse on design methods, with roots in different design fields such as human-centered design, graphic design, and engineering design. Many academic fields have undergone critical phases, which is true for geography and cartography. Within the discipline of geography, critical discussions were initiated as early as the 1980s and 1990s; scholars such as John Brian Harley, Denis Wood, and John Fels published works of self-examining orientation within the domain, referred to as critical cartography (Harley, 1988, 1992; Fels and Wood, 1992). These works questioned the built-in power relations in a map, how maps are designed and how the visual image of a map could be said to represent a reality outside of it—or failing to represent it—as a postcolonial critique of maps and mapping constitutes a vivid field of research.

A similar critical discussion within mapping and navigational design within information design has not been as noticeable. A younger, cross-disciplinary subject area like information design can learn something from such a discourse. Information design tends to be rather uncritical, perhaps due to its relative youth and interdisciplinarity. Perhaps a critical discourse presupposes an interest in the history and development within the discipline when a critical mass emerges. Finally, Information design has a strong practical tradition, focusing on the production of design artefacts before a critical discussion.

This chapter aims to provide a critical overview of different concepts that are used when designing wayfinding and navigational aids in information design—and their historical underpinnings. I will discuss the concepts of wayfinding, environmental graphic design, cognitive maps, getting lost, and also touch upon accessibility arguments in these areas, as well as standardisation and regulations in wayfinding design. This overview examines the wayfinding theory and practice on a conceptual and historical level exploring how different fields of knowledge have disseminated into it, and how with time, different influences have come to dominate or decline in empirical examples and handbooks. This overview sees wayfinding in information design as an inherently interdisciplinary area, both in its academic manifestations and its practical applications.

The tentative timeline of literature on wayfinding design reviewed here spans from c. 1960 into the 2020s. There are clear tensions between standardised guidelines versus deep contextual and conceptual explorations of space and place. To address some of these tensions, the focus here is on a few, not too distinct, nodes in the literature on wayfinding design, starting with the legacy of Kevin Lynch (1918–1984), American urban planner and professor at MIT, primarily his influential book, *The Image of the City* (1960), moving on to the renewed interest in wayfinding design in the late 1970s and 1980s, with Canadian professor in architecture and environmental psychology, Romedi Passini's writings on wayfinding as the main example.

The early 2000s saw the rise of the concept of Environmental Graphic Design or as the mainly North American community SEGD (Society for Experiential Graphic Design), started naming it in 2013, Experiential Graphic Design (Calori, 2015). This discourse produced a few richly illustrated handbooks and guidelines, still using "wayfinding" in the title. These are all readily available publications, often used in first cycle education or used as handbooks for practitioners. They represent a surge in handbooks on wayfinding design from the 1990s and early 2000s. This is also when Danish designer Per Mollerup coined the term *wayshowing* in 2005 (Mollerup, 2005), developed in the book's second edition (2013). This term is now widely spread, distinguishing between users' navigational strategies and design on a more systemic level, versus the design of artefacts of navigational aids such as signage and maps.

Apart from handbook literature, the dissemination of wayfinding ideas into a Swedish context of policy documents, reports, and guidelines from institutes and agencies is used to exemplify a more practical implementation and design of wayfinding systems, also *avant la lettre*. The main examples are design manuals for hospitals and healthcare institutions, since this area is the most well documented in public records when it comes to wayfinding in complex, public buildings. In the policy documents and standards for healthcare facilities, the focus is also on elements that have to do with signage in a formal manner: hierarchies, typefaces, and standardised pictograms. These exemplifications of wayfinding and signage in public buildings and mainly hospital and healthcare facility design provide visual examples when it comes to principles of wayfinding. These examples from reports and guidelines from the 1970s and 1980s emanate from a background that was heavily standardised and centralised. There are a few examples from later dates just to show differences in attitudes towards centralisation and standardisation.

For much of the past decade the author of this chapter, holding a background as a student and practitioner in exhibition design, has been teaching Information Design for undergraduate students, mainly in Spatial Design. The area of wayfinding in design appears to be so instrumental and rational in the handbook and design manual literature from the 1990s and 2000s. When reading Lynch, the impression is that it conveys a more holistic view on the city and its inhabitants, and it can almost be read as a poetics of the city. In the project leading to the publication of *The Image of the City* in 1960, Lynch collaborated with artist, designer, and visual theorist György Kepes, also at MIT. Lynch, at least rhetorically, grounded more of his findings in psychology and anthropology rather than making aesthetic claims (Raynsford, 2011). *The Image of the City* is far from a handbook on efficient navigation from start to goal where detours must be eliminated and the feeling of getting lost is seen as a disaster. But its place among its contemporaries in urban design research is also complex, and for instance, Lynch does not always acknowledge the ideas of other researchers in the field (Raynsford, 2011). The complexity and contradictions in *The Image of the City* are not often discussed, and it's tempting to use its more impactful terms as labels rather than as a starting point for discussion. The use and adaptations of Lynch's terminology are one of the incentives to go through some of the succeeding literature in the field.

Another issue that wayfinding literature seldom addresses is that designing for wayfinding in general is designing for an individual activity. Facilitating wayfinding by design means addressing one single user at a time. She or he must travel alone with the potential fear of getting lost or being late. The individual has no personal guide or group to stick to, the individual must manage on their own with instructions given by signs and at best, the layout of buildings. The handbook version of wayfinding design promises smoothness and ease in personal navigation and movement. It rests on a rationalist paradigm of viewing any individual resident or visitor to a city as a user whose needs can always be summarised by taking the shortest and most efficient route from point A to point B. Eliminating wrong turns and hesitation saves time and avoids potential fear and frustration. This is not the case with Lynch and Passini, who both write about wayfinding in cities as more than a one- dimensional experience, and that in Lynch's words, a legible and imageable city is an agreeable environment to be in. Somehow, the art of getting lost got lost along the way in the 1990s and 2000s. The potential space for being and dwelling in a city, one fundamental feature of urban life, may be a question for politicians and city planners and not for information designers. But it is also relevant to consider what effect wayfinding systems have on the way we behave and function in urban spaces.

Discussing how digitalisation and digitisation have affected the wayfinding field has a more secondary place in this text. There are many reasons for this. A mobile GPS application typically ends when entering a building or building complex, so there is a persisting need for physical signage in public and semi-public indoor environments. The details on how to find the right floor and rooms still rely on some kind of wayfinding design for first time visitors to a building, even if there may be dedicated digital apps and devices or info kiosks with screens. Such a user is dependent on some form of physical signage. Digitalisation also affects the design process. The examples from the Swedish context of wayfinding in hospitals and healthcare facilities tell a story about a rather analogue working process. From the 1990s, the design tools and the data used are digital, and a designer working with wayfinding may have used CAD files and GIS files as a raw material earlier than that. Various design applications are also used to produce print originals.

Rather than giving digital and interactive media a special position, what is more interesting is the ways in which digitalization in the form of GPS maps and instructions may have affected the way people move and their decisions about wayfinding in more seamless ways. The use of digital devices for navigation enhances "turn-by-turn navigation", meaning that the user follows a route without having a contextual view of the area. The user can follow written, spoken, or visual cues, digital or analogue, with the common trait that only decision points and their intended behaviour are of interest. Such navigation works momentarily, but deprives the user of getting an overview. Here the concept of the cognitive map, which will be discussed later, is in play: the process of building up an impression of an area, not just at the decision points, is part of human navigation and helps future navigation in the same or similar milieus. Landmarks are important for more spontaneous wayfinding, in contrast to turn-by-turn navigation. A cognitive scientific and geoinformatics study by Schwering et al. (2017) criticises that wayfinding support does not incorporate what is known about cognitive processes involved in wayfinding. Computerised models for wayfinding, not tested by users in a physical setting, have a tendency to simplify, to only include the information that is needed and nothing more. Instead, wayfinding is to be seen as a situated learning activity, recognising that people won't always take the optimal route. The authors are proposing a model of "Wayfinding Through Orientation"—to enable users not to be completely dependent on wayfinding systems and be able to make their own choices (Schwering et al., 2017).

To summarise: The outlook towards digital navigation is not particularly developed in this chapter, simply since digital maps and interactive applications so far play a lesser role in indoor wayfinding. However, the implicit results of movement and mobility—such as a close-sighted focus on decision points rather than a more holistic view—may be greater than just using a digital app in a new place.

Wayfinding

The concept of way-finding (with hyphen) derives from Lynch's *The Image of the City* (1960). The legacies of this book are vast, so the focus here is on the concepts that have become instrumental in the emergence of wayfinding principles. *The Image of the City*, aims to discuss the appearance of cities and how they are perceived by people dwelling and moving around there. It aims to provide a model for understanding and analysing the qualities of a city based on how it's perceived and understood—its *legibility*—via a handful of concepts. It not only takes into account what is discernible to the eye. Lynch writes: "At every instant, there is more than the eye can see, more than the ear can hear, a setting or a view waiting to be explored" (1960: 1). The focus is on how a human city dweller, titled the observer, perceives the unfathomable scale of a city, with their senses. It embraces that a city is experienced from a perspective of movement in time and space in instances that, together with bird's-eye views and maps, all interact like a mosaic to form a coherent image in the individual observer's mind. Lynch's narrative about this city image seems to be quite a poetic combination of memories, impressions, history, and sentiments personal for the individual observer. He was influenced by his peer, Kepes' interest in visual and graphic design, and of city views as a filmic succession of images that constitute the observer's image of the city (Raynsford, 2011). On the other hand, the central concept of legibility appears as a strong quality in cities: that they can be perceived and understood as coherent, clear, comprehensive and identifiable. By analysing a few American cities, Lynch offers tools to improve future urban design, a legible city is perceived as a "beautiful" city, bringing clarity and well-being, ease of movement into complexity for its inhabitants and visitors. An in-depth explanation of how to use a set of methods to understand imageability and use of a city is given in an appendix. Here methods to get to know how inhabitants of a city perceive it are presented. These resemble some of the ethnographic or anthropology inspired methods used in design research as of today, such as in-depth interviews, map sketching, photo sorting, and field trips (Lynch, 1960).

The principal elements that have become well-known trademarks in creating the image of the city are paths, landmarks, edges, nodes, and districts. In short: paths are the routes that people use for movement and transportation and can be anything from narrow alleys to broad streets and routes for public transportation. Edges are the demarcations of an area such as walls, fences, and roads. Districts are areas that are perceived as finite two-dimensional units out of which a city is structured, and they can go against administrative divisions. Nodes are junctions, hubs and crossings, or other perceived centres of gravity. Landmarks may be the most well-known concept as being visible physical signs such as public art, building parts that stand out in terms of striking height and size, but can also be elements of nature and smaller details as long as it is something that people notice and value as a mnemonic point in the city (Lynch, 1960). These elements, with slight variation, exist in some form in the predominantly indoor wayfinding discourse.

Early on in the *The Image of the City* introduction, the word *way-finding* appears for the first time, in the context of how humans use different sensory cues for orientation, leaning on research publications in psychology on orientation in space. Here, *way-finding devices* are also introduced, which are described as the aids in the form of map signs, street numbers and the like (Lynch, 1960). The impression is that Lynch did not conceptualize "way-finding", it seems to appear as a convenient description of the act of finding one's way in an environment—the activity in focus throughout the book is orientation. The conceptualization and use of the word *wayfinding* in other contexts appears to have occurred later on.

Cognitive Mapping

Wayfinding revolves around orientation and navigation and the individual's ability to form a mental image of a place within their minds. Geographer Robert Kitchin explains variations in the definitions of the concept by stating that cognitive maps and mapping are multidisciplinary in nature. The concept appears in writings from many different subject domains, from psychology and cognitive science to the social sciences, the humanities, architecture and planning (Kitchin, 1994). A cognitive map is composed of different perceptual inputs from vision and movement and from what Lynch calls "way-finding devices" such as maps and plans. Together, these sensory impressions, from different points in memory are woven together into an overview, helping the individual to navigate. The established term is cognitive map (Kitchin, 1994). In Kitchin's comprehensive cross-disciplinary survey of cognitive maps and mapping, he refers to the first use of the term by Edward C. Tolman in 1948, "Cognitive maps in rats and men", referring to observations of rats using labyrinths. Passini (1992) also refers to these experiments in a full chapter about labyrinths.

The concept of cognitive maps attracted new interest in the 1970s, not least within geography (Kitchin, 1994). Kitchin cites and refers to quite a few definitions of the concept: the most general definition by Kaplan: "At its most general, a cognitive map is a mental construct which we use to understand and know the environment" (Kitchin, 1994: 3). What constitutes a cognitive map, and if it can be said to be a representation is also at stake. In the research that Kitchin has surveyed there are different positions, from "a cognitive map is a map" to a metaphoric or an analogical connection between cartographic and cognitive maps (Kitchin, 1994). Whichever model you use, it may be confusing to call a cognitive map a map, or even an image, since these intrinsic mental structures may not even have a distinct visual form, like the image of a map on paper or screen.

Cognitive maps matter in the wayfinding discourse since the idea of a successful wayfinding strategy is to merge the cognitive map with the design of the environment and vice versa. In Lynch's *The Image of the City*, that would be a city environment with a high level of legibility. Lynch uses the concept of *environmental image* instead of cognitive map, referring to a personal image in the mind of the individual observer, based both on past and current sensory impressions. A good environmental image is a sharp and clear one (Lynch, 1960). The concept of the environmental image in human orientation is also further discussed in an appendix. Here, Lynch writes about some of the sources that have formed the concept, from fiction, history and news articles, with a leaning towards anthropology (Lynch, 1960). It is also likely that Lynch found inspiration for the legibility of cities and how well they are suited for the visitor to form a mental image of them from his contemporaries in urban theory (Raynsford, 2011). But Lynch omits mentioning Tolman's experiments with rats in mazes.

Kitchin places Lynch in a category of geographers and planners who had an instrumental attitude towards the use of cognitive maps in urban planning where correspondence between people's cognitive maps and the design of cities and environments was something to strive for (Kitchin, 1994). To compare Lynch and Passini in terms of their respective use of cognitive maps, Lynch also represents a cognitive turn in the conceptualisation of the cognitive map. This means that human internal cognitive capacities are seen as the main driving forces in the activity of navigation. Cognitive mapping is considered to be involved in decision-making and behaviour in spatial settings. Cognitive maps' role in human decision-making on where to go link them to the concept of wayfinding. At the same time, both Lynch and Passini are seen as representatives of a more instrumental use of cognitive maps in Kitchin's survey of the concept (Kitchin, 1994). The underlying assumption is that if the cognitive map and the physical environment correspond, it makes navigation easier and thus environments can be designed in a better way.

Passini also discusses what is known about the constitution of cognitive maps in cognitive research. They can be structured like an image of a map, or as a more linear sequential route, a more abstract structure or as a more gestalt-like representation. This becomes interesting if you compare maps sketched by people from their memories of a place. People then tend to structure their conceptions of a place around landmarks, and then extend the sketch from there (Passini, 2002).

Cognitive maps and cognitive mapping are recurring concepts in the writings on wayfinding, tentatively appearing before it was an established concept in Lynch's *The Image of the City*, to become an expected ingredient in manuals on wayfinding and signage design in the 2000s.

Avoidance of Getting Lost: At All Costs

The opposite of wayfinding is not finding your way at all. Not reaching your goal, perhaps even getting lost. The fear of getting lost becomes an important argument for implementing wayfinding design, and this issue is addressed in different ways. Passini (1992) uses the labyrinth as a counter-example of wayfinding. A labyrinth's whole aim is to make people lose their direction. Mollerup also includes a short section about getting lost, also using the metaphor of the maze. He lists that such places have in common that they lack transparency, organisation, landmarks, accessibility, hierarchy, among other features (Mollerup, 2013). Lynch (1960), in a passage where he introduces the concept of way-finding, writes that it quite seldom happens that someone becomes lost in a modern city with all the help from "way-finding devices: maps, street numbers, route signs, bus placards", but then continues: "But let the mishap of disorientation once occur, and the sense of anxiety and even terror that accompanies it reveals to us how closely it is linked to our sense of balance and well-being. The very word 'lost' in our language means much more than simple geographical uncertainty; it carries overtones of utter disaster" (Lynch, 1960: 4). Lynch acknowledges that the feeling of being lost in a modern urban setting can evoke a deep, almost fundamental feeling of existential fear, as opposed to the harmonious sense of being able to read and understand the surroundings. The legibility of a city, combined with the wanderer being able to form a coherent environmental image (cognitive map) is necessary to feel safe and secure. Lynch does acknowledge the "value in mystification, labyrinth, or surprise in the environment" (Lynch 1960: 5). But this must happen in "a visible whole", where the wanderers are reassured that they will not become completely lost (Lynch, 1960). In *The Image of the City* there is room for mystery, the unexpected, and detours along the route, as long as the milieu as a whole is legible. As a predecessor to the later wayfinding discourse, it does not have the purpose of efficiency in navigation and transporting oneself towards the intended destination.

In Passini's *Wayfinding in Architecture* (1992), a thorough review of navigation with essays on labyrinths, mazes, and maps which open the book, Passini in the introduction states that: "Being lost is one of the most frustrating and destabilizing experiences." (Passini 1992: vii) and "What is more annoying than the feeling of having wasted time and energy in a futile attempt to reach a destination; what more exasperating than searching for information on displays giving everything but what is needed? What is more bizarre than the twinge of fear or excitement and the second of panic when momentarily lost?" (Passini 1992: ix) "The consequences for individual users may be minimal or they may be harmful, costly or, at the extreme, disastrous" (Passini, 1992: ix). Frustration and stress are also in focus in the later book *Wayfinding: People, Signs, and Architecture* (2002) in the introductory chapters on "Wayfinding difficulties". But here the important issue of accessibility is addressed in a list of "wayfinding difficulties", the problems that can occur for a wheelchair user, a person with low eyesight or low hearing, and more. Passini emphasises that having the user in mind when designing for good wayfinding also means not only designing for a "super-user", but for users with potential challenges (Passini, 2002).

Comparing Passini's statements on getting lost with Lynch's is of course not entirely reasonable, *The Image of the City* did not have the purpose of being a design manual, but to understand cities and their use by their inhabitants, in the next step improving urban design and planning. Passini's two books that I refer to here also have an open and creative tone, but have more explicit purposes to serve as a knowledge base and support in wayfinding design. Therefore, the tone is more instrumental and the opposite of wayfinding—being lost—needs to be spelled out as something both frustrating and time-consuming, and the aspect that lost time also means costs is emphasised. On the other hand, in the later book, *Wayfinding: People, Signs, and Architecture* (2002), users with difficulties to access wayfinding information are addressed, which makes the discussion of the difficulties of getting lost more valid.

The arguments on the avoidance of getting lost bear witness to how exposed the individual subject is when wayfinding fails. They are then left to their own devices when the impersonal directions are missing or happen to be incorrect. As individual wanderers in a city of today, comfort is often to be found in the smartphone and many people use mapping apps on a daily basis. However, these apps don't always work in areas with poor connection or indoors. That sort of security can quickly be replaced by uncertainty.

Case Study: Guidelines for Signage in Public Buildings

In 1972, the National Board of Public Building (KBS, Kungliga Byggnadsstyrelsen) published an illustrated handbook on building signage (KBS, 1972). The institution was a Swedish government agency between 1918–1993 responsible for managing state buildings and the state's provision of premises and building development. The responsibilities have since been taken over by other governmental bodies. The handbook, intended for internal use, in its bibliography, refers to some systemic graphic design for airports and hospitals, but also to graphic design in architecture and graphic and corporate design in general. There is for instance no reference to Lynch, nor any of his well-known terms or concepts in the main handbook text, but in an appendix on environmental psychology, referred to below. The motivation for the handbook is stated in the preface: "The environment around us changes character in a way that makes it increasingly difficult for people to orient themselves in their surroundings" (author's translation, KBS 1972: 10). The handbook's scope is to cover directions and signs within a site or a building. There is generally an emphasis on accessibility, stressing the importance of a well-designed wayfinding system for people with disabilities. There is a dedicated text in the appendix called, "The views of the disabled on signs" (author's translation, KBS, 1972) testifying that this was an issue that was taken seriously. It is said that signs are more important to people with different kinds of disabilities (mobility, hearing, eyesight) and it is underlined that the guidelines in the handbook have the ambition to result in better signage for everyone, including people with disabilities. It is also highlighted that not only signage promotes accessibility, but also floor plans with simple, unambiguous connections provide good conditions for a useful wayfinding system (KBS, 1972). The striking and humorous formulation "sign forest" is used for poor wayfinding with a multitude of ambiguous signs, difficult to distinguish from each other, which of course hinders accessibility (KBS, 1972).

The visual design of the handbook is consistently friendly and playful. The illustration on the cover shows a couple drinking coffee at a café table with an oversized sign saying, "Maja's patisserie: Coffee and tea" with squiggly letters behind them (Figure 1). This illustration and many more in the report are drawn by the widely popular and prolific Swedish artist and author Stig Claesson (1928–2008). In a humorous way, the illustrations address different kinds of problems that can arise when we humans need to find our way in new places with the help of signs. The style of drawing is a contrast to the bureaucratic formal content of the handbook such as the standardisation of pictograms and typography in signs.

In a text like this, it is also interesting to read about the professional roles involved in implementing wayfinding programmes. The handbook's target group is mainly architects and civil engineers and the planning instructions are very hands-on. There seems to be no graphic designer or similar profession acting as an intermediary between the architects and the sign producer (KBS, 1972).

Ideas and concepts about wayfinding from Lynch are more or less absent from the main handbook, but they appear in one of the appendices called, "Psychological Views on Signs" (author's translation) by Rikard Küller from what was later to become a department of Environmental Psychology at Lund University (KBS, 1972). Here, Lynch's *The Image of the City* is referred to via the terms imageability and legibility, stating that a milieu as a whole ideally contributes to readability, but that signs remain indispensable for people to find their way in new environments. In the more concrete examples the dissemination from Lynch comes down to

Fig. 1 Cover design of Kungliga Byggnadsstyrelsen 1972. *Vägvisning: Handbok i skyltning av byggnader (Wayfinding: Handbook on Building Signage)*.

signposting, at least in this publication. The text by Küller also highlights that in order for us to notice a sign, it is recommended that the sign should neither stand out too much nor too little from its background, since excessive visual stimulation could cause stress for the individual (KBS, 1972). Accessibility is once again important: it is emphasized that the driving force must be that people with low vision and mental disabilities are able to understand signs. The text contains instructions and criteria for the placement and design of signs and also methods for calculating reading distance, sharpness, movement, motion tolerance, and, among other things, various observation methods where new technology for measuring people's view point is briefly described (KBS, 1972). The handbook is an interesting record that oscillates between standardisation and hands-on instructions in detail on the one hand, to the presentation of environmental psychology for an architectural context on the other.

Seventeen years later, in a report from the same institution, the development has come to more standardisation, and the report is also presented as a proposal for a standard. It contains concrete descriptions of responsibilities in the construction process of wayfinding systems. Regarding new construction, the house architect is responsible for the overall planning of wayfinding and signage and the roles and stages in the construction process are listed (KBS, 1989). Lynch's terminology is noticeable here: paths, nodes, landmarks, districts, and the more general "places" and "spaces" appear. Instead of "cognitive map", or Lynch's "environmental image", a concept named "memory map" is used to describe the overview people form in their minds of an environment (KBS, 1989).

It is underlined that people's orientation is facilitated if the building is well designed from the outset with clear "main thoroughfares" (corridors) and "secondary thoroughfares", where the first has a more public function as opposed to the secondary which has a more private character (KBS, 1989). Hierarchy is key, and the fewer opportunities for route choices, the better, is the general message.

The photography on the cover exemplifies what the earlier handbook from 1972 named "sign forests" for too many and ambiguous signs in one place. The photo is taken on Skeppsholmen in Stockholm and depicts no less than 15 directional signs on two signposts, with the addition of street signs and road signs in the background (Figure 2). The signs appear to be pointing out directions to the different locations, but differences in angle are so small that a visitor would find it very difficult to have any use for them. The visual material in the report as a whole consists of highly standardized, schematic examples of signage, room distribution of building plans and typography.

When it comes to signage, there is a discussion about a communication model, also represented graphically, where signs are spoken of as a kind of one-way communication where the sender cannot really know whether the message reaches the recipient and therefore it must be designed as clearly as possible from the start (KBS, 1989). There are many instructions on how the design of a building that can facilitate or hinder orientation. For example, it is said that a person can get a

Byggnadsstyrelsens rapporter 165
1989-06

Vägvisning, orientering och skyltning

Fig. 2 Cover design of Kungliga Byggnadsstyrelsen 1989. *Vägvisning, orientering och skyltning* (*Wayfinding, orientation and signage*).

"picture of the building" from the outside of it, which could then facilitate interior orientation when the visitor enters the building. The report offers terminology for different types of rooms and corridors, hierarchically arranged with abbreviations. There are also various illustrated examples of room layouts with names and whether they facilitate orientation or not (KBS, 1989). A distinction is also made between so-called active wayfinding, i.e., personal guidance, or instructions, for instance to be met at the entrance and being accompanied to the destination—versus passive wayfinding, referring to signage with the help of which the visitors can navigate on their own. There are often mixed forms where visitors can ask for directions at a reception desk or ask staff. Instructions written on paper are also considered to be active wayfinding. In 1989, the relative novelty of movable electronic signage was also included (KBS, 1989).

Both these publications from the National Board of Public Building, 1972 and 1989, present a basic model of a wayfinding or signage system consisting of "building orientation": a map overview or a model that should be in the same direction as the building, with a "you are here" indication. Furthermore, signs are categorized into "refer to" signs—signs with arrows, "identify" signs that name certain parts of a building. "Orientation" signs to identify parts of a building, floor, corridor, corridor section or room (KBS, 1989). The guidelines for graphic design include symbols and pictograms (KBS, 1989). The colour scheme is set according to a SIS (Swedish Institute for Standards) standard of distinguishable colours in the NCS palette (described further below). There are also strict rules for sign terminology, hyphenation and abbreviations, instructions for choice of materials and detailed drawings of ways to mount and hang signs (KBS, 1989). There are also accessibility instructions in the form of Braille and relief maps for people with poor eyesight (KBS, 1989).

In 1989 there had been quite some development in the field of literature on wayfinding and orientation. In the bibliography, Lynch is represented and there are also references to, among others: Donald Appleyard, professor of urban design and author of *Livable Streets* (1982), Jan Gehl, Danish architect and professor of urban planning who published extensively on cities and public spaces favouring pedestrians and cyclists; and Arne Branzell, Swedish architect and professor at Chalmers University of Technology who developed a notation system for spatial experiences.

This means that there was competence and knowledge of the theories behind wayfinding in urban planning in the team behind the publication in 1989, which is reasonable as it was a growing field. At the same time, the concept of indoor wayfinding had not yet begun to be used. As with Lynch, it was mainly theories about outdoor environments and city planning that formed the theoretical basis for wayfinding within and in close proximity to the public buildings in the report.

In terms of time, in between these two publications, there is a report from the state agency Health and Social Care Planning and Rationalization Institute (SPRI)

(1968–1989), *Wayfinding within hospitals* 1977. The terminology and names of units within hospitals and healthcare facilities are in focus here. Wayfinding (*vägvisning* in Swedish) has a slightly different definition in this publication: "The concept of wayfinding can be defined as a conscious organization of signals to help the individual reach a desired goal" (author's translation, SPRI, 1977: 14). There is a clear orientation towards a general communication model and the wayfinding focus is on signage, not on architectural guidance in the form of room layout and design, as recommended by the National Board of Public Building. However, the technical and practical guidelines of sign hierarchies, placement, colours, and installation are similar.

The inner cover of the report shows a hand-drawn illustration, which is one of few in the report as a whole. It depicts a person on crutches with a bandaged leg approaching a large hospital building (Figure 3). The person seems to have reached the destination: the emergency ward. On the right hand of the drawn sign is a plan of the hospital and on the left hand side a list of departments. The hospital building is typical of Swedish hospital architecture of the 1960s and 1970s and the entrance has large glass doors. Inside, two figures in medical outfits can be glimpsed. The person on crutches is about to reach their goal to get help! The illustration shows an educational approach and an intention to include the human user in the theme of the report. The four squares on the lower end of the page are likely examples from the standard colour palette presented in the report.

The report contains a set hierarchy of sign types where orientation signs, direction signs and position signs are the main categories, with illustrations (SPRI, 1977). There are also principles for numbering of floors and rooms as well as guidelines for technical execution and materials, placement, typography, and pictograms (SPRI, 1977). The question of sign colour is given particular attention in an appendix written by the colour researcher and physicist Gunnar Tonnqvist, one of the originators of NCS, mentioned earlier, the internationally used Natural Colour System, which became a Swedish standard in 1979. Consequently, the bibliography is much more focused on signage, signage systems, and terminology than the earlier described KBS publications.

In a Swedish governmental context of the 1970s and 1980s, represented by these three publications, it is clear that signs and signage systems were seen as important for people's orientation and navigation, and that there was a need for standardisation of terminology, hierarchy, design, and placement. Extra attention was also given to people with disabilities or poor eyesight, so questions of accessibility were clearly on the agenda. Looking at the dissemination of the most well-known concepts in wayfinding design, that is terminology originating in Lynch's *The Image of the City*, it is discernible in some parts, most notably in a text from the emerging field of environmental psychology (KBS, 1972) and in the report from 1989. Generally, however, there seems to be a conflation between wayfinding as a navigational activity that is facilitated by the design and layout of a building and what can be achieved with signs and signposting.

Fig. 3 Sjukvårdens-och socialvårdens planerings-och rationaliseringsinstitut 1977. *Vägvisning inom sjukhus och andra sjukvårdsanläggningar med termer för vägvisningsskyltar (Wayfinding within hospitals and other healthcare facilities with signage terminology).*

Wayfinding Beyond Signage

In Paul Arthur's and Romedi Passini's book, *Wayfinding: People, Signs, and Architecture* (2002), first published 1992, signage is only a small part of wayfinding in spatial contexts. The overall planning of buildings, room sequences, and milieus is the primary target, and signage is secondary. It relies quite heavily on Lynch and wayfinding design is seen as something strictly bound to space and architecture. Secondly, there is a strong focus on the cognitive functions of human navigation. According to Passini, wayfinding is "spatial problem solving", consisting of "decision- making, decision execution, and information processing" (Arthur and Passini, 2002: 25). The concept of the cognitive map is at the centre of this process to make these navigational decisions. The decision making in the wayfinding process is also broken down into a decision plan, illustrated as a graphic chart (Arthur and Passini, 2002).

Passini expressed reluctance towards too much focus on signs and graphic design in wayfinding design, stating that "signs should be plain, not fancy" (Arthur and Passini 2002: 151). They discuss a vision about a new professional role in "wayfinding design"—a new speciality alongside architects and graphic designers. In an essay in the anthology *Information Design* 2001, Passini sees potential in the then relatively new knowledge area of information design and wayfinding design. He hopes that if graphic design had become too preoccupied with surface qualities, information design could take on a cognitive approach and take into account how people's decision strategies and processes operate when orienting in unknown places (Passini, 1999).

The wayfinding solutions in buildings and environments are based on the design of buildings: and the book offers a multitude of illustrated spatial planning solutions. Here some of Lynch's concepts are transferred to suit indoor spaces, such as "circulation systems": they could be shaped in different ways: linear, centralised, composite and networked, which in turn are exemplified with different sub types. There is a huge variety and a very visual compilation, different from the more sign-focused writings that would emerge later (Passini, 2002). The focus is on the overall planning of built structures, both indoors and outdoors.

Wayfinding, Wayshowing and Environmental Graphic Design

In the first decades of the 2000s, the use of the concept, *wayshowing*, found its way into the wayfinding and navigation discourse, coined by Danish designer Per Mollerup. "Wayfinding and wayshowing relate to each other as do writing and reading, teaching and learning, or cooking and eating. One activity deals with sending, the other receiving," according to Mollerup (2005). Wayfinding is the problem that the user needs help with, and facilitation in the form of designed signs can indeed help. It is even said that the former use of the concept of wayfinding to denote designed navigation systems and devices such as signs was imprecise

(Mollerup, 2013). It is reasonable to assume that wayshowing comes from a direct translation of the Danish word for navigation: *vejvisning* into English. Wayfinding does not have an unambiguous translation from English to the Scandinavian languages (Danish, Norwegian, Swedish), but for instance, in the KBS report *Vägvisning, orientering och skyltning* (1989) referred to earlier, it is quite clear that for instance *wayfinding systems* have been translated into *vägvisningssystem,* and that there was an awareness of the international concept of wayfinding (KBS, 1989).

Lynch and Passini in different ways refer to the processes and active participation involved in wayfinding. A multitude of impressions and stimuli make the urban wanderers aware of their surroundings. There are differences in how they handle signals and stimuli compared to what can be seen in Mollerup's handbook. Mollerup refers to communication models and to semiotics as the theories behind wayshowing (Mollerup, 2013). Mollerup's book is beautifully illustrated with lots of photographs of signs of different kinds. The concept of wayfinding is referred to as "a spatial problem solving process", similar to Passini's definition but without a reference (Mollerup, 2013).

There is also a short section about cognitive maps referring to Lynch's concepts paths, landmarks, nodes, edges, and districts as important entities for facilitating memory of the surroundings (Mollerup, 2013). A difference is that since Mollerup's handbook is not about urban planning, there are no claims about the importance of making environments legible in order to facilitate cognitive mapping. Rather, it refers to the concept and then attempts to link cognitive maps to communication theory and how "receivers", meaning people, understand messages according to their culture and internal cognitive maps (Mollerup, 2013). Accessibility is also addressed in a section focusing on people with visual impairments (sic.) with illustrated examples from design projects (Mollerup, 2013).

With Mollerup's handbooks from 2005 and 2013 as a bridge, the first decades of the 2000s saw a surge in publications about signage and wayfinding systems with lots of visual presentations of inspiring projects, but less text. The same era also saw the establishment of books about environmental graphic design with a more pronounced ambition to function as design handbooks.

Designer David Gibson's, *The Wayfinding Handbook* (2009) is written from a practitioner's perspective and is a typical hands-on handbook. A benefit of this publication is that it lists the early practice of wayfinding design in the English speaking world—private companies and designers working for them—in implementing and developing wayfinding systems. It was even a field where female designers from the 1960s and onwards could make a career since it was a relatively new design field (Gibson, 2009). A wayfinding design project in this handbook's perspective is purely bespoke for the commercial client, underlining the understanding of three variables: the client, the client's clients, and the environmental context in which the wayfinding system is to be installed (Gibson,

2009). Compared to, for instance, Passini, it says very little about cognitive processes and decision hierarchies. As in Passini, a discussion about the role of a specific wayfinding designer is continued: on how to become a wayfinding designer, which kind of teams are needed and the types of organisations that could hire a wayfinding designer (Gibson, 2009). It is even said that there should be a dedicated signage design team in medium-sized and large projects (Gibson, 2009). The design process seems to have few frameworks and standards to relate to, each project is framed as unique and with a unique research process (Gibson, 2009). On the positive side, the unique status of each design project makes them situated in space with carefully defined user needs. On the other hand, tailor-made projects are attractive for a design company, and handling as much of the process as possible means more project time. Needless to say, from a design perspective, such projects, with good collaboration between design firm and client, can result in innovative and beautiful design solutions. There is not a great focus on accessibility in this book, but The Americans with Disabilities Act (former American Disabilities Act) is mentioned (Gibson, 2009).

Despite the title, designers Chris Calori and David Vanden-Eynden's designer handbook, *Signage and Wayfinding Design* 2009/2015, has a strong focus on signage, or Environmental Graphic Design (EGD). EGD is presented as the renewed Experiential Graphic Design, and encompasses signage, interpretation, and placemaking (Calori and Vanden-Eynden, 2015). The professional organisation SEGD (the Society for Experiential Graphic Design), changed its name from "Environmental" to "Experiential" graphic design in 2013. Signage and wayfinding are used interchangeably in this handbook, and a simplified reference to the cognitive processes involved in orientation and navigation is presented. It is also said that signage is more than just signage, it is the design of built environments, roads, corridors and sequences of rooms. Still, the physical environment is seen as a given factor to adapt the design to. Again, the handbook promotes a client-driven process, where every project is unique. The design process is described in a concrete way like a working plan, and it is generously illustrated and transparent with lots of detail. It attempts to divide the wayfinding design process approach into three different systemic parts in the so-called "signage pyramid model", dividing the design and production process into content, graphic design, and sign production (Calori and Vanden-Eynden, 2015).

Designer Craig M. Berger's, *Wayfinding: Designing and Implementing Graphic Navigation Systems* (2009) springs from the same school of Environmental Graphic Design. It has a holistic approach, but is less of a handbook, even if there are tips and tricks from the trade. It is accessible to read with many image examples divided by genres such as museums, health care facilities, universities, and sports facilities among others. The book also has a dedicated chapter on "legibility for users with disabilities" and symbols for universal design, making this book the one that does have a more comprehensive approach to accessibility in its genre (Berger, 2009).

Wayfinding in Hospitals and Healthcare Facilities in the 2000s

To sum up, there is room for a brief follow-up on the guidelines on signage and wayfinding design in Swedish public buildings from the 1970s and 1980s referred to earlier. To give a true picture of the field in the 2020s would of course have required a more extensive study. But there is some documentation on institutional responsibilities regarding hospitals and healthcare buildings on the government authority Boverket, the Swedish National Board of Housing, Building and Planning, which in the 2020s had a strategic and regulatory role on a national level, even though hospitals and healthcare facilities are under the responsibility of regions and municipalities in Sweden. The first striking difference from the 1970s and 1980s is decentralisation and that standardisation and regulations in indoor design, such as signage and wayfinding design have decreased in the 2000s. There may be many reasons for this, but regional and municipal autonomy in planning construction and reconstruction is one, and deregulation with a diversity of healthcare providers may be another.

In the 1960s and 1970s, a large number of hospitals and healthcare facilities were built in Sweden. Many of them are today in need of renewal and some have already undergone major changes (Boverket, 2022). The National Board of Housing, Building, and Planning does not issue any guidelines, but refers to other actors, such as the Swedish Association of Local Authorities and Regions, an employers' organisation for local governments. In 2013, they issued a report on hospital entrances where a schematic picture on navigation to and within a hospital is taken from Locum, a property manager owned by Region Stockholm (Bergström and Ljungkvist, 2013). Another example for understanding the field in the past decades is architect Anna Barne's overview from her own experience as an architect specializing in wayfinding in hospitals (Barne, 2006). In turn, this means that to get a picture of how wayfinding design was implemented in hospitals and healthcare facilities in the 2000s in Sweden, one would have to visit each region and hospital. This goes in tune with the trend in the international literature on wayfinding design, which in the 1990s and early 2000s reflects a more bespoke and particular approach to such projects. This probably gives more freedom to designers, but also creates greater challenges. The knowledge of well-designed wayfinding is to find with many different actors: property managers and owners, architects, graphic designers, and sign producers, and this makes it harder to navigate the field.

Summary and Conclusions

This chapter has attempted to track and compare some main ideas and concepts in wayfinding design. Mainly Lynch, Passini, Mollerup, and finally the Environmental/Experiential Graphic Design discourse have served as examples. For more concrete examples of the dissemination of these ideas, a few handbooks and reports from Swedish government agencies on wayfinding design in public buildings from the

1970s and 1980s have been discussed. To summarise, there is no straight diachronic development in the timeline of wayfinding design. Instead a few tensions can be discussed.

The first tension is the one between architecture and planning versus graphic design. Wayfinding is ideally an inherently multidisciplinary field of knowledge both in theory and in practice, but looking at the different positions reviewed here it is clear from which profession they originate. Lynch was a professor in urban planning and deeply interested in cognitive psychology and the communicative aspects of human orientation and navigation in urban settings. Passini is professor in architecture and environmental psychology. Both Lynch and Passini were, in different ways, favouring the cognitive psychological aspects of wayfinding. The wayfinding design that they promoted or indirectly gave rise to was more focused on spatial disposition and layout than on facilitating artefacts of wayfinding design such as signage. The concept of the cognitive map lies at the core of how to understand and facilitate wayfinding. The more recent writings from the 2000s are partly collected under the umbrella term of Environmental/Experiential Graphic Design. In the same era, designer Mollerup coined the term "wayshowing" for the facilitating artefacts in wayfinding design. These voices mainly come from the extended field of graphic design. The focus lies in adding wayfinding design to existing environments rather than influencing for instance room layout in new and rebuilt environments. Information design, being a relatively young discipline, has an advantage in this tension to understand user needs and take advantage of different positions and reconcile between professional roles.

The second tension is that between standardisation versus tailor-made design and is the one that becomes most clear when discussing wayfinding in hospitals and healthcare facilities in Sweden. The practice has clearly moved away from standardisation and has become decentralised where every design project is seen as unique and is made bespoke to each client. This can of course generate innovative design projects. But one goal for design must also be efficiency and sustainability. Clients, public and private, need to be able to implement wayfinding solutions that are useful, transparent and can be easily updated. Accessibility is also an important reason for standardisation. Users with limited mobility and poor eyesight have the right to be able to navigate under the same conditions as everyone else. Working within and developing useful standards in design is an important consideration for information designers.

The third tension regards the balance between exploration and efficiency in wayfinding. For instance, Passini lingers on the negative aspects of getting lost and structures the wayfinding process around decisions and execution. This is not to say that his writings on wayfinding lack aspects of human exploration and aesthetic aspects of spatial experiences, but the later interpretations in the era of Environmental/Experiential Graphic Design are more overtly promoting efficient routes, however with innovative graphic design. At the end of the day, users are multi faceted actors with different needs and abilities and that is commonplace in design with a human centered perspective. It is important that efficiency in spatial information does not eliminate space for users to experience environments, for detours, wrong-turns, and hesitation since human users learn by multi sensory experiences. Going back to Lynch's *The Image of the City*, the multi sensory complexity of spatial experience, orientation, and navigation is always present in descriptions of experiencing the city.

Lastly, the promises of digital devices and platforms in navigational design have been vast, but in indoor settings, users are still dependent on functional wayfinding via architecture and signage. Such site-specific design also provides space to experience and understand the environment with all senses. Information design would benefit from opening up to a critical discussion on the theories, practices, and concepts that form the basis of what we mean when we work with wayfinding design and related activities in information design education, research, and practice.

References

Arthur, P. and Passini, R. 2002. *Wayfinding: People, Signs and Architecture.* Focus. Oakville.

Barne, A. 2006. *Skyltning av sjukhus,* Stiftelsen för arkitekturforskning. Gothenburg.

Berger, C. 2009. *Wayfinding: Designing and Implementing Graphic Navigational Systems*. RotoVision, Mies.

Bergström, A. and Ljungkvist, S. 2013. *Sjukhusentréer,* Sveriges Kommuner och Landsting. Stockholm.

Boverket. 2022. *Styrning av vårdmiljöer*. https://www.boverket.se/sv/samhallsplanering/arkitektur-och-gestaltad-livsmiljo/arbetssatt/vardens-miljoer/manniska-och-vardmiljo/styrning-av-vardmiljoer/ (Retrieved 2025-10-09)

Calori, C. and Vanden-Eynden, D. 2015. *Signage and Wayfinding Design: A Complete Guide to Creating Environmental Graphic Design Systems*. John Wiley & Sons, Hoboken.

Gibson, D. 2009. *The Wayfinding Handbook: Information Design for Public Places*. Princeton Architectural Press, New York.

Fels, J. and Wood, D. 1992. *The Power of Maps*. Guilford Press, New York.

Harley, J.B. 1988. Maps, Knowledge and Power. In: D. Cosgrove and S. Daniels (Eds.), *The Iconography of Landscape: Essays on the Symbolic Representation, Design and Use of Past Environments* (pp. 277–312). Cambridge University Press, Cambridge.

Harley, J.B 1992. Deconstructing the Map. In: T. Barnes and J. Duncan (Eds.), *Writing Worlds: Discourse, Text and Metaphor in the Representation of Landscape* (pp. 231–247). Routledge, London and New York.

[KBS] Kungliga Byggnadsstyrelsen. 1972. *Vägvisning: Handbok i skyltning av byggnader*. Byggnadsstyrelsen, Stockholm.

[KBS] Kungliga Byggnadsstyrelsen 1989. Vägvisning, orientering och skyltning. *Rapport, 165*. Byggnadsstyrelsen, Stockholm.

Kitchin, R. 1994. Cognitive Maps: What Are They and Why Study Them? *Journal of Environmental Psychology*, *14* (1): 1–19.

Klein, J.T. 2023. Boundary Discourse of Crossdisciplinary and Cross-Sector Research: Refiguring the Landscape of Science. *Minerva*, *61*(1): 31–52.

Lynch, K. 1960. *The Image of the City*. MIT Press, Cambridge, Mass.

Mollerup, P. 2005. *Wayshowing: A Guide to Environmental Signage: Principles and Practices*. Lars Müller, Baden.

Mollerup, P. 2013. *Wayshowing > Wayfinding: Basic and Interactive*. BIS Publishers, Amsterdam.

Passini, R. 1992. *Wayfinding in Architecture*, Van Nostrand Reinhold, New York.

Raynsford, A. 2011. Civic Art in an Age of Cultural Relativism: The Aesthetic Origins of Kevin Lynch's Image of the City. *Journal of Urban Design*, *16*(1): 43–65.
Schwering, A., Krukar, J., Li, R., Anacta, V.J. and Fuest, S. 2017. Wayfinding through Orientation. *Spatial Cognition and Computation*, *17*(4), 273–303.
[SPRI] Sjukvårdens- och socialvårdens planerings- och rationaliseringsinstitut. 1977. *Vägvisning inom sjukhus och andra sjukvårdsanläggningar med termer för vägvisningsskyltar*. SPRI, Stockholm.

10 Creating Space for the Unspeakable

Max Kleijberg[1*]

Alex (9 years old): "*My mother says that my grandmother has gone on a trip, but she's dead. She still says that she's gone on a trip and I don't believe her.*"
(Kleijberg, 2021)

Introduction

In this chapter information design is approached from the perspective of topics and experiences that are often considered sensitive, taboo, marginalized, or hard to verbalize—in other words: *unspeakable.* Examples of such topics are severe illness, death, sexuality, racism, discrimination, violence, and abuse. Information design can create space for meaningful interaction regarding such unspeakable topics. "Creating space", here, refers to designing environments in which the unspeakable can be acknowledged and engaged with sensitively with empathy and respect. This often involves creating settings where individuals feel safe (or at least feel safer) to broach and address these potentially difficult and sensitive issues.

In this chapter, I will first define what I mean with *unspeakable.* I will then draw from my own research to discuss approaches, methods, and theories that can help designers to create spaces for people to interact and engage with unspeakable topics. I will focus on three projects in particular: Studio DöBra[#], a project in which children and older adults met in a series of intergenerational arts-based workshops to engage with topics related to dying, death, and loss (Kleijberg, 2021). DöBra Rainbow Café, a project in which I collaborated with LGBTQ+ organizations to create community places by and for LGBTQ+ people who want to discuss topics related to aging, dying, death, loss, and care at the end-of-life. And the Peer Advisor Project, an initiative by the Regional Cancer Centre Stockholm Gotland to promote cancer prevention in communities affected by challenging socioeconomic conditions.

I do not claim that the approaches, methods, and theories I discuss here are universally applicable, as the nature of what is considered unspeakable and how it is experienced, varies significantly across cultural, social, and individual contexts.

[#] *DöBra* is a Swedish pun literally meaning dying well, but figuratively meaning "awesome".

[1] Ph.D. Senior Lecturer in Information Design, Department of Health Science, Innovation and Design, Mälardalen University, Box 325, 631 05 Eskilstuna, Sweden.
[*] Corresponding author: max.kleijberg@mdu.se

DOI: 10.1201/9781032642598-10

Rather, my aim is to contribute to a conversation within information design about how to design environments for meaningful interaction regarding sensitive topics, and invite others to adapt, reimagine, and build further on these ideas in ways that are responsive to their own topics, users, and contexts.

Exploring Unspeakable Topics

The quote above comes from Studio DöBra, a research project in which children and older adults met in a series of five intergenerational arts-based workshops to engage with topics related to dying, death, and loss (Kleijberg, 2021). The children and older adults were unrelated but lived in the same neighborhood. The project was called Studio DöBra and took place in two iterations in different Swedish cities. One workshop was about the question "where do we end up after we die?" Children and older adults worked in small groups to answer this question by visualizing various afterlife destinations to form a kind of travel guide for the afterlife. After that, they together built a large rocket ship with which they played that they traveled to these different afterlife destinations, Figure 1. In another workshop, the children and older adults discussed how people can express that they are grieving, for example, by wearing black clothes or wearing a black ribbon. They then created their own symbols for grief from fabric and other material which

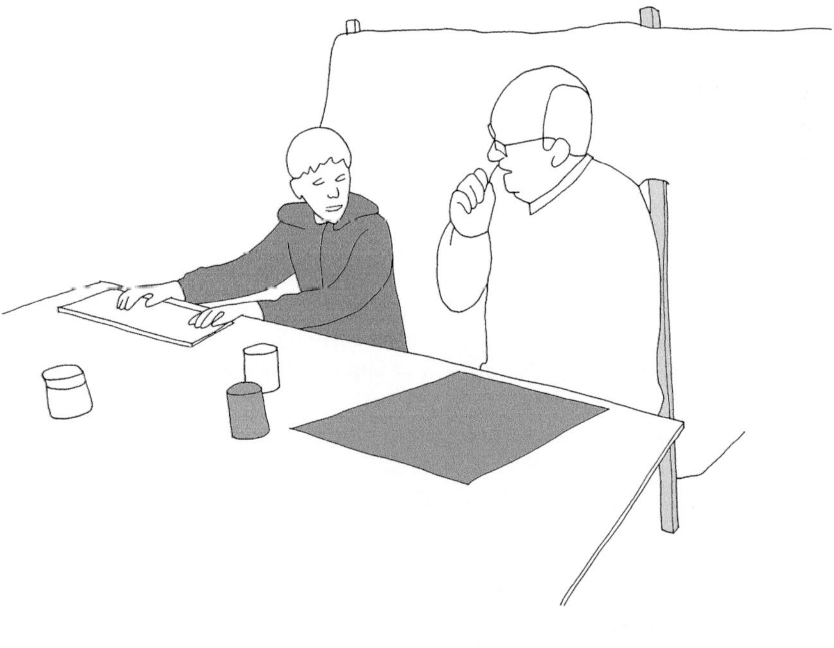

Fig. 1 Traveling to the afterlife, (illustration based on a photograph of the workshop first published in Kleijberg et al. (2021), adapted and shared under CC BY 4.0 license).

they could wear on their clothes as an expression for grief. In the subsequent session, the participating children and older adults held a funeral for a dead bird, wearing their own grief symbols.

During the workshop about afterlife destinations, Alex, the 9-year-old boy from the quote above, explained that he did not believe his mother when she told him that his grandmother had gone on a trip—"she's dead", he said. We can only speculate about why Alex' mother was not straightforward with him about the death of his grandmother. Perhaps it was too painful or sensitive for her, or maybe she just did not know how to put this loss into words. Maybe she tried to protect her child from becoming sad and upset by the death of his grandmother. Or maybe there was a miscommunication. In Swedish, it is common to say that someone has "gone away" (gått bort) when someone has died. Similarly, the English phrase "has passed" does not directly mention death either. A child, who might be unfamiliar with the ambiguity of such phrasing, might take this to mean that the deceased person has gone on a trip and may come back.

With the unspeakable, I refer to *information, topics, and experiences that may be difficult to broach or put into words due to personal sensitivities, emotional intensity, and/or societal and cultural norms.* Death can be considered an unspeakable topic as it may be sensitive on a personal level and come with emotional intensity that can be hard to put into words, perhaps caused by prior painful experiences or a fear of death. Furthermore, societal and cultural norms shape what is appropriate to talk about and with whom. For example, parents may be reluctant to bring up death with their children based on societal norms implying that adults should protect children from feeling sad and upset (Kleijberg et al., 2019). Paradoxically, in our culture, death is talked about often on the news (for example, in relation to crimes, wars, and pandemics) and in popular culture (for example, through fictional deaths in movies, games, and literature).

Societal and cultural norms change over time. For example, in many contemporary societies, dying, death, and grief are managed by professionals such as healthcare staff, funeral directors, and psychologists, while in the past, families, social networks, and communities played a much larger role in caring for the dying and those affected by grief (Sallnow et al., 2022). It has been argued that this change has made us less prepared when we encounter death and loss in our own lives (Sallnow et al., 2022). Therefore it is important to create spaces to engage with death and loss for those who want to, as this may help people to, for example, come to terms with their own mortality, make informed decisions regarding care at the end-of-life, and support people who are grieving (Sallnow et al., 2022).

Above, I mentioned severe illness, sexuality, racism, discrimination, violence, and abuse as other examples of topics and experiences that can be considered unspeakable. These are all topics and experiences that can be charged with personal sensitivities and the ways they are spoken about (or avoided in conversations and societal discourse) are directed by societal and cultural norms. It is important to recognize that norms also often determine whose voices are amplified and whose voices are marginalized. Norms typically favor dominant narratives, people in power positions, and the perspectives of the majority, while the voices of minority groups, those with less power, or those with non-conforming perspectives can be silenced, ignored, or not given space. For example, a lesbian woman who lost her same-sex life partner might lack the space to express and deal with her grief in contexts with low acceptance of LGBTQ+ people and/or a lack of equal rights (M. Kleijberg & L.E. Eriksson [in manuscript]). Another example is that people with a migration or refugee background may lack space to engage with relevant health and care information when healthcare systems do not provide culturally adapted and inclusive information (Efremius et al., 2023). These issues contribute to inequalities in society. I suggest, therefore, that information design should counteract these inequalities by critically engaging with societal and cultural norms, especially norms that marginalize certain voices or create silence around certain topics. In doing so, information design can amplify underrepresented perspectives and create inclusive spaces for dialogue.

Challenging Norms and Navigating Power Dynamics

As described above, topics and experiences can be unspeakable partly due to societal and cultural norms. Norms are often implicit expectations about how we should behave which we have learned through partaking in society and social interaction. Not only do norms affect individuals and human interaction, but they also shape how society is organized because norms are embedded in systems such as politics, education, and healthcare. For example, norms around culture and socioeconomic contexts influence healthcare systems through a design of healthcare services around the needs and contexts of the majority population. In doing so, healthcare services often fail to provide information and care practices that respond to the needs and contexts of people with migration or refugee backgrounds, people with financial challenges, and/or people with lower education levels (Efremius et al., 2023; Horrill et al., 2024). Therefore, designers who want to create space for the unspeakable, need to navigate and challenge those norms (Kleijberg et al., 2019). To clarify this, I draw from the Studio DöBra project in which norms and power structures were challenged to create space for intergenerational interaction about death. I will also discuss the Peer Advisor project which challenges norms and power structures to create space for conversations about cancer prevention with people affected by challenging socioeconomic conditions. While Studio DöBra predominantly challenged norms to influence interpersonal interaction, the Peer Advisor project challenges norms embedded in the healthcare system. Both examples illustrate that challenging norms implies the need for designers to navigate power structures (Efremius et al., 2023; Kleijberg et al., 2019).

Studio DöBra, Challenging Norms to Engage with Death and Grief

The aim of Studio DöBra was to create space for children and older adults to engage with end-of-life topics through intergenerational arts activities. I initiated the first iteration of Studio DöBra by establishing a project group which included representatives from a children's library, an artistic organization for children, an activity center for older adults, and an after-school center, all based in an urban muti-ethnic neighborhood in a large city in Sweden (Kleijberg et al., 2019; Kleijberg et al., 2022). We developed the first Studio DöBra iteration through a co-design process. My role was to facilitate this process. Together we decided to invite eight children (nine years old) and eight older adults (mostly over 80) to a series of five 2-hour arts-based workshops dealing with end-of-life related topics. Project group members who worked with children planned to invite children to Studio DöBra, and the representative of the activity center for older adults would invite the older adults. However, these project group members found this more difficult than expected due to the sensitivity of the topic. In a follow-up research interview, the representative of the organization for older people said:

> "*I didn't really expect this. But there was a fear of how to bring it up, because I know that there are many who are widows and widowers. Maybe they've experienced a lot of loss not too long ago either, so how would they react when you bring up the topic?*" (Kleijberg et al., 2019)

Thus, even though this project group member was initially positive about involving older adults in end-of-life related topics, she realized that she was afraid of raising the topic with older adults as she felt responsible for them and did not want to make them upset. However, eventually she began approaching older people who attended the activity center, talked with them about the project and found that many were indeed interested in talking about death. Similarly, the project group members who planned to invite children found it difficult to broach the topic with children due to the sensitive nature of death and loss. They asked me, as the initiator of the project and facilitator of the co-design process, to invite children. To this end, I presented the project for a large group of nine-year-olds at an after-school center. The children seemed engaged and when I asked who wanted to participate almost everyone raised their hand. Research based on follow-up interviews with participants found that the older participants were motivated to participate in Studio DöBra as they lack spaces for discussing end-of-life related experiences, and children talked about wanting to participate as they were curious about the topic (Kleijberg et al., 2020).

Thus, both older adults and children wanted to participate and engage with topics related to death and grief, but we (the project group) would have not found out about this if our fear of broaching these sensitive topics led us to avoid the topics. As project group members, we had to navigate societal and cultural norms regarding what is appropriate to talk about with children and older adults (Kleijberg et al., 2019). This also shows that project group members were in a position of power due to their professional role in relation to the children and older adults. We had the power to create space for children and older adults to engage with end-of-life related topics by raising the topic and opening up for conversations, but we also had the power to hinder children and older adults from engaging in these topics by avoiding the topic and silencing conversations (Kleijberg et al., 2019). The project group members, including myself, can be conceptualized as *"adults-in-between"* the children and older people—literally in terms of age, and figuratively in terms of power (Kleijberg et al., 2019). As adults-in-between, we were "gate keepers" for the children and older people to meet each other and we were gate keepers for their engagement with topics related to death and loss (Kleijberg et al., 2019).

In order to create space for intergenerational conversations about death in Studio DöBra, we first needed to reflect on our own position of power over the children and older adults, as well as our personal perspectives on death and loss (Kleijberg et al., 2019; Kleijberg et al., 2022). One pragmatic way to think about power in collaborative and participatory design processes is by making a distinction between "power over", "power to act", and "power to act in concert with others" (Gaventa and Cornwall, 2015). A *power-over relationship* indicates that one person or group has the ability to control or dominate others, often in a hierarchical manner (Gaventa and Cornwall, 2015; Kleijberg et al., 2019). In the case of Studio DöBra, the project group members had power over the children and the older adults. In many contemporary societies, age-related norms are embedded in legal and institutional frameworks, shaping how power is distributed across age groups. For example, children need the consent of their legal guardians for many decisions which places the guardians in a place of *power over* children. Similarly, older adults often lose autonomy when they become dependent on elder care, where staff make decisions about their daily lives.

Power to act can be characterized by the agency and ability a person has to take action, for example to change a situation or to influence circumstances (Gaventa and Cornwall, 2015; Kleijberg et al., 2019). This form of power often requires having knowledge, resources, and a mandate to act, for example a professional position that allows or requires a certain action. In Studio DöBra, this power to act was exclusive to the adults-in-between, as we had the resources, mandate, and professional agency to invite children and older adults through our professional roles (Kleijberg et al., 2019). Furthermore, our power to act was evident in the way that we were able to shape the intergenerational engagement with death and loss by designing the arts-activities for the Studio DöBra workshop series (Kleijberg et al., 2019; Kleijberg et al., 2021).

Power to act in concert with others refers to a collective ability to make decisions and take actions, building on strengths, resources, and capabilities to create a greater impact (Gaventa and Cornwall, 2015; Kleijberg et al., 2019; Kleijberg et al., 2022). An important research finding from Studio DöBra was that power to act in concert with others within the project group was limited at the start of the collaboration but increased as the project evolved and progressed (Kleijberg et al., 2019). For example, even though my aim was for the design process to be collaborative and participatory, at the beginning, project group members felt little ownership over the project and relied on me to make decisions (Kleijberg et al., 2019). As I had initiated the project and collaboration and facilitated the co-design process, I was seen as the owner of the project and had a position of *power over* the other project group members. Therefore, I made efforts to increase a sense of shared ownership to shift power dynamics towards power to act in concert with others. I did this primarily by initiating a collective reflective practice on power dynamics, both within the project group and in relation to the children and older adults. We did this through conversations during which we discussed roles and expectations, our view of age-related societal norms and how this affected how we approached children and older adults, and also our personal views and beliefs of death and how this affected how we designed the workshops for children and older adults (Kleijberg et al., 2019). Findings from Studio DöBra research indicate that this collective reflective practice helped to increase a sense of shared ownership and power to act in concert with others within the project group (Kleijberg et al., 2019). Furthermore, a sense of shared ownership was increased as project group members invested resources (e.g., time, money, ideas) and power to act in concert with others increased as project group members became increasingly confident in facilitating intergenerational arts activities about death and loss through trial and error (Kleijberg et al., 2019; Kleijberg et al., 2022).

Thus, a main learning from this first iteration of Studio DöBra was that, before the project group could create space for children and older adults to engage with the unspeakable topic of death, we had to first individually and as a group reflect on our own perspective on these topics, our power over the children and the older adults, and power dynamics within the project group (Kleijberg et al., 2022). Furthermore, as initiator of the co-design process, it was my role to facilitate this process. These lessons were applied in the co-design process of the second iteration of Studio DöBra, which took place in another city in Sweden. The project group there consisted of myself, an artist, representatives from the municipal organizations for culture, elder care, and an after-school center (Kleijberg, 2021). Prior to inviting children and older adults to participate, I shared lessons learned from the first iteration and initiated a collective reflective practice in which we discussed our own views and experiences related to death and grief, and our power relation to children and older adults. Studio DöBra research indicates that this practice helped to develop partnerships among project group members, interest and confidence, as well as skills to facilitate the intergenerational arts-based workshops about death (Kleijberg et al., 2022).

Peer Advisors, Challenging Norms to Discuss Cancer Prevention

The above-described Studio DöBra project challenges societal norms in a community-based setting to influence interpersonal behavior. However, as noted above, norms regarding unspeakable topics are also embedded in systems such as healthcare. The Peer Advisor project is an example of how norms in the healthcare system can be challenged to create space for conversations about cancer with people facing challenging socioeconomic conditions.

Cancer can be a severe illness and, for many, a sensitive topic. However, it is important to discuss issues related to cancer, such as informing people about healthy behaviors and cancer screening programs that can help prevent the disease or detect it early. One such screening method is mammography, which plays a key role in the early detection and prevention of breast cancer. In Sweden, the healthcare regions are supported by regional cancer centers to improve the quality of cancer care. One goal is to provide everyone, regardless of socioeconomic context and geographical location, with the same high quality cancer care. However, the Regional Cancer Centre Stockholm-Gotland (RCC) noted that in some areas and neighborhoods in the region of Stockholm, more people get cancer diagnoses and more people die from cancer compared to other neighborhoods (Regionalt Cancercentrum Stockholm Gotland, 2024). One reason for this may be that there is a lower participation in cancer screening programs among people in these neighborhoods (Regionalt Cancercentrum Stockholm Gotland, 2024). These geographical areas are generally characterized by a population with challenging socioeconomic conditions and a higher percentage of people with a migration background from outside the European Union. To address this inequality, the RCC began the Peer Advisor project. Peer Advisors are people who live in these neighborhoods and received an education from the RCC to inform people in their own social networks and communities about behaviors that can prevent cancer and help to detect it early-on, such as not smoking, reducing alcohol intake, exercising, choosing healthy food options, and participating in cancer screening programs. The Peer Advisors often have migration backgrounds themselves and speak various languages (Efremius et al., 2023).

This Peer Advisor project challenges norms related to the provision of information about health and care in the Swedish healthcare system in order to diminish inequality. The healthcare system often relies on a unidirectional form of information provision, for example a website with information regarding the mammography procedure that addresses the entire group that is recommended to participate in this type of screening. However, such one-size-fits-all solutions tend to overlook the diverse needs, cultural contexts, and lived experiences of minority groups, which can lead to misunderstanding and exclusion. In contrast, research investigating the strategies of Peer Advisors, shows that the Peer Advisors aim to create space for a *dialogue* regarding cancer prevention and early detection. In a research interview, one Peer Advisor said:

> "[The community] *wants to have a dialogue, not just a paper… we try to create a dialogue.* [A lot of what we do is to] *talk with people and try to understand, try to get more information from them and help them in a way… it helps* [me] *to connect.*" (Efremius et al., 2023)

For example, Peer Advisors approach people in shopping malls, on squares, or neighborhood events, and in these spaces broach discussions about cancer prevention and early detection, see Figure 2. Some Peer Advisors organize group activities, such as language cafés during which peer advisors discuss issues related to cancer prevention and the Swedish Healthcare system with people who are

Fig. 2 A peer advisor working in a shopping center, creating space using a rollup, a table with information leaflets, a sweater from the RCC and a name tag (illustration based on a photograph).

learning Swedish (Efremius et al., 2023). In meeting people in these settings, Peer Advisors continuously assess these individuals' need for information regarding cancer prevention and early detection by combining their first impressions of these individuals, with their contextual knowledge about diversity, and their own unique experience and knowledge (Efremius et al., 2023). One Peer Advisor explained this in a research interview, saying:

> "*I add things up... I connect various things,* [for example] *this person* [I can see] *that she's shy and then I approach her and* [begin] *to talk.* "*Hi, how are you?*" *in her language. Eventually she'll tell me things. Then I tell her what I know, what I want to help* [with]. *It's important that you don't disturb people... they come from a different country that doesn't have democracy, or has war the whole time, women don't have rights like here in Sweden, they are shy, they are afraid too. So that's why they need extra, extra explanation.*"
> (Efremius et al., 2023)

Thus, rather than relying on a one-size-fits-all and unidirectional approach for information provision, Peer Advisors create dialogues and continuously adjust the way they inform people based on the various needs and questions that people in their communities have (Efremius et al., 2023).

In researching the Peer Advisor project, we found that the education that the Peer Advisors receive from the RCC increases their agency and mandate to inform people about cancer prevention and early detection (Efremius et al., 2023). This can be seen as increasing their power to act, i.e., their power to create dialogues with people in their communities regarding these potentially sensitive issues. These dialogues in turn increase a power to act within the target groups, as they are empowered with relevant and culturally adjusted information that allows them to make informed decisions about their own health in the Swedish cancer care system (Efremius et al., 2023). The research also indicates that Peer Advisors often work together to address challenges they encounter in their work (Efremius et al., 2023), which can be seen as a form of power to act in concert with others. The research found that the RCC also benefits from the Peer Advisor project. An RCC representative said in a research interview:

> "[The coordinator of the Peer Advisors] *had a lot of information for us, which we would never have been able to capture otherwise. So that was awesome, I thought. She has so much knowledge about what works and what does not work, and she hears what is being talked about on the street right now.*"
> (Efremius et al., 2023)

Thus, it could be argued that the power to act at the RCC is also increased as the Peer Advisors provide the organization with relevant information regarding specific target groups. For example, Peer Advisors learned from the community that one important reason for lower mammography screening rates in these areas on the outskirts of Stockholm is because screening is only offered in the city center. This makes participation difficult for those with limited socioeconomic resources, as travel is both time-consuming and costly, and taking time off work may not be feasible. Therefore, in a spin-off project, the RCC currently offers free public transportation tickets to people who live in certain geographical areas, to increase participation in screening. This system-based innovation is made possible through the flow of information between the Peer Advisors and the target group and between the Peer Advisors and the RCC.

While the Peer Advisor project addresses norms on a system level, research showed that it is challenging to integrate this organization wide. One RCC representative said in a research interview:

> "*Sometimes I think we work too little systematically* [...] *It's about the connection between us at RCC and the* [peer advisors]. *This direct contact... we would need to tighten that, because then I think we could improve the* [project] *even more* [...] *Their work should be more integrated in RCC's other work. I think we would benefit from that.*" (Efremius et al., 2023)

This quote highlights the need for leadership to challenge established norms and drive sustainable organizational change. This supports the argument that challenging norms involves navigating power dynamics, pointing out the importance for designers to closely work with those in position of power.

The examples from Studio DöBra and the Peer Advisor project illustrate that information design, when applied to unspeakable topics such as death, grief, and cancer, must consider both societal norms and power dynamics that affect what is appropriate to talk about, who is allowed to speak, and who is included. Designers should reflect on their own positions of power and how they relate to societal norms in order to understand how to challenge societal norms and navigate power dynamics rather than unconsciously uphold them. The examples also illustrate the importance of working with relevant leaders and stakeholders to shift from a *power over* dynamic to *power to act* and *power to act in concert with others*, particularly to empower marginalized voices (e.g., children, older adults, people facing challenging socioeconomic conditions) and increase their agency and mandate to express their needs, set the agenda, and help create the space. In this way, information design can be seen as a political and ethical practice that creates space for dialogue and meaning making rather than reinforcing silence.

A Magic Circle

To explore how spaces can be created for people to engage with unspeakable topics, I have drawn on play theory in research related to Studio DöBra. This analysis primarily builds on Huizinga's original description of play theory from 1938 (Huizinga, 1949) as well as Frissen et al.'s later development of the theory (Frissen et al., 2015). Huizinga described that a central feature of play is its separation from "ordinary life", occurring within what he termed a "magic circle", i.e., the temporal and spatial boundary that defines the play context (Huizinga, 1949). Think, for example, about spaces such as playgrounds, sports arenas, theatres, and the boards of board games. When (temporal aspect) players enter such spaces (spatial aspect), they understand that the rules of the game/play apply, distinct from the "rules" of everyday life (Kleijberg et al., 2021). Thus, within a magic circle, the norms of everyday life no longer count (Huizinga, 1949). It should be noted that such spaces are not only physical but can also be interpersonal. As discussed above, engagement with unspeakable topics is shaped by social and cultural norms, i.e., the rules of everyday life. Thus, the concept of the magic circle offers a framework for creating spaces (physical and/or interpersonal) where alternative rules can apply, potentially enabling more open engagement with the unspeakable.

Within the Studio DöBra magic circle, different rules apply from everyday life. In everyday life, we often live in age-segregated ways, meaning that opportunities for interaction between different age groups are limited as most people tend to socialize in their own age group (Kleijberg et al., 2020). Additionally, as noted above, end-of-life issues are largely managed by professionals with a lack of engagement from communities and social networks (Kleijberg et al., 2020; Sallnow et al., 2022). But in Studio DöBra, children (9 years old) and older adults (mostly

80+ years old) interacted across generations and engaged with topics related to dying, death, and loss.

Research indicated that the Studio DöBra magic circle was established by the time and space during which the arts activities were held (Kleijberg et al., 2021). In terms of time, the arts activities in both Studio DöBra iterations, were held for 2 hours on Friday afternoons, for five consecutive weeks. In terms of space, the arts activities were held in various places, for example, a children's library, activity centers for older adults, and an artistic workshop for children (Kleijberg et al., 2021). While the places varied, a commonality was that these were closed for the general public during the arts activities and thus exclusively accessible to the participating children, older adults, and the project group, which created spatial and temporal boundaries from everyday life (Kleijberg et al., 2021).

The "rules" of Studio DöBra, i.e., engaging with end-of-life topics across generations, were established and communicated throughout the project in various ways. At the stage of inviting children and older adults to the project, project group members explained that they would interact with older people and engage with end-of-life topics through arts activities, early-on setting the stage for the magic circle (Kleijberg et al., 2021). The children's parents and the older adults signed informed consent forms as part of the research project. To enhance the children's sense of participation and agency, they signed an age-adapted informed consent form. This further contributed to participants' understanding of and commitment to the Studio DöBra rules (Kleijberg et al., 2021). Furthermore, at the start of each arts workshop, project group members explained to the participants what the various activities would be about, for example, that participants would explore the question "where do we end up after we die" through making visualizations of afterlife destinations. This further clarified what was expected within the Studio DöBra magic circle (Kleijberg et al., 2021). Thus, the project group members played an important role in establishing the magic circle, because they had the power to decide the time, space, and content of the arts activities (Kleijberg et al., 2021). The Studio DöBra magic circle provided participants with permission and encouragement to interact with people they would not usually interact with and engage with topics that are often considered unspeakable in everyday life.

The magic circle concept was applied in another project, called DöBra Rainbow Café, which focused on creating space for LGBTQ+ community support regarding aging, dying, death, and loss. Older LGBTQ+ people in Sweden have lived in times when their identities were criminalized, pathologized, and often socially not accepted (Bromseth et al., 2023). While the situation for LGBTQ+ people has improved in recent decades, current global movements increasingly question and limit LGBTQ+ rights. Research shows that experiences of such marginalization raise concerns among LGBTQ+ people regarding being able to live authentically at the end-of-life (Bromseth et al., 2023; M. Kleijberg and L. E. Eriksson [in manuscript]). I therefore partnered with LGBTQ+ community organizations throughout Sweden to develop DöBra Rainbow Cafés, i.e., meeting places by and for LGBTQ+ people who want to engage with end-of-life related topics and interact across generations. The spatial and temporal boundaries of the magic circles for these cafés were the time-limited events held in well-known LGBTQ+ meeting places. In the advertisement of the cafés, it was explained that the social event focused on end-of-life topics. This advertisement contributed to providing the permission for participants to engage with these issues. At the start of each DöBra Rainbow Café, a set of guidelines was shared, for example, maintaining confidentiality, showing respect, and welcoming all feelings and emotions. Clearly stating these "rules" helped to create a magic circle in which all participants are aware of what behavior was expected. Participants reflected afterwards that this contributed to an intimate setting where they could discuss topics that they would not usually talk about and with people that they would otherwise not interact with. Participants also said that the DöBra Rainbow Cafés contributed to a safer place for LGBTQ+ people to interact authentically regarding sensitive issues, without having to take in account dominant heteronormative societal discourses (M. Kleijberg and L. E. Eriksson [in manuscript]).

The concept of the magic circle as providing a space for engaging with unspeakable topics is also evident in the Peer Advisor project. Research investigating the strategies of peer advisors found that they aim to create a safe space for dialogues with people in their communities about cancer prevention and early detection (Efremius et al., 2023). This safe space was described as "an interpersonal space in which both the peer advisor and the community seem comfortable in engaging with potentially sensitive and personal cancer-related issues" (Efremius et al., 2023). Whereas people are unlikely to discuss cancer prevention and early detection with strangers in everyday life, Peer Advisors seemed successful in creating temporary interpersonal spaces for such conversations in various public contexts, including shopping malls, squares, or community events. Thus, this interpersonal space can be seen as existing within a magic circle.

Building trust seemed to be an important way to create an interpersonal safe space (Efremius et al., 2023), or magic circle. One peer advisor said in a research interview:

> "*I give myself, like I am your friend, you can trust me, no-one will know. And I have met many women and I... their trust is important...* Max: *How do you create that trust?* Peer advisor: *Often, I have met those women many times, not just once, they are active* [in the same] *association* [as I am]."
>
> (Efremius et al., 2023)

This peer advisor mentions several strategies to create trust, such as connecting with people based on shared characteristics, having a pre-existing relationship, and basing dialogue on confidentiality (Efremius et al., 2023). Another peer advisor explained that he created trust by consistently working at the same place and time, for example, every Wednesday afternoon at a specific spot in a shopping mall. In this way, the local community came to expect him to be there, which contributed to creating a magic circle; the community knew that at that time and place they could meet the Peer Advisor to discuss cancer prevention and early detection.

A challenge some peer advisors mentioned in research interviews was that people sometimes tried to avoid them as they seemed to perceive peer advisors as salespeople in shopping centers (Efremius et al., 2023). Rather than trying to sell something, peer advisors aim to empower people with information so they can make decisions regarding healthy behaviors. To emphasize this and differentiate themselves from salespeople, peer advisors used materials such as role-up banners, information materials (e.g., leaflets and cards), name-tags, and clothes with the RCC logo (Efremius et al., 2023). Peer advisors often use portable tables and chairs in combination with these visual tools to mark the space within which they work, for example, in a well-visited and visible part of a shopping center, square, or community event (Figure 2). The arrangement of these visual tools and furniture seemed to help the peer advisors to signal who they are, what they do and to create a magic circle for dialogues about cancer-related issues.

Research found that peer advisors often tried new ways of creating safe spaces for cancer dialogues in their communities. Recently, peer advisors offered Zumba classes (dance-based fitness classes often with Latin music), free of charge both outdoors and in public indoor spaces. One peer advisor said in a research interview:

> "*Everyone appreciated it and stayed* [after the class] *and started to talk with each other* […]. *We had a group with just women, and we informed them about* [cancer screening like] *mammography, and they had questions, and the men who were on the other side, they had questions about how you can quit smoking, those kinds of things. So it was a very successful activity.*"
>
> (Efremius et al., 2023)

Thus, the Zumba class spontaneously attracted people who were passing through the public spaces. As explained in the quote above, people who did not originally know each other, stayed after the Zumba class upon which the peer advisors struck up conversations about cancer screening and healthy behaviors (Efremius et al., 2023). To start these conversations, Zumba was used as an example of a healthy behavior. This illustrates how an event out of the ordinary, can create a magic circle within which the community feels safe to discuss potentially sensitive issues such as mammography.

As the above examples illustrate, the magic circle concept, originally a component of play theory (Huizinga, 1949; Kleijberg et al., 2021), can be used in information design to create spaces for the unspeakable. The concept can help designers to think about how they can create time and space within which different rules apply from ordinary life and within which the target group can feel safe (or safer), respected, and confident to engage with potentially sensitive topics. To create spaces that are considered safe(r) for the target group, it is important that designers involve the target group in designing these spaces. For example, the DöBra Rainbow Cafés were developed and facilitated in collaboration with LGBTQ+ stakeholders, and the peer advisors are part of the communities they inform.

Modes of Communication

Throughout the chapter, I have intentionally used the word *engage* in relation to the unspeakable, rather than for example *talking about* or *discussing*. This choice reflects the nature of unspeakable topics and experiences, which are often difficult to articulate, whether due to societal and cultural norms or emotional complexity. While some people may feel comfortable to express themselves verbally, others may prefer alternative modes of communication. By using the term engage, I aim to encompass a broad spectrum of communicative and interactive possibilities. When designers create spaces for engaging with the unspeakable, it is essential to consider the diverse ways individuals may prefer to participate. This requires a deep understanding of the various needs and contexts within the target group, and a flexible and responsive design approach that accommodates various forms of expression.

In Studio DöBra, we explored alternative modes of communication about death and loss. As end-of-life related experiences and beliefs can be sensitive and personal, the project group tried to design the arts activities in a way that could foster diverse and intergenerational perspectives (Kleijberg et al., 2021). Studio DöBra research shows that this was done through approaching end-of-life topics through a combination of imagination and real-life experiences (Kleijberg et al., 2021). One way this was done was to incorporate elements of play in the arts activities. An example of this is the workshop mentioned above about the question "where do we end up after we die", in which participants were asked to visualize after-life destinations and to build a vehicle with which they could play that they traveled to these after-life destinations. This playful activity led to playful communication regarding this sensitive topic. For example, one of the older adults said to the participating children that she would soon send a postcard to them from the afterlife. By doing so, she seemed to convey her awareness of nearing the end of her life in a playful manner (Kleijberg et al., 2021).

There were also activities that focused on communicating through visual materials. For example, in the workshop following the "traveling to the afterlife" session, participants worked in small intergenerational groups to create abstract compositions of flowers to express the feeling of grief and loss (Kleijberg, 2021). Participants used fresh flowers in various colors that they arranged, deconstructed, and composed on a piece of paper. These compositions were later dried and framed. Conversations shifted gently between sharing experiences of loss, the arts activity, and everyday topics like football. However, I observed that participants were noticeably quieter during this workshop compared to the previous, more playful one. It seemed that participants let the materials and compositions largely speak for themselves. (Kleijberg, 2021). Oftentimes, research relies on oral or written data to analyze communication. However, as these examples illustrate, information design can facilitate forms of expression that relies on creative expression rather than words. Therefore, arts and design-based methods need to be considered and developed to understand and facilitate such communication processes and products.

During the DöBra Rainbow Cafés, various modes of communication were explored to support engagement with end-of-life issues in LGBTQ+ communities. For instance, in some cafés, participants were invited to create memorial cards for someone close to them who had died. Using collage materials and printed photos from their mobile phones, they crafted personal tributes. At the end of the café, the cards were displayed, and a moment of silence was held for those who wished to participate. Throughout the creative process, participants responded in different ways. Some seemed deeply immersed in the process of making, experimenting with different materials, reflecting on old photographs, and sometimes sharing stories about the person who died. Others, however, expressed discomfort, saying they did not feel creative or skilled in making something beautiful. I encouraged them by emphasizing that the focus was not the quality of the product, but the process of making, an embodied engagement that engages all senses without relying on verbal expression. This was helpful for some, for others it was not. This highlights the importance of offering multiple modes of engagement, being responsive to the diverse needs of individual participants. Thus, inclusive design for unspeakable topics requires flexibility, empathy, and a commitment to supporting different ways of interacting and engaging with the topic.

It is important to recognize that preferred modes of communication may be culturally determined. Cultural aspects played an important role in the Peer Advisor project, as target groups are often people with migration backgrounds with various cultural backgrounds. Peer advisors described needing to deal with various culturally-based sensitivities to cancer in their dialogues. For example, in some cultures there is a sense of shame in living with cancer and in other cultures using the word cancer brings bad luck (Efremius et al., 2023). In these situations, peer advisors often focused on discussing healthy lifestyles rather than directly mentioning cancer. Other modes of communication involved using information artifacts, for example, showing a video about mammography screening. In this way, the topic was broached through an information artifact rather than primarily relying on the peer advisor. Another strategy was to embed cancer-related information in materials for Swedish language classes. In this way, individuals indirectly learned about cancer prevention and early detection through learning Swedish words about for example exercise, food, and healthcare services. This allowed individuals to engage with the information on their own terms. These various modes of communication highlight how information design can facilitate culturally sensitive engagement, enabling individuals to access and interact with cancer-related information in ways that aligns with their personal needs and cultural backgrounds.

Conclusion

In this chapter I have drawn from my own work with Studio DöBra, the DöBra Rainbow Café's and the Peer Advisor Project. These deal with unspeakable topics related to severe illness, death, and loss, but it is important to recognize that there are many more unspeakable topics as noted in the introduction. My hope is therefore that this chapter contributes to an ongoing conversation in information design about creating space for unspeakable topics. Information design that addresses unspeakable topics must extend beyond the immediate design of information artifacts. It needs to consider and challenge norms and power dynamics that influence what topics are considered speakable or unspeakable and whose voices are prioritized or marginalized. Information design in this area also involves creating the physical and interpersonal environments that foster trust, safety, and meaningful engagement among individuals in the target group. Additionally, designers must consider appropriate modes of communication in response to the needs and context of these individuals. It is important to recognize design as a political and ethical practice, with designers having power to either reinforce structures that marginalize topics and voices or create space for dialogue and meaning- making. To become conscious of this power, a strong self-reflective practice is required. This practice should involve examining personal assumptions, expectations, biases, and power positions. Such practice can increase awareness of our own positionality, and can help to prioritize the needs, values, and context of the target group.

References

Bromseth, J., Siverskog, A. and NIKK Nordisk information för kunskap om kön. 2023. "He went back into the closet". Older LGBTI people's interactions with health and social care in the Nordic countries.

Efremius, S., Eriksson, L.E. and Kleijberg, M. 2023. Investigating Peer Advisors' Strategies to Promote Cancer Prevention and Early Detection in Swedish Communities with Challenging Socioeconomic Conditions. *Health and Social Care in the Community*, *2023*: 6692655. https://doi.org/10.1155/2023/6692655.

Frissen, V., Lammes, S., De Lange, M., De Mul, J. and Raessens, J. 2015. Homo ludens 2.0: Play, media and identity. In: V. Frissen, S. lammes, M. De Lange, J. De Mul and J. Raessens (Eds.), *Playful Identities: The Ludification of Digital Media Cultures*. Amsterdam University Press.

Gaventa, J. and Cornwall, A. 2015. Power and knowledge. In: H. Bradbury (Ed.), *The Sage Handbook of Action Research: Participative Inquiry and Practice* (pp. 465– 471). Sage.

Horrill, T.C., Bourgeois, A., Kleijberg, M., Linton, J., Leahy, K. and Stajduhar, K.I. 2024. Services, models of care and interventions to improve access to cancer treatment for adults who are socially disadvantaged: A scoping review protocol. *PLoS One*, *19*(2): e0296658. https://doi.org/10.1371/journal.pone.0296658.

Huizinga, J. 1949. *Homo ludens: A Study of the Play-element in Culture* [*Homo ludens. Proeve eener bepaling van het spel-element der cultuur, 1938*]. Angelico Press.

Kleijberg, M. 2021. *Studio DöBra: Creating Spaces for Engaging with End-of-life Issues and Interacting across Generations through Community-based Arts Activities*. Karolinska Institutet. Stockholm.

Kleijberg, M., Ahlberg, B.M., Hilton, R. and Tishelman, C. 2020. Death, loss and community: Perspectives from children, their parents and older adults on intergenerational community-based arts initiatives in Sweden. *Health and Social Care in the Community*, *28*(6): 2025– 2036. https://doi.org/https://doi.org/10.1111/hsc.13014.

Kleijberg, M., Ahlberg, B.M., Macdonald, A., Lindqvist, O. and Tishelman, C. 2019. Navigating power dynamics in engaging communities in end-of-life issues: Lessons learned from developing community-based intergenerational arts initiatives about death and loss. *Death Studies*, *45*(8): 651–664. https://doi.org/10.1080/07481187.2019.1671547.

Kleijberg, M. and Eriksson, L.E. ([in manuscript]). *Developing LGBTQ+ community Spaces for Social Support Regarding End-of-life Issues*.

Kleijberg, M. and Eriksson, L.E. ([in manuscript]). *Queering the Death System: LGBTQ+ Perspectives on Dying, Death, Loss and Care at the End-of-life*.

Kleijberg, M., Hilton, R., Ahlberg, B.M. and Tishelman, C. 2021. Play Elements as Mechanisms in Intergenerational Arts Activities to Support Community Engagement with End-of-Life Issues. *Healthcare*, *9*(6): 764. https://doi.org/https://doi.org/10.3390/healthcare9060764.

Kleijberg, M., Hilton, R., Ahlberg, B.M. and Tishelman, C. 2022. Conceptualizing impact in community-based participatory action research to engage communities in end-of-life issues. *Palliative Care and Social Practice*, *16*: 263235242210951. https://doi.org/10.1177/26323524221095107.

Regionalt Cancercentrum, Stockholm, Gotland. 2024. *Sjukvårds Regional Cancerplan Stockholm Gotland 2024–2027*.

Sallnow, L., Smith, R., Ahmedzai, S.H., Bhadelia, A., Chamberlain, C., Cong, Y., Doble, B., Dullie, L., Durie, R., Finkelstein, E.A., Guglani, S., Hodson, M., Husebø, B.S., Kellehear, A., Kitzinger, C., Knaul, F.M., Murray, S.A., Neuberger, J., O'Mahony, S., Rajagopal, M.R., Russell, S., Sase, E., Sleeman, K.E., Solomon, S., Taylor, R., Tutu Van Furth, M. and Wyatt, K. 2022. *Report of the Lancet Commission on the Value of Death: Bringing Death Back into Life*. The Lancet. https://doi.org/10.1016/s0140-6736(21)02314-x.

11 Presence in the Hybrid State

Ulrika Florin[1*]

Introduction

Where people meet, the modalities through which they communicate, and the type of presence and engagement these interactions require have changed significantly. Parameters related to the physical body—as well as the understanding of place and space—are being challenged by the possibilities introduced through digital and hybrid communication. While most individuals appear to have adapted to these relatively new interaction modes, this adaptation has often occurred without fully considering its broader implications. One could even argue that the shared and deeply rooted human understanding of place and space has been disrupted.

These shifts in communication practices with the changed and expanding relationship between body and space are central concerns in the field of Information Design, that explore how information is formed, mediated, and experienced across both physical and digital environments.

As digital and hybrid modes of communication challenge traditional notions of presence and spatiality, a reconsideration of core design principles may be necessary. For instance, what do accessibility and user-friendly mean in hybrid contexts, and are established design principles still sufficient when designing for increasingly fluid and dynamic spaces. It is now necessary to reflect not only on what has been gained, through recent technological leaps, but also on what may have been lost.

The rapid digital advancements of recent years—accelerated by the necessity of remote work during the COVID-19 pandemic—have compelled employees across sectors to quickly adapt to hybrid modes of working. The pandemic also prompted broader acceptance of New Ways of Working (NWW), an organizational approach that emphasizes flexibility and autonomy, particularly for knowledge workers. Despite its name, this approach is not new. Its roots trace back to the 1960s and 1970s, when experimental mobile and paperless office projects were launched, incorporating videoconferencing systems and flexible workplace design.

---

[1] Ph.D., Senior Lecturer and Associate Professor in Information design, Department of Health Science, Innovation and Design, Mälardalen University, Box 325, 631 05 Eskilstuna, Sweden.
* Corresponding author: ulrika.florin@mdu.se

DOI: 10.1201/9781032642598-11

NWW was formally introduced as a concept in the Netherlands in the early 1990s (Kingma, 2018). Today, it is described as a multidimensional organizational strategy that integrates flexible work arrangements—especially in terms of spatial and temporal autonomy.

This way of working "has changed the experience of belonging at work" (Urrila et al., 2025). Closely tied to this sense of belonging is the notion of presence—the feeling of being seen, heard, and engaged. In hybrid work, presence takes on new dimensions, becoming less about physical location and more about meaningful participation and interaction. However, humans, as a species, have only limited experience with digital and hybrid forms of presence, especially when compared to the long-standing familiarity with physical co-presence that has shaped generations of social and professional interaction. This shift is easy to overlook, even as our understanding of presence continues to evolve. Physical parameters—those linked to the body—are being redefined and challenged.

So, how do human bodies and minds function in this new hybrid landscape? And what should be considered when designing hybrid environments that aim to nurture presence, engagement, and well-being? This chapter sheds light on how *presence* can be understood in today's hybrid world, grounded in the notion that human spatial understanding begins with the upright bodily position and its connection to orientational parameters in physical space.

Body, Space, and Place

Human existence is grounded in physical space, and the understanding of *space* begins with the upright posture of the body and the extent to which limbs can be stretched relative to the body's mid-axis, all while maintaining balance. Sight, balance, and sensory perception collectively contribute to spatial awareness (Twersky, 2001; Florin and Eriksson, 2020). *Place*, on the other hand, can be understood as a consolidation of value in its most essential form—a setting where fundamental human needs such as food, water, rest, and protection are met. In line with the reasoning of human geographer Yi-Fu Tuan, *place* may be defined as the more anchored, home-like point of orientation, whereas *space* represents the dynamic relationship between bodily position and time (Tuan, 1977/2011).

Moreover, the human way to move their limbs, with recurrent phases of tension and ease are a cue to the experiential relations between space and time. The movement itself gives a sense of space; when stretching the limbs, humans experience space and time simultaneously: with "space as the sphere of freedom from physical constraint and time as duration in which tension is followed by ease" (Ibid.: 118).

The phenomenologist and philosopher Maurice Merleau-Ponty suggested in the 1940s that the understanding and experience of environments are intrinsically tied to the body, emphasizing the significance of bodily presence in a given location. Such bodily experiences, that entails physical presence, are both linked to how memories are formed, and to the human capacity to move beyond immediate perception toward anticipated—or projected—futures (Merleau-Ponty, 1949/1978).

Let us consider a few concepts that can be abstracted from the human body. Firstly, the upright body in physical space is prepared to act in the world. In contrast, when a person is sleeping, lying down, and perhaps dreaming, spatial organization shifts entirely, transcending gravity. Some people dream that they can fly. However, spatial orientation terms such as vertical–horizontal, up–down, top–bottom, front–back, on–off, center–peripheral, near–far, high–low, and right–left all relate to the awake, upright human body and extrapolate to physical space. Tuan articulates that "space opens out before [us]" and is "immediately differentiable into front–back and right–left axes" according to the structure of the human body (Tuan, 1977/2011: 35) (Figure 1).

Secondly, these body–space relational coordinates are deeply reflected in verbal language, particularly through metaphors, which add intangible qualities to spatial concepts. For instance, expressions as *something has been left behind*, *there is no point in looking back*, and *something lies ahead* are time-related and link to a horizontal axis.

Spatial metaphors are also used to express social or hierarchical distinctions—for example, phrases such as *this is over my pay grade,* or *this is beneath my level.*

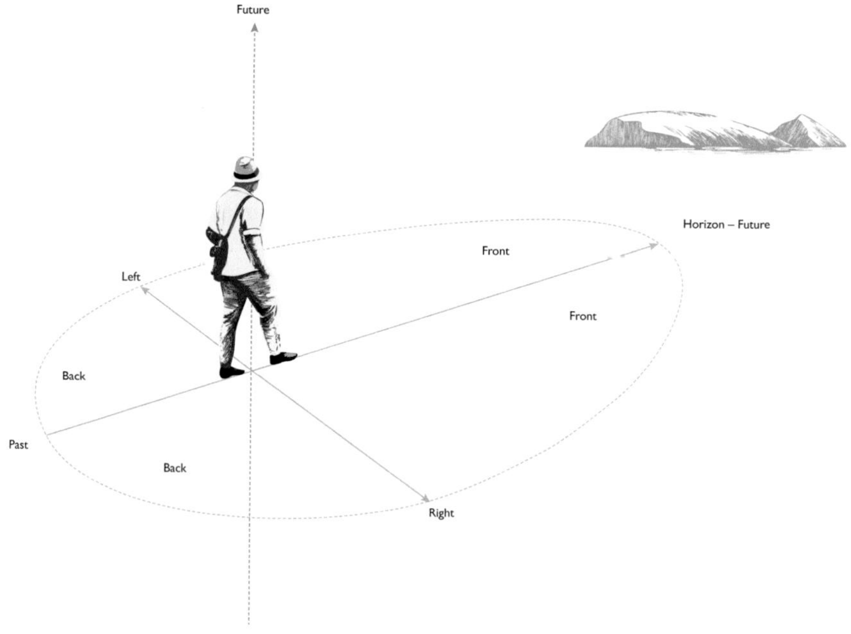

Fig. 1 Body–space–time–horizon. Shows the most significant body–space relational coordinates. This illustration builds further to Yi-Fu Tuan´s schematic illustration from 1977 (Illustration, Ulrika Florin 2025).

In these latter examples, the relationship is oriented along the vertical axis. Both sets of expressions—those related to time and those reflecting hierarchy—are grounded in the upright human posture (Figure 1). Everyday language is rich with such cogent metaphors, which not only reflect spatial thinking but also embody thought in action (Lakoff and Johnson, 1980/2003; Tuan, 1977/2011).

In Figure 1, a human is depicted moving forward toward the future, gazing at the horizon. The future is conceptualized as ahead, above, up, and in front; the past, conversely, is understood as behind and back, but also down and below. These body–space relational terms in verbal language form the foundation of orientational metaphors, which have emerged from both physical and cultural experiences of space.

Although orientational metaphors may vary across cultures, they tend to remain relatively stable in Western contexts. For example, expressions such as "happy is up" and "sad is down"; "more is up" and "less is down"; "good is up" and "bad is down" reflect consistent metaphorical mappings (Lakoff and Johnson, 1980/2003). Furthermore, some metaphors of this kind are both time- and culture-bound, and their meanings can shift over time. One such example is the metaphor pair "rational is up; emotional is down", which has undergone a shift in connotation. Historically, humans have perceived themselves as "in control over animals, plants, and their physical environment" (Ibid.: 17), but this notion is arguably no longer viewed as a positive (upward) attribute, particularly in light of contemporary concerns such as climate change.

Presence and Memory

As outlined in the previous section, embodied experience is closely linked to the way humans build memories. This connection enables individuals to relate past observations to future expectations—an ability that plays a vital role in navigating the world. Experience, therefore, entails the capacity for continuous learning, in an evolving interplay of emotions and thoughts situated at either end of what Tuan described as "an experiential continuum and way of knowing" (Tuan, 1977/2011: 10).

In addition to embodied experience, humans rely on verbal and visual representations of their surroundings, from which personal records are gradually constructed—spatial experiences that can be recalled almost instinctively. These representations include physical maps, signs, and written instructions, as well as auditory cues, such as the voice in elevators announcing the destination floor. In this context, the widespread use of smartphones equipped with Global Positioning System (GPS) technology presents an intriguing development. Such use may diminish spatial experience, as attention is directed toward the screen rather than the surrounding environment. Consequently, individuals may become less attuned to the spatial context they are navigating, and long-term implications for memory and spatial understanding might occur.

Regarding the frequent use of GPS in everyday navigation, there is a surprisingly limited body of research investigating its potential impact on individuals' ability to navigate physical space. However, a recent study on how GPS technologies are used and perceived in relation to users' spatial abilities concludes that a greater reported use of GPS is associated with "a poorer sense of direction and higher spatial anxiety", and that "dependency on GPS is associated with poorer spatial knowledge acquisition" (Topete et al., 2024).

Until this point, this text has built on theories from human geography, philosophy, phenomenology, linguistics, and art—which have long explored how humans navigate the world. Even so, emerging technologies offer new avenues for investigating how memory is created and stored.

The neurologists and philosophers Antonio Damasio and Hanna Damasio, share a similar perspective on this section's theme, emphasizing that the body—all its systems and cells—is involved in both the creation and storage of memories. They do not separate the body from the brain; instead, they describe them as coexisting within the same organism, where "the relations between the two can be entirely circular" (Damasio and Damasio, 2006: 20–21). Historically, neuroscience experiments have focused either on the body—referred to as the body proper, excluding the brain—or solely on the brain, often excluding the rest of the body due to the complexity and volume of data involved.

However, modern technologies such as Computed Tomography (CT) and Magnetic Resonance Imaging (MRI) now enable real-time, image-based data collection through experiments that treat the body and brain as an integrated whole. Artificial Intelligence (AI) further supports the clustering and interpretation of this data. These advancements have made it possible to conduct well-designed neuroimaging studies on healthy individuals. Revealing that concepts are represented in the brain in an abstract form, and show how a specific idea:

> "[--] is 'represented' in the cerebral cortex, independently of the sensory channels that contributed to forming such a 'representation' and of the word that denotes the particular idea, in whatever language"(Damasio, H. et al., 2023).

Even more fascinating is the brain's continuous capacity to represent ongoing bodily states in detail. This functional ability extends beyond mere observation; the brain can actively modify its depiction of current physiological conditions and, most strikingly, simulate bodily states before they actually occur (Damasio and Damasio, 2006; Damasio, A. 2003).

In Hybrid Modes

"Presence in the hybrid state", as this chapter is titled, refers to how individuals engage and feel connected in work environments that blend physical and digital spaces. In traditional office settings, presence is primarily defined by physical proximity—being on-site, interacting face-to-face, and participating in in-person conversations. In hybrid work environments, however, presence extends beyond physical space to include virtual interactions, digital communication, and engagement through online platforms. Here, presence is not only about location but also about participation—such as responsiveness in digital meetings, active involvement in discussions, and the ability to build meaningful connections despite physical separation. A key challenge is ensuring that both remote and on-site employees feel equally involved, valued, and visible within the organization.

Today, regular work meetings conducted in digital and hybrid formats are widely regarded as the new normal. However, expectations regarding the number of days employees must be physically present in the office often remain fluid and subject to change. The nature of meetings—and how we perceive and understand workplace presence—has undergone a significant transformation. Research on workplace technology highlights the effects of this rapid hybridization of human activity and socio-technical systems, examining its impact on employees' social interactions (Baptista et al., 2020).

Regarding hybrid work, as of 2024, no universally accepted definition has been established. Work and organizational psychologist Matti Vartiainen (2024) notes that it remains "an open question precisely what the elements, content, and implications of hybrid work are in practice" (p. 1). Later in his text, Vartiainen offers the following definition:

> "[--] any type of work arrangement where a worker operates in a sustainable manner alone or with others, as agreed upon by the worker and organization, based on the latter's purpose, the former's needs and tasks, and the context, with flexibility regarding the time and place of the work–on the employer's premises or—default location or remotely at home, other locations or on the road—using digital technologies such as laptops, mobile phones, and the Internet" (Vartiainen, 2024: 139).

According to Vartiainen, hybrid work is a flexible and sustainable arrangement shaped through mutual agreement between the worker and the organization. It considers the organization's purpose, the employee's needs and tasks, and the broader context. It allows autonomy in choosing when and where to work—whether on-site, at home, or elsewhere—supported by digital tools.

Within the research project Concepts for the Sustainable Office of the Future, which aimed to develop concepts for future office environments, promote healthy working life, and contribute to sustainable development, a photo-voice study was conducted. The material consisted of 63 photographs, supplemented with written comments, taken by 17 office workers from five different companies. Participants were asked to photograph what they perceived as supporting or hindering their health while performing office work (Florin et al., 2025). The study identified nine aspects that influence health in office settings, organized into two clusters:

1. **Environmental perspective:** *spaces, physical ergonomics, technical, aesthetic-sensuous*
2. **Behavioral perspective:** *flexibility, focus, breaks-recovery, physical activity, eating habits*

Among these aspects, *spaces* and *flexibility* were considered bridging and connecting aspects between the environmental and behavioral perspective.

The *aesthetic-sensuous* aspect refers to specific environmental qualities—objects and spaces with appealing characteristics—that can lift individuals out of their everyday routines. Examples from the study include how light enters a room, a scenic view from an office window, a plant, or other elements that evoke feelings of calmness or happiness. Regarding *technical* aspects, the results suggest that unstable internet connections are a stressor, as today's office work largely depends on reliable connectivity—whether working from home or at the office. Moreover, poor technical solutions and lack of technical equipment cause frustration (Florin et al., 2025).

Flexibility, on the other hand, is often seen as more of an enabler for individuals to adjust their life puzzles—relating to place, space, and time—when working from home or other locations.

In contrast, flexibility within the office was described quite differently, examples such as, having access to various collaborative environments (spaces), being allowed to bring a dog to work, or using one's backpack as a mobile workspace. The latter example illustrates a mobile working life and that one can perceive something as small as a backpack as the workplace—a stretched space unbounded in time and space (Florin et al., 2025).

However, flexibility may also complicate the balance between work and personal life (Lyzwinski, 2024; Zapata, 2023), potentially leading to distractions and reduced focus. This duality was conveyed by several participants, particularly in relation to working from home. On an organizational level, flexibility is also part of a broader value system, that is tied to spatial design, which should support personalized solutions. The degree of flexibility and autonomy afforded by spatial arrangements depends on organizational culture and management style. As a result, office design has become a strategic tool in managing contemporary—and increasingly hybrid—work environments. Likewise, organizational flexibility is part of a broader interplay of values that can be associated with the environmental perspective and hybrid work. Spaces should be designed and organized with this interplay in mind, offering flexibility that allows for personalized solutions (Bodin Danielsson, 2019; Florin et al., 2025).

The term *hybrid work* gained prominence in the context of NWW during autumn 2020, shortly after the first phase of the pandemic (Vartiainen and Vanharanta, 2024). However, concepts often associated with hybrid work—such as telecommuting, telework, information and communication technology (ICT) mobile work, and online platform work—had already evolved decades earlier (Vartiainen, 2024).

Hybrid work is commonly defined as a combination of remote and on-site work, and this shift has fundamentally altered employees' experience of the workplace. As prior noted this this way of working has reshaped employees experience of belonging. (Urrila et al., 2025) In hybrid settings the notion of *presence* shifts ergo focus from physical attendance to quality of engagement and interaction. However, as stated in the introduction, humans have limited experience with digital and hybrid presence compared to the longstanding familiarity with physical presence. Moreover, managerial structures such as self-leadership and autonomous work teams have emerged. These models are increasingly recognized for their potential to democratize workplace governance and enhance transparency within organizations (Taskin et al., 2017). Additionally, the pandemic accelerated the development of new and improved online collaboration tools, which have reshaped workplace dynamics and made adaptability a key professional trait.

Furthermore, the pandemic accelerated the development of more advanced digital tools, enabling work from home or other locations beyond traditional office spaces. Meeting platforms evolved, and individuals learned to use them. This enhanced flexibility and autonomy in working life, granting employees greater freedom to determine their work conditions. Particularly for knowledge professionals, this shift has significantly increased workplace adaptability. Moreover, flexible arrangements and remote work have opened the door to asynchronous collaboration across time zones (Urrila et al., 2025). At the same time, challenges remain, such as maintaining team cohesion, avoiding burnout, and ensuring equitable access to digital resources.

Flat Screen—Medium or Bounded Artifact

New systems and technologies have consistently influenced how individuals perceive the world and engage with space. In the context of hybrid work, the flat screen has become the central medium through which people are expected to be present and interact with others—whether in physical meeting rooms at the workplace, on mobile devices while on the move, or from home offices located in living rooms, kitchens, or small nooks within spaces originally designed for other purposes.

This dispersed mode of presence reflects what the philosopher Michael de Certeau has described as a hidden poiesis[#]—a form of production that is scattered across domains dominated by systems of production exemplified by "television, urban development, commerce" (de Certeau, M., 1984, 2009), to which we can add digital meeting platforms. As such systems expand, they increasingly limit the ability of users (or "consumers" with de Certeau´s word) to express what they make or how they engage with the products and environments shaped by these systems. Even though the digital interfaces of today may prompt users with questions like: Are you satisfied with this new version? such built-in mechanisms rarely offer meaningful opportunities to articulate use, experience, or agency.

Furthermore, de Certeau's notion of "trajectory"—as a movement that becomes flattened into a graph or transcription—that also can highlight how digital systems often reduce complex, temporal, and embodied actions into (simplified) representations (Ibid.) These flattening risks obscuring the lived experience of

[#] From the Greek *poiein* 'to create, invent, generate'.

users, replacing acts with traces. To counter this reduction, a distinction might be proposed, in de Certeau spirit, between tactics and strategies, where tactics represent the everyday, situated practices of users navigating systems not of their own making. In hybrid work environments, these tactical engagements become increasingly relevant. Users adapt, appropriate, and reconfigure spaces and technologies in ways that may not be visible or acknowledged by the systems themselves, yet remain central to how work and presence are actually performed.

The flat screen has become the central medium for hybrid work, enabling interaction across physical and digital spaces. Drawing on de Certeau, this dispersed presence reflects a hidden form of production, where users' actions and creativity are shaped—and often constrained—by expanding technological systems. Digital platforms rarely allow users to meaningfully express their experiences or agency, instead reducing complex activities to simplified traces. In response, users develop everyday tactics to navigate and adapt these systems—practices that remain essential yet often invisible within hybrid work environments.

On a physical level, it is worth considering how this mode of working may affect memory. Individuals sit in static positions, gazing into square, flat screens, encountering flattened representations of those they interact with. The body is physically separated from the image and the gaze, diminishing the sense of embodied presence. Earlier in this chapter, examples from human geography, philosophy, phenomenology, linguistics, art, and neurology illustrated how humans navigate the world and emphasized that bodily movement plays a crucial role in both the creation and storage of memories.

Recent research within workplace design and information environments highlights the need for co-creative and adaptive approaches to hybrid work, and emphasizes the value of design science and co-design methods in developing hybrid workplaces (Lahti and Nenonen, 2021). That to support both organizational goals and individual well-being. Their study demonstrates how participatory design processes can help identify spatial and informational needs, enabling the creation of environments that are responsive to the dynamic nature of hybrid work. From an Information Design perspective, this means that design strategies must not only address usability and accessibility but also incorporate user-driven insights into how space, technology, and information interact. Co-design becomes a tool for uncovering how presence, orientation, and engagement are experienced across physical and digital contexts. By involving users in the design process, environments can be tailored to support cognitive clarity, emotional connection, and spatial coherence—key factors in hybrid work settings.

Concluding Reflections

This chapter has illuminated the relationship between bodily presence and memory, emphasizing that embodied experience shapes the human capacity to learn, understand, and navigate the world. The aim has been to explore how presence may be understood within today's hybrid realm, grounded in the notion that spatial understanding begins with the upright posture and orientational parameters that connect body and space.

Bodily presence in physical space is central to how environments are perceived and interpreted. Moreover, spatial relationships are embedded in verbal language through metaphors that reflect both physical orientation and cultural values. Embodied experience is thus fundamental to both spatial understanding and the formation of memory.

The concept of presence in hybrid work environments goes beyond physical attendance, encompassing digital engagement and meaningful participation in both virtual and on-site settings. Hybrid work, which became prominent after the pandemic, lacks a universally accepted definition. However, it generally refers to flexible work arrangements that combine remote and on-site work, supported by digital technologies. This shift has transformed workplace dynamics, affecting how employees experience presence—both as a sense of belonging and as being physically present at work. While hybrid work offers greater flexibility, autonomy, and opportunities for asynchronous collaboration, it also presents challenges such as maintaining team cohesion and ensuring equitable access to resources.

In hybrid work, the flat screen has become the central medium for presence and interaction—across offices, homes, and mobile settings. Drawing on Michel de Certeau, this dispersed presence reflects a hidden poiesis: a form of everyday production shaped by dominant systems such as digital platforms. These systems often reduce complex, embodied actions into simplified traces, limiting users' ability to express experience or agency. De Certeau's concepts of trajectory and the distinction between strategies (system-driven structures) and tactics (user-driven adaptations) help illuminate how users navigate and reconfigure technologies and spaces. In hybrid work, such tactical engagements—like adapting home spaces or creatively using digital tools—are essential, yet often invisible.

As digital and hybrid modes of communication continue to reshape traditional understandings of presence and spatiality, it becomes increasingly clear that established design principles must be critically re-evaluated. In hybrid contexts, accessibility extends beyond physical access to include digital inclusivity, cognitive clarity, and the ability to navigate between virtual and physical environments with ease. Similarly, user-friendliness must account for diverse modes of interaction, varying levels of digital literacy, and the need for intuitive transitions across platforms and spaces.

The described shift—from embodied, physical interaction to screen-based, disembodied communication—has significant implications. In hybrid environments, individuals engage with colleagues through digital platforms that flatten presence into two-dimensional representations. This separation from the body affects cognitive processes such as memory, movement, and spatial orientation, all of which play a vital role in how memories are formed and retained.

Designing for fluid and dynamic hybrid environments requires principles that are adaptable, context-sensitive, and responsive to both individual and collective needs. From an Information Design perspective, this includes integrating flexibility, sensory engagement, and embodied interaction into design strategies. Rather than discarding established principles, the challenge lies in expanding and reinterpreting them to support presence, engagement, and well-being in hybrid work and communication settings.

Established design principles remain valuable but must be reinterpreted through the lens of hybrid interaction—where presence, accessibility, and usability are no longer limited to physical space but extend across fluid, dynamic environments. This means creating adaptive environments that respond to fluctuating user needs and contexts, designing spatially aware interfaces, and integrating temporal and spatial cues into design to support memory, orientation, presence, and engagement.

References

Baptista, João, Mari-Klara Stein, Stefan Klein, Mary Beth Watson-Manheim, and Jungwoo Lee 2020. Digital work and organisational transformation: Emergent Digital/Human work configurations in modern organisations. *The Journal of Strategic Information Systems*, *29*, (2): 101618. ISSN 0963-8687, https://doi.org/10.1016/j.jsis.2020.101618.

Bodin Danielsson C. 2019. The shaping factors forming contemporary and future office designs. Nord J Archit Res. 2019 10/21;Oct(2):53–80.

de Certeau, M. 1984. *The Practice of Everyday Life*. Berkeley, CA: University of California Press.

de Certeau, M. 2009. The Practice of Everyday Life. In: S. Thornham, C. Bassett, and P. Marris (Eds.), *Media Studies: A Reader* (pp. 76–88). Edinburgh University Press. http://www.jstor.org/stable/10.3366/j.ctvxcrv1h.12.

Damasio A., 2003. *På spaning efter Spinoza: glädje, sorg och den kännande hjärnan*, In: Translation by Björn Nilsson, Original title: *Looking for Spinoza. Joy, Sorrow, and the Feeling Brain*. Natur och Kultur.

Damasio, A. 2025. The new science of consciousness. *Prospects*, Springer.

doi: https://doi.org/10.1007/s11125-025-09716-0.

Damasio, Hanna, et al. 2023. *Neuron, 111* (21): 3337–3340.

Damasio, A. and Damasio, H. 2006. Minding the body. *Daedalus*, *135*(3): 15–22.

Florin, Ulrika, and Yvonne Eriksson. 2020. Visual Awareness Aiding Communication. *The International Journal of Visual Design*, *14* (2): 21–33. doi:10.18848/2325-1581/CGP/v14i02/21-33.

Florin, Ulrika, Christina Bodin Danielsson, Katarina Bälter, and Susanna Lehtinen-Jacks. 2025. *Aspects Influencing Health in Office Work Environments: A Photo-voice Study*. Manuscript under publication.

Kingma, S. 2018. New ways of working (NWW): Work space and cultural change in virtualizing organizations. *Culture and Organization*, *25*(5): 383–406. https://doi.org/10.1080/14759551.2018.1427747.

Lahti, M. and Nenonen, S. 2021. Design Science and Co-Designing of Hybrid Workplaces. *Buildings*, *11*(3): 129. https://doi.org/10.3390/buildings11030129.

Lakoff, G. and Johnson, M. 2008. *Metaphors We Live By*. University of Chicago Press.

Lyzwinski L.N. 2024. Organizational and occupational health issues with working remotely during the pandemic: a scoping review of remote work and health. *Journal of Occupational Health*, *66*(1): uiae005. https://doi.org/10.1093/joccuh/uiae005.

Merleau-Ponty, Maurice. (1949) 1979. Phenomenology of Perception. New York: Routlethat the walkingdge.

Taskin, L., Ajzen, M., Donis, C. (2017). New Ways of Working: From Smart to Shared Power. In: Muhlbauer, V., Harry, W. (eds) Redefining Management. Springer, Cham.

Topete, A., He, C., Protzko, J. et al. 2024. How is GPS used? Understanding navigation system use and its relation to spatial ability. *Cogn. Research*, *9*: 16. https://doi.org/10.1186/s41235-024-00545-x.

Tuan, Yi-Fu. 1977/2011. *Space and Place: The Perspective of Experience*. University of Minnesota Press, Seventh printing 2011, Minneapolis, USA, ISBN 978-8166-3877-2.

Tversky, B. 2001. Spatial schemas in depictions. In: *Spatial Schemas and Abstract Thought* (Vol. 79, p. 111). The MIT Press.

Urrila, Laura, Aija Siiriäinen, Liisa Mäkelä and Hilpi Kangas. 2025, March. Sense of belonging in hybrid work settings, *Journal of Vocational Behavior*, *157*: 104096. https://doi.org/10.1016/j.jvb.2025.104096.

Vartiainen, M. 2024. *Flexible Hybrid Work.* Edward Elgar Publishing.

Vartiainen, M. and Vanharanta, O. 2024. True nature of hybrid work. *Frontiers in Organizational Psychology, 2*:1448894. doi: 10.3389/forgp.2024.1448894.

Viererbl, B., Denner, N. and Koch, T. 2022. "You don't meet anybody when walking from the living room to the kitchen : Informal communication during remote work". *Journal of Communication Management*, *26* (3): 331–348. https://doi.org/10.1108/JCOM-10-2021-0117.

Zapata, L.E. and Ibarra, G. 2023. Engaging new ways of work: The relevance of flexibility and digital tools in a post-COVID-19 era. *Journal of Organizational Effectiveness.* https://scispace.com/papers/engaging-new-ways-of-work-the-relevance-of-flexibility-and-3l4hv8eo.

12 Design Dimensions for Interactive Information Artefacts

Adam Palmquist[1*] and William Hagman[2]

Introduction

The rapid proliferation of digital systems in domains such as healthcare has increased the demand for design approaches that integrate clarity of information with quality of interaction. In such contexts, success is no longer defined solely by technical performance or aesthetic appeal, but by an artefact's ability to support meaningful, context-sensitive exchanges between humans and technology, which is often conducted through intentional design choices. For interactive information artefacts, this challenge lies at the intersection of interaction design, which focuses on crafting the dynamic, situated qualities of engagement, and information design, which structures data content to improve comprehension, decision-making, and action. Such challenges include navigating complex information landscapes, digital transformations, and evolving interactions between humans and technologies (Benyon, 2019; Löwgren and Stolterman, 2004).

Interaction without Information, and vice versa

While it is debatable whether interaction design should be considered a subset of information design or the reverse, this is not the central concern of this chapter. Instead, our focus is on how these two fields increasingly converge in practice, particularly around shared design concerns such as meaning-making (Krippendorff, 2006), context (Suchman, 2007), and human agency (Stolterman and Wiberg, 2010).

Information and interaction are inseparable yet oriented in different directions. Information structures provide meaning and clarity, whereas interaction affords exploration, verification, and action. Together, they constitute the dual affordances that designers must reconcile when shaping what we propose to term *Interactive*

[1] Ph.D. Senior Lecturer in Information Design and affiliated to *ExcITEd – Norwegian Centre for Excellence in IT Education (SFU)*, Department of Health Science, Innovation and Design, Mälardalen University, Box 325, 631 05 Eskilstuna, Sweden.

[2] Ph.D. Senior Lecturer in Information Design, Department of Health Science, Innovation and Design, Mälardalen University, Box 325, 631 05 Eskilstuna, Sweden.
Email: william.hagman@mdu.se

* Corresponding author: adam.palmquist@mdu.se

DOI: 10.1201/9781032642598-12

Information Artefacts. The concept of Interactive Information Artefact denotes a digital artefact deliberately designed both to communicate information and to structure interaction. It is presented here as a constructed category developed to emphasise the interdependence of informational content and interactive form.

The theoretical grounding of Interactive Information Artefacts lies in Human–Computer Interaction (HCI) (e.g., Norman, 1986; Dourish, 2001), design theory (e.g., Nelson and Stolterman, 2014), and human-centred design (e.g., Norman, 2013). Within these traditions, artefacts are not conceived as passive carriers of data but as mediators of meaning. The concept of interactive information artefacts highlights the underexplored intersection of interaction design and information design, two traditions that have primarily developed in parallel. By situating artefacts in this shared space, the concept provides both a rationale for its introduction and a framework for addressing the dual demands of informational clarity and interactive engagement. Positioned in this way, interactive information artefacts functions both as a descriptive category, encompassing artefacts such as dashboards, simulations, visualisations, and gamified interfaces, and as an analytical framework for examining how informational and interactive dimensions co-produce experiences, engagement, and interpretation.

Although this intersection has not been extensively theorised, earlier efforts such as Shedroff's (1994) notion of information interaction design show that the ambition to unify these domains is not new. The challenge lies in reconciling distinct orientations: information design has historically prioritised clarity, structure, and communicative precision (Pettersson, 2002), while interaction design has foregrounded dynamics, human experience, and agency (Winograd, 1996; Löwgren and Stolterman, 2004). As Buchanan (1992) observes more broadly, design domains often operate with divergent logics, making cross-domain convergence inherently complex. To understand why such convergence is complex, we must return to the conceptual foundations of each domain.

Information

For information design, the key distinction is between data, information, and meaning. Information is not equivalent to data. By data, we mean facts or measurements such as figures, names, places, or numbers. To construct information, these must be interpreted and structured with a clear purpose. As Silver (2012) notes, numbers gain meaning only when assigned a "voice". The value of data, therefore, emerges from the context of its interpretation. Shannon's information theory (Shannon and Weaver, 1998) highlights unpredictability as a measure of information; however, design must also account for human interpretive contexts (Krippendorff, 2006).

Information can thus be seen as actionable communication, distributed across time and space. However, information alone is rarely sufficient to prompt action. As Pontis and Babwahsingh (2023) define, information design is the practice of helping people understand something unfamiliar or complex so they can achieve their goals. We emphasise that information contributes to action, but does not guarantee it.

Interaction

For interaction design, the conceptual foundation lies in distinguishing interaction from mere communication. Interaction is not reducible to the transfer of messages but instead involves engagement with and exploration of the context. Winograd (1997) defines interaction design as "designing spaces for human communication and interaction" (p. 160), thereby framing interaction as a means of enabling users to access, explore, and act upon information while verifying their understanding through use. Within this framework, designers can orient interaction towards different experiential goals. Utilitarian design emphasises functionality, usability, and task efficiency, aiming to support users in achieving specific objectives with minimal effort or cognitive load. In contrast, hedonic design foregrounds experiential pleasure, stimulation, and personal meaning, focusing on the emotional and motivational aspects that make an interaction engaging and enjoyable (Hassenzahl, 2008; van der Heijden, 2004). Together, these two orientations illustrate that interaction design operates across a continuum, from the instrumental to the experiential, requiring designers to balance pragmatic effectiveness with affective appeal.

Cognitive Foundations

The need for design in both domains becomes clearer when viewed through the lens of human cognitive limitations. Humans operate under bounded rationality (Simon, 1955). One "blade" of the scissor is human heuristics; the other is the structured presentation of information in context (Gigerenzer and Selten, 2001). In practice, humans satisfy rather than optimise, settling for "good enough" solutions. The cognitive system, oriented toward efficiency, often relies on fast and intuitive processing (System 1) while reserving effortful, analytical reasoning (System 2) for more demanding situations (Kahneman, 2011; Evans, 2003; Evans and Stanovich, 2013). Designers must therefore either accommodate System 1 processing or deliberately encourage System 2 engagement.

Cognitive load theory adds further nuance. Functional state (e.g., alertness), intrinsic load (e.g., task difficulty), and extraneous load (e.g., time pressure) interact to shape workload (Galey et al., 2012). While good design seeks to reduce extraneous load, intrinsic complexity may be necessary to achieve the intended goals of an interactive information artefact. Wicked problems exemplify this condition: complex, ill-structured challenges without clear solutions (Buchanan, 1992; Simon, 1969/1996). In such cases, artefacts may need to embody sufficient complexity to manage the problem space while still supporting the user in handling unavoidable intrinsic load. As Ashby (1956) argues, only variety can absorb variety.

Toward Synthesis

While information and interaction design share the overarching goal of enabling effective and satisfying experiences, they differ in emphasis. Information design prioritises semantic clarity, visual hierarchy, and strategies that optimise cognitive processing. Interaction design focuses on how people engage with artefacts over time, how systems respond, and how experiences evolve in context (Rogers et al., 2019). Building on these perspectives, we propose the *Structural–Behavioural–Experiential* (SBE) dimensions, adapted from the *Mechanics–Dynamics–Aesthetics* (MDA) framework, as a means of bridging information design and interaction design by positioning meaning-making as the central design responsibility in interactive information artefacts. The SBE dimensions for analysing and designing Interactive Information Artefacts rest on four foundations:

1. *Data* needs to be structured since raw facts only turn into information through interpretation and intentional organisation.
2. *Information* is actionable because it represents communication that facilitates, but does not dictate, human action.
3. *Interaction* fosters engagement by going beyond mere communication, shaping how individuals explore, verify, and act within a given context.
4. *Cognition* is bounded; thus, all design must consider human limits, balancing representation and interaction to support both intuitive and analytical processing.

Taken together, these foundations highlight the importance of recognising that information and interaction do not simply emerge; they must be shaped through effort, structure, and time. Situated Interactive Information Artefacts at the intersection of information and interaction design clarified why this convergence is conceptually challenging and outlined the cognitive constraints that make deliberate design essential. By articulating the SBE dimensions as guiding principles, we establish both a rationale and a foundation for analysing Interactive Information Artefacts as artefacts that simultaneously communicate information and structure interaction. The subsequent section further develops this foundation by adapting the MDA framework to Interactive Information Artefacts, thereby operationalising the SBE dimensions.

A Possible Framing of Interactive Information Artefacts

To operationalise the SBE dimensions introduced above, we draw on the MDA framework (Hunicke, LeBlanc, and Zubek, 2004). Developed initially within game design, MDA provides a structured way of linking formal design elements, emergent behaviours, and experiential outcomes. This makes it particularly suitable for adaptation to Interactive Information Artefacts, which likewise depend on the interplay between structural configurations, behavioural patterns, and experiential effects. By reframing MDA into Structural–Behavioural–Experiential dimensions, we position our proposal within an established lineage while extending its applicability beyond games to the wider category of Interactive Information Artefacts. The MDA framework formalised a vocabulary that had long been used informally by both practitioners and scholars (Zubek, 2020).

We argue that design artefacts beyond games can benefit from a similar approach. Employing both holistic and component-level lenses allows designers of Interactive Information Artefacts to make more transparent design choices and conduct systematic analyses. The adaptation highlights how rules, structures, and informational scaffolds shape emergent behaviours, and how those behaviours, in turn, generate experiential outcomes. Drawing inspiration from MDA, the SBE framework thus provides a means to analyse and design Interactive Information Artefacts as systems that simultaneously communicate information and structure interaction.

The MDA Framework in Detail

Hunicke et al. (2004) introduced the MDA framework in their influential paper, "MDA: A Formal Approach to Game Design and Game Research." At the time, terms such as mechanics, dynamics, and aesthetics were used informally by practitioners to describe different aspects of games, but without shared, formalised meanings. The MDA framework offers a game design lens that conceptualises games as artefacts in which structures generate behaviours that, in turn, shape experiences. It provides both definitions and an organising structure for understanding how rules and components (mechanics) interact with emergent patterns of behaviour (dynamics) to produce experiential outcomes for players (aesthetics). MDA's three categories also function as distinct, yet interdependent, analytical lenses (Zubek, 2020). Importantly, MDA highlights the dual perspective inherent in game design: designers work forward from mechanics toward aesthetics, whereas players experience the game in reverse, beginning with aesthetics and moving toward mechanics. From the designer's standpoint, mechanics give rise to dynamics, which produce aesthetic outcomes; from the player's standpoint, aesthetic expectations shape the perception of dynamics and ultimately structure interaction with mechanics (Hunicke et al., 2004). To examine MDA in greater detail, it is helpful to unpack each of its three categories separately. Following the designer's perspective, we begin with mechanics, since these establish the foundational structures and rules from which dynamics and aesthetics emerge.

Mechanic:s Structures, Rules, and Information Architecture

Mechanics refers to the game elements, their organisation, and the formal rules that govern the functioning of the game system. Mechanics comprise the foundational design elements, their organisation and frames deliberately implemented to define what is possible within the game system.

In Interactive Information Artefacts, mechanics may include database schemas, filtering functions, navigation menus, input constraints, or algorithms controlling how information is retrieved and displayed. From an Interactive Information Artefact perspective, mechanics shape the information architecture, its arrangement, accessibility, and categorisation. Designers must ensure that these structures allow users to locate, retrieve, and act on relevant information efficiently, while managing cognitive load without oversimplifying necessary complexity.

Dynamics: Emergent Behaviour and Designer-Influenced Sense-Making

Dynamics refers to the player behaviours and system responses that emerge through interaction with the mechanics during play. Dynamics describe the patterns of interaction that emerge when players engage with mechanics. These patterns are not scripted, but arise from the interplay between system rules and players' adaptive practices, as framed by the game designer's affordances and constraints.

In Interactive Information Artefacts, dynamics may include browsing patterns, decision-making processes, or the adaptive use of filters and search tools, behaviours that designers must anticipate and support. Dynamics are central to sense-making: they reflect how users interpret, adapt to, and navigate the system's structures considering their goals, prior knowledge, and situational context.

Aesthetics: Targeted Experiential Qualities

Aesthetics refers to the player's subjective experience of gameplay, which emerges as the mechanics generate dynamics that shape and influence emotional, cognitive, and sensory engagement. Aesthetics refers not to surface-level visual design but to the experiential qualities that the designer aims to evoke through the mechanics–dynamics relationship.

In Interactive Information Artefacts, relevant aesthetic goals may include trust (confidence in the reliability of information), clarity (ease of interpreting content), and assurance (confidence in decision-making based on system outputs). These qualities arise from the interaction of mechanics and dynamics: well-structured mechanics coupled with supportive dynamics foster reliability and ease of use, whereas poor alignment may generate confusion or mistrust. Designers must therefore intentionally align mechanical and dynamic layers to achieve the desired experiential outcomes.

The MDA framework links structures, behaviours, and experiential qualities into a coherent model of design. This makes MDA a powerful starting point for analysing and guiding the design of Interactive Information Artefacts. However, as we shall see, transferring MDA directly from games to Interactive Information Artefacts exposes certain limitations. To illustrate this, we first apply the unmodified framework to a Digital Healthcare System, an example of an Interactive Information Artefact. As shown in Table 1, this baseline mapping clarifies how mechanics, dynamics, and aesthetics can be aligned with structures, behaviours, and experiential qualities. At the same time, it reveals the limits of an unadapted MDA when applied to Interactive Information Artefact.

Table 1 Baseline MDA Applied to a Digital Healthcare System.

Lens	*Digital Healthcare System*	*Design Implication*
Mechanics	Patient records database; lab results integration; appointment booking modules	Ensure accurate data structures, interoperability between subsystems, and reliability of access
Dynamics	Patients booking appointments, doctors entering notes, and administrators generating reports	Support efficient workflows, minimise redundant input, and enable adaptive decision-making
Aesthetics	Patients' trust and clarity; doctors' confidence in workflow; administrators' sense of accountability	Reinforce interpretive accuracy, reduce cognitive load, and support reliability across roles

While the MDA framework provides a lens for linking mechanics, dynamics, and aesthetics to game design, applying it directly to Interactive Information Artefacts risks overlooking several key design considerations. Unlike games, which are often centred on hedonic outcomes such as sensation, challenge, or fantasy (Hassenzahl and Tractinsky, 2006; Juul, 2010; Salen and Zimmerman, 2004; Zubek, 2020), Interactive Information Artefacts are typically oriented toward interpretive accuracy, contextual relevance, and trust. This divergence suggests that MDA, although highly influential, requires contextual adaptation to function as a diagnostic and planning tool beyond the game domain. The following section outlines the limitations of applying MDA unmodified to Interactive Information Artefacts and introduces the rationale for our proposed adaptation.

Need for Adaptation

Applying the unmodified MDA framework directly to Interactive Information Artefacts risks overlooking several key design considerations. While MDA provides a clear and structured approach to linking system components with experiential outcomes, its direct application to information-centred systems presents limitations. These arise because design priorities in non-game contexts often emphasise interpretive accuracy, contextual relevance, and temporal alignment, which are outcomes that require intentional (re)structuring by Interactive Information Artefacts designers.

First, in Interactive Information Artefacts, the primary objective is not entertainment but the accurate interpretation and practical application of information to achieve situated goals (McCarthy and Wright, 2004; Benyon, 2019). Meaning-making thus becomes a core design responsibility, with aesthetic outcomes such as trust, clarity, and confidence taking precedence over game-centric qualities like challenge or fantasy (Hassenzahl and Tractinsky, 2006).

Second, these systems typically unfold through iterative action–feedback cycles, echoing Donald Norman's (1986) seminal model of the stages of action, which conceptualises interaction as a sequence of goal formulation, execution, and evaluation. Norman's work, foundational to user-centred design, emphasises that effective interaction depends on users being able to perceive, interpret, and act upon feedback in ways that support their intentions. This perspective, further elaborated within the broader HCI literature (Rogers et al., 2019), highlights that designers must determine not only what information is delivered, but also when and in what form, so that comprehension builds cumulatively over time. Without this temporal lens, MDA risks overlooking the development of understanding in information-rich contexts.

Third, semantic framing is a central design responsibility. Klaus Krippendorff (2006), whose semantic turn profoundly influenced contemporary design theory, argues that meaning is not inherent in artefacts but constructed through users' interpretive engagement with them. From this perspective, design is not merely about creating functional or aesthetic forms but about shaping communicative artefacts that enable sense-making and meaningful interaction. Accordingly, choices of representational form, terminology, and communicative cues determine how users interpret and act upon information (Krippendorff, 2006; Stolterman and Wiberg, 2010). An adapted MDA must therefore integrate semantic and contextual considerations into mechanics (data structures and representations), recognise their influence on dynamics (interpretation and use), and assess their impact on aesthetics (trust, clarity, confidence). This aligns with Krippendorff's (2006) semantic turn, which foregrounds meaning as a deliberate design outcome.

Fourth, games, as modelled in MDA, are typically voluntary; players may disengage at will if the experience fails to sustain their interest (Juul, 2010; Salen and Zimmerman, 2004). By contrast, many Interactive Information Artefacts are mandatory, such as patient databases or compliance-driven HR systems. This distinction shifts the design mandate: while games often pursue fun or challenge, Interactive Information Artefacts must prioritise trust, clarity, efficiency, and interpretive accuracy (Hassenzahl and Tractinsky, 2006; Benyon, 2019). Dynamics likewise differ, often manifesting as compliance, adaptation, or workaround behaviours in response to constraints (Bødker, 2006; Dourish, 2001). Designers must anticipate these and actively design for resilience, reliability, and intelligibility.

Fifth, Interactive Information Artefacts are deployed across diverse contexts and heterogeneous audiences, meaning identical mechanics can produce very different dynamics and aesthetic outcomes depending on prior knowledge, situational constraints, and external conditions (Bødker, 1991; Dourish, 2001). Designers must, therefore, anticipate variability and adopt adaptive strategies that maintain accessible pathways for meaning-making across contexts.

For these reasons, we propose an adapted MDA framework for interactive information design that extends the original model to address the structural, temporal, and semantic demands of these systems. In this adaptation, the designer's forward perspective becomes primary, ensuring the framework functions as a planning and diagnostic tool for structuring meaning-making over time. Some scholars suggest that the MDA framework has limitations when applied to artefacts outside of entertainment games, particularly for gamified systems or experience-oriented design contexts, arguing that its emphasis on mechanics and its emotional targets can be less valuable or prescriptive in non-game settings (Walk, Görlich, and Barrett, 2017; Junior and Silva, 2021).

At the same time, research on *serious games*, games whose primary purpose extends beyond entertainment, as well as gamification research has expanded and refined the MDA framework (Kusuma et al., 2018; Pistono et al., 2024), demonstrating its broad applicability across diverse contexts. We contend that contextual adaptation of MDA is both possible and necessary. Applying the framework directly to Interactive Information Artefacts risks overlooking three critical aspects. Accordingly, we propose three required adaptations: (1) incorporating both utilitarian and hedonic design views, (2) addressing the perspectives of multiple stakeholders rather than a single monolithic "user", and (3) embedding meaning-making, semantics, and contextual/temporal framing as primary design responsibilities. Together, these adaptations extend MDA's original tripartite structure while preserving its core value as a design and analytical lens. The preceding discussion highlights recurring challenges when applying MDA to information-intensive contexts. These can be synthesised into three required adaptations, which together reposition MDA as a robust planning tool and analytical lens for the design of Interactive Information Artefacts.

Multiple Stakeholder Perspective in Interactive Information Artefacts

Interactive Information Artefacts seldom address a single, homogeneous audience. Instead, they mediate between multiple stakeholder groups, each with distinct goals, interpretive frames, and evaluative criteria. This multi-stakeholder orientation is central to understanding the challenges of information-intensive contexts. In contrast, the original MDA framework foregrounds two perspectives: designers working forward from mechanics to aesthetics, and players experiencing the process in reverse from aesthetics to mechanics (Hunicke, 2004). While valuable for game design, this dual orientation does not capture the complexity of artefacts embedded in organisational and social infrastructures, where divergent stakeholder perspectives must be reconciled through design.

Over the past few decades, the design field has increasingly foregrounded the stakeholder concept (Palmquist, 2023a). Krippendorff (2006) placed stakeholders at the centre of semantic design, famously arguing that "the end-user is a myth, or at best a statistical artefact" (p. 63). He distinguishes between *end-users*—those who directly interact with technology—and stakeholders, who include all actors with a vested interest in an artefact. *End-users* are thus only one stakeholder group among many, alongside those who finance, approve, design, or are indirectly affected by the artefact's use. In this chapter, we emphasise operational stakeholders (patients, doctors, administrators, and policymakers) to illustrate how design provisions play out in specific contexts.

Stakeholder groups usually hold different perspectives on artefact design when implemented within digital ecosystems (Nancy, Currie, and Whitley, 2016;

García-Sánchez et al., 2018). Aligning these perspectives is essential, since the failure or disengagement of even one stakeholder group can compromise the entire implementation and ripple across others (Krippendorff, 2006; Palmquist, 2023b). Considering the heterogeneous conditions and expectations of stakeholders is therefore not peripheral, but central, to the design of Interactive Information Artefacts. Healthcare offers a compelling case to illustrate this point. For instance, a digital healthcare system is simultaneously experienced by patients, doctors, administrators, and policy makers, each of whom depends on the same artefact, but evaluates its adequacy differently.

Patients use digital healthcare system interfaces to book appointments, review medical records, and interpret diagnostic results. Their primary evaluative concern is trust: that the information provided is accurate, accessible, and communicated in comprehensible terms. Poor semantic framing can cause anxiety, misinterpretation, or even harm.

Doctors rely on the digital healthcare system to enter patient notes, retrieve lab results, and coordinate care. Their perspective foregrounds efficiency and reliability. Cognitive (over)load from redundant input, delays in retrieving information, or unclear system signifiers directly reduce the time available for clinical decision-making. For them, trust is not only about the correctness of information, but also about the system's ability to sustain complex workflows under time pressure.

Administrators approach the digital healthcare system as a compliance and oversight tool. Their concerns are transparency, accountability, and standardisation. Audit trails, permissions management, and reporting functions must be designed to ensure systemic integrity and regulatory compliance. For administrators, the aesthetics of the system are not about reassurance or usability flow, but about confidence that the system is reliable, secure, and legally defensible.

Policy makers interact with the digital healthcare system less directly, but their stake in the system is no less central. Their perspective foregrounds population-level oversight, systemic integrity, and evidence-based decision-making. Essential aesthetic qualities are systemic trust, transparency, and legitimacy. Thus, their evaluative concern is not whether the system supports an individual's workflow, but whether it produces reliable, intelligible, and actionable insights at scale to inform policy, allocate resources, and maintain accountability across institutions.

These perspectives demonstrate how the same mechanics (e.g., medical record structures, lab data integration, appointment workflows) can produce distinct dynamics (interpretive actions, diagnostic practices, compliance oversight) and yield varying aesthetics (trust, clarity, efficiency, accountability) depending on the stakeholder. The implication for designers is clear: Interactive Information Artefacts cannot be understood through the monolithic figure of "the end-user". Their success depends on anticipating, provisioning, and aligning meaning-making across multiple stakeholder groups. By embedding stakeholders as a central design category, MDA is repositioned for information-rich, organisationally embedded contexts, situating meaning-making as a distributed and negotiated process. To address this limitation, we introduce the first adaptation: a multi-stakeholder perspective. Table 2 illustrates how patients, doctors, administrators, and policy makers experience the same digital healthcare system differently, with distinct evaluative concerns and design implications. This highlights the inadequacy of treating "the end-user" as a homogeneous category, and foregrounds the need to design for stakeholder heterogeneity.

Table 2 MDA with Multi-Stakeholder Adaptation in a DHS.

Stakeholder	*Mechanics*	*Dynamics*	*Aesthetics*	*Design Implications*
Patients	Patient dashboards, appointment booking modules, and lab result integration	Reviewing test results, booking or rebooking appointments	Trust in accuracy, reassurance through clarity, and reduced anxiety	Ensure accuracy and accessibility by providing precise terminology and a visual hierarchy, and reduce anxiety through semantic framing
Doctors	Clinician input interfaces, prescription tools, and lab data retrieval	Entering notes, cross-checking prescriptions, coordinating care	Confidence in reliability, efficiency in workflow	Minimise redundant input; streamline data retrieval; ensure reliability of real-time updates; reduce cognitive overload
Administrators	Compliance dashboards, reporting modules, and permissions management systems	Generating audit reports, managing permissions, and monitoring compliance	Assurance of transparency, accountability, and systemic reliability	Provide robust audit trails, ensure security and legal defensibility, and support oversight through standardisation
Policy Makers	Aggregated data repositories, anonymisation protocols, public health dashboards	Monitoring population-level data, issuing directives, and auditing systemic compliance	Confidence in systemic integrity, trust in data accuracy, and evidence-based policy legitimacy	Ensure secure and ethical data aggregation; provide meaningful visualisations for large-scale monitoring; maintain compliance with privacy and data protection laws (e.g., GDPR); support decision-making at the societal level

In the example in Table 2, mechanics determine the availability, accuracy, and integration of medical information; dynamics describe how patients, doctors, and administrators interact with and adapt to this information in real-world contexts; and aesthetics capture the experiential qualities—trust, clarity, confidence, and reliability—that define whether the system supports effective decision-making and safe-care delivery. This account of the MDA framework parts reflected in the multistakeholder digital healthcare system example highlights its core strength as a design lens, linking the organisation of design elements to behaviours and emotional outcomes, while operating along multiple interwoven dimensions rather than a dual one, as in the original framework.

In the next section, we will discuss applying MDA directly to Interactive Information Artefacts through the psychological aspect of utilitarian and hedonic design traits. While games are usually linked with hedonic qualities, Interactive Information Artefacts often require a balance of both.

Utilitarian and Hedonic Design Traits in Interactive Information Artefacts

The design practice of Interactive Information Artefacts often requires balancing two complementary yet distinct orientations: utilitarian and hedonic design. In information systems research, van der Heijden (2004) conceptualises these orientations as functional versus enjoyment-related evaluations, while in HCI and design research, similar distinctions have long been made between pragmatic and hedonic attributes (Hassenzahl, 2003; Hassenzahl and Tractinsky, 2006). Utilitarian dimensions appeal to functional and conscious evaluations, such as purpose, usefulness, quality, and safety.

In contrast, hedonic dimensions address experiential and affective concerns, including aesthetics, stimulation, pleasure, and comfort (Jordan, 2000; Norman, 2004). These qualities are not fixed but are subject to ongoing psychological negotiation shaped by context, audience, and technological affordances (Palmquist, 2023c). For this reason, design requires an adaptive mindset attuned to when utilitarian precision or hedonic engagement should be foregrounded (Forlizzi and Battarbee, 2004).

The dominant design intent largely determines a technology's affinity: utilitarian design seeks to improve performance and task efficiency (productive use), whereas hedonic design aims to create pleasurable experiences and enhance engagement (prolonged use) (Hassenzahl, 2008; van der Heijden, 2004). Nevertheless, many successful artefacts integrate both orientations. A spreadsheet application that is accurate and efficient (utilitarian) may also be visually satisfying and rewarding to use (hedonic). Conversely, a smartphone designed for aesthetic appeal (hedonic) may also excel in reliability and functionality (utilitarian).

From a disciplinary design perspective, entertainment-oriented game design has traditionally aligned with hedonic goals, maximising enjoyment, engagement, and affective resonance. Information design, in contrast, has pursued primarily utilitarian purposes, emphasising clarity, precision, and functional effectiveness. Interaction design often occupies an intermediate position, combining usability and utility with experiential qualities of engagement and appeal (Hassenzahl and Tractinsky, 2006), as illustrated in Figure 1.

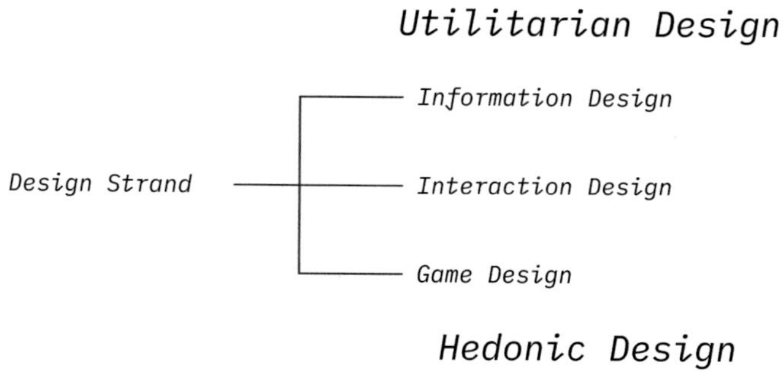

Fig. 1 Utilitarian and hedonic design traits provide different interpretive frames for various stakeholders.

The second MDA adaptation incorporates utilitarian and hedonic design traits. As Table 3 demonstrates, stakeholders evaluate Interactive Information Artefacts not only in terms of utility and efficiency but also in terms of affective and experiential qualities, such as reassurance, comfort, and confidence. Integrating these dual dimensions of evaluation deepens the analysis of how artefacts support or hinder meaningful engagement.

Table 3 MDA with Multi-Stakeholder and Utilitarian/Hedonic Adaptation in a Digital Healthcare System.

Stakeholder	*Mechanics*	*Dynamics*	*Aesthetics: Utilitarian vs Hedonic*
Patients	Dashboards, appointment booking, and lab results	Reviewing records, interpreting diagnostic results	**Utilitarian:** accuracy, safety, clarity **Hedonic:** reassurance, comfort, sense of control
Doctors	Input interfaces, prescription tools, and lab data integration	Entering notes, cross-checking, and interpreting lab data	**Utilitarian:** efficiency, reliability, error prevention **Hedonic:** reduced frustration, increased pleasure, ease of use
Administrators	Compliance dashboards, reporting tools, and permission management	Monitoring compliance, generating reports, and managing system access	**Utilitarian:** accountability, transparency, standardisation **Hedonic:** Confidence in use, Error Prevention, and reassurance of data integrity
Policy Makers	Population dashboards, anonymisation protocols, aggregated data	Monitoring trends, auditing compliance, issuing directives	**Utilitarian:** systemic accuracy, evidence-based policy **Hedonic:** trust in system legitimacy, confidence in systemic transparency

For Interactive Information Artefacts, the balance between these orientations depends on context and stakeholder group. In some cases, functional requirements dominate; in others, eliciting positive affect and sustaining engagement is decisive. Recognising when a context calls for predominantly utilitarian or hedonic qualities provides a foundation for design decisions that are both effective and appealing. This balance is central to creating Interactive Information Artefacts that remain desirable and relevant over time, underscoring the adaptive strategies discussed in the following section.

Meaning-Making, Semantics, and Context in Interactive Information Artefacts

Besides utilitarian and hedonic aspects and the concern of multiple stakeholders in Interactive Information Artefacts, applying MDA without adaptation presents specific challenges. Three interrelated aspects must be foregrounded: the centrality of meaning-making, the temporal sequencing of interaction, and the semantic and contextual framing of information.

Meaning-making refers to interpretive accuracy and trust as primary design outcomes (Krippendorff, 2006; McCarthy and Wright, 2004; Dourish, 2001; Benyon, 2019). Temporal sequencing highlights how designers deliberately structure action–feedback cycles to support comprehension over time (Norman, 1986; Rogers, Sharp, and Preece, 2019; Hassenzahl and Tractinsky, 2006).

Semantic and contextual framing concerns how representational choices, such as terminology, iconography, and cues, shape interpretation and guide subsequent action (Krippendorff, 2006; Forlizzi and Battarbee, 2004; Stolterman and Wiberg, 2010). Together, these aspects expand the diagnostic capacity of MDA while retaining its core tripartite structure.

In practice, this adaptation allows designers and researchers to connect the structural what of system elements with the temporal when of interaction and the interpretive how of meaning-making. In Interactive Information Artefacts, success is determined less by entertainment or challenge and more by whether design supports accurate interpretation, timely action, and coherent usability in real-world environments. Four emphases summarise this adaptation:

1. Meaning-making as a primary design outcome (Krippendorff, 2006; McCarthy and Wright, 2004; Dourish, 2001; Benyon, 2019).
2. Temporal sequencing of interaction through action–feedback cycles (Norman, 1986; Rogers et al., 2019; Hassenzahl and Tractinsky, 2006).
3. Contextual framing of information presentation (Bødker, 2006; McCarthy and Wright, 2004; Forlizzi and Battarbee, 2004).
4. Design semantics determine how form, structure, and signifiers communicate intended meaning (Krippendorff, 2006; Norman, 2013; Stolterman and Wiberg, 2010).

Together, these emphases ensure the framework accounts for both the structural where of design elements and the temporal when of interaction moments, while foregrounding the communicative role of design choices in shaping interpretation.

At the heart of both information and interaction design lies meaning-making: the intentional shaping of how stakeholders interpret and act upon information. This unfolds as a tacit dialogue (McCarthy and Wright, 2004; Suchman, 2007):

> The designer provisions the artefact with clear goals, structured content, and actionable pathways, informed by principles of information architecture (Horn, 1999; Tufte, 2001) and interaction heuristics (Rogers et al., 2019).
>
> The system responds to stakeholder actions at specific, intentionally chosen points in the interaction cycle (Norman, 1986).
>
> The value of the response depends on whether its timing, framing, and semantic clarity align with the stakeholder's expectations, situational demands, and cognitive capacity (Sweller, Ayres, and Kalyuga, 2011).

When these elements are misaligned, for example, through inconsistent signifiers (Norman, 2013), interactional latency (Hassenzahl and Tractinsky, 2006), or poorly framed cues (Bødker, 2006), cognitive friction arises, undermining both usability (Nielsen, 1993) and experiential coherence (Forlizzi and Battarbee, 2004). Meaning-making is therefore not a decorative layer but a structural and temporal design responsibility. As discussed in the previous section, Interactive Information Artefacts seldom address a single homogeneous audience, but must instead accommodate the distinct perspectives of patients, doctors, and administrators. However, many systems are also embedded in broader organisational and societal infrastructures. In healthcare, this includes policymakers and regulators, who rely on aggregated and anonymised data to inform decisions, monitor compliance, and maintain public trust. Their perspective illustrates how meaning-making extends beyond immediate interactions to encompass population-level insights and systemic accountability.

Table 4 presents the Interactive Information Artefact-adapted MDA framework, which embeds meaning-making, semantics, and contextual/temporal framing as core design responsibilities. This table illustrates how structural provisions, behavioural dynamics, and experiential qualities must be linked with semantic clarity, contextual relevance, and temporal sequencing to sustain interpretive accuracy and trust across stakeholder groups. In this way, the complete adaptation operationalises our proposed extensions to MDA, providing a comprehensive lens for Interactive Information Artefacts. By considering these four groups together, Table 4 also illustrates how mechanics, dynamics, and aesthetics facilitate distinct forms of meaning-making for patients, doctors, administrators, and policymakers.

Table 4 MDA Incorporating Meaning-Making, Semantics, and Context in a DHS.

Stakeholder	*Mechanics (Structural Lens)*	*Dynamics (Behavioural Lens)*	*Aesthetics (Experiential Lens)*	*Meaning-Making Dimensions*
Patients	Dashboards for appointments, prescriptions, and results	Interpreting lab results, rebooking appointments	**Utilitarian:** Accuracy, safety, clarity **Hedonic:** Reassurance, comfort, sense of control	Semantic framing of medical terms; timely release of results; accessible visual hierarchy
Doctors	Clinician input interfaces, lab integration, prescription tools	Cross-checking results, entering notes, coordinating care	**Utilitarian**: Efficiency, reliability, error prevention, and reduced work or cognitive load **Hedonic:** Reduced frustration, increased pleasure, ease of use	Contextual alignment with workflows; error-prevention cues; semantic clarity in data presentation
Administrators	Compliance dashboards, audit trails, ease of documentation and reporting	Managing permissions, generating compliance reports	**Utilitarian:** Accountability, transparency, systemic reliability **Hedonic:** Confidence in use, error prevention, and reassurance of data integrity	Consistent signifiers for oversight; standardised terminology; intelligible reporting
Policy Makers	Aggregated data repositories, anonymisation protocols, population dashboards	Monitoring population health data, auditing compliance, issuing directives	**Utilitarian:** Systemic trust, legitimacy, assurance for policy **Hedonic:** Trust in system legitimacy, confidence in systemic transparency	Data aggregation protocols; semantic clarity in reporting; contextually relevant indicators for decision-making

Adapting MDA for Interactive Information Artefacts requires more than a direct transfer of concepts from game design. We propose three complementary adaptations: integrating utilitarian and hedonic design traits, embedding the perspectives of multiple stakeholders, and foregrounding meaning-making, semantics, and contextual/temporal framing as primary design responsibilities. Together, these adaptations extend the original tripartite structure of MDA while preserving its analytical clarity. They position the framework not only as a tool for linking mechanics, dynamics, and aesthetics, but also as a lens capable of diagnosing breakdowns and guiding design decisions in information-rich, socially embedded systems. By doing so, the adapted MDA offers information and interaction designers a structured yet flexible approach for aligning structural provisions, behavioural patterns, and experiential outcomes across diverse contexts of use.

Structural, Behavioural, and Experiential Dimensions for Interactive Information Artefacts

Building on the three proposed adaptations of the MDA framework: (1) hedonic–utilitarian balance, (2) multi-stakeholder perspectives, and (3) meaning-making through semantic and contextual framing—outlined in Tables 2 to 4, we introduce a new framework tailored explicitly for Interactive Information Artefacts, the SBE dimensions.

The SBE dimensions extend the analytical and design benefits of MDA by addressing the specific demands and values of interactive information artefacts where meaning-making, semantic clarity, stakeholder heterogeneity, and the balance between utilitarian and hedonic traits are decisive. By translating the MDA layers into the categories of structural, behavioural, and experiential, we retain continuity with established theory while making explicit the interpretive and contextual demands that characterise Interactive Information Artefacts.

From *MDA* to *SBE*

The shift from MDA to SBE reflects a transition from the game design context, where mechanics, dynamics, and aesthetics were primarily linked to enjoyment and engagement (Hunicke et al., 2004), to information design contexts, where accuracy, clarity, trust, and contextual relevance become central outcomes. Each of the three adaptations we have proposed contributes to this reframing of the issue. First, the utilitarian and hedonic dimensions ensure that Interactive Information Artefacts are not reduced to functional tools but are understood as artefacts that must balance pragmatic efficiency with experiential qualities such as reassurance, comfort, or confidence. Second, the multi-stakeholder perspective makes explicit that Interactive Information Artefacts serve heterogeneous and often divergent audiences, rendering the notion of a single "user" insufficient. Third, embedding meaning-making, semantics, and contextual/temporal framing ensures that Interactive Information Artefacts are analysed not only for what they provide, but also for when, how, and in what context provisions occur. The SBE dimensions crystallise these three adaptations into a coherent lens for both design planning and diagnostic analysis.

The Structural Dimension

The ***Structural*** dimension corresponds to what MDA termed *mechanics*: the deliberate provisions and rules encoded into the system. In Interactive Information Artefacts, this includes data models, information architecture, filtering functions, dashboards, and protocols for input and output. Structural design choices determine what is possible within the artefact and set the conditions for stakeholder engagement. Importantly, structural design must be attuned to semantic clarity, terminology, representational forms, and organisational logics, which are not neutral but directly shape interpretive accuracy (Krippendorff, 2006). For example, whether a healthcare system labels test results with technical abbreviations or plain-language terms has immediate consequences for patient comprehension and anxiety. Structurally, Interactive Information Artefacts therefore require not only robustness and reliability but also alignment with the heterogeneous knowledge bases and expectations of different stakeholders.

The Behavioural Dimension

The ***Behavioural*** dimension parallels MDA's *dynamics*: the emergent patterns of interaction that arise as stakeholders engage with the system's structural provisions. In Interactive Information Artefacts, these behaviours include patients booking or rebooking appointments, doctors adapting workflows under time pressure, administrators generating compliance reports, or policy makers interpreting population-level dashboards. Unlike in entertainment games, these behaviours are often mandatory rather than voluntary (Juul, 2010), making design for resilience and reliability essential. The behavioural dimension also highlights how utilitarian and hedonic qualities intertwine: a system that enables efficiency (utilitarian) while reducing frustration or offering reassurance (hedonic) fosters more sustainable patterns of use. Designers must therefore anticipate behavioural adaptations, workarounds, and sense-making strategies across diverse contexts, ensuring that the artefact supports rather than constrains stakeholders situated practices.

The Experiential Dimension

The ***Experiential*** dimension builds on MDA's *aesthetics*, but with a reoriented focus. In Interactive Information Artefacts, experiential outcomes extend beyond enjoyment or challenge to include trust, confidence, clarity, transparency, and systemic legitimacy. These qualities are not mere by-products but core design responsibilities. They emerge from the alignment (or misalignment) of structural provisions and behavioural patterns with semantic, contextual, and temporal framing. For instance, timely and comprehensible feedback fosters reassurance for patients, efficiency for doctors, accountability for administrators, and legitimacy for policy makers. Conversely, vague signifiers, inconsistent terminology, or poorly timed outputs create cognitive friction, reduce trust, and compromise both individual and systemic outcomes (Norman, 2013; Forlizzi and Battarbee, 2004).

The experiential dimension thus foregrounds meaning-making as the evaluative anchor of Interactive Information Artefacts, ensuring that interaction outcomes resonate with stakeholders' goals and contexts.

Integrative Value of SBE Dimensions

The SBE dimensions offer integrative value in two ways. First, as a planning scaffold, it enables designers to ensure that structural provisions, behavioural patterns, and experiential qualities are coherently aligned, balancing utilitarian and hedonic traits and anticipating stakeholder heterogeneity. Second, as a diagnostic tool, it helps researchers and evaluators identify where breakdowns occur whether in the structure (ambiguous terminology, weak interoperability), the behaviour (inefficient workflows, workarounds), or the experience (loss of trust, reduced clarity). By mapping SBE across the stages of interaction (Norman, 1986; Rogers et al., 2019), designers can also account for when and how provisions support or hinder meaning-making, bridging the structural "what" with the temporal "when" and interpretive "how". The SBE dimensions thus represent the outcome of adapting MDA for information-rich, multi-stakeholder, and semantically sensitive contexts. By retaining the tripartite logic of SBE, it provides continuity with game design theory while reframing its categories to meet the specific challenges of Interactive Information Artefacts. SBE positions meaning-making, stakeholder diversity, and the utilitarian–hedonic balance as central, not peripheral, design concerns. In doing so, it offers designers and researchers a lens that is both analytically precise and practically actionable, capable of guiding design decisions and diagnosing interactional breakdowns across diverse real-world contexts.

To consolidate the three adaptations and make them analytically actionable, we present the SBE dimensions. Table 5 summarises the dimensions definition, design responsibilities, and targeted outcomes, illustrating how structural provisions, behavioural patterns, and experiential qualities together capture the central design concerns of Interactive Information Artefacts. By situating meaning-making, stakeholder diversity, and the utilitarian–hedonic balance within this tripartite framework, the table provides a concise overview of how SBE can serve as both a design planning scaffold and a diagnostic tool for evaluating interaction in practice.

Six Actionable Stages of SBE Dimensions

We present a six-stage process that serves as a practical tool for both planning and analysing the design of Interactive Information Artefacts, which integrates temporal, contextual, and semantic factors by aligning the three SBE dimensions, structural provisions, behavioural patterns, and experiential qualities, with the six actional stages. In doing so, these stages operationalise our argument that Interactive Information Artefacts must be understood not only as theoretical

Table 5 The Structural–Behavioural–Experiential Dimensions

Dimension	*Definition*	*Design Responsibilities*	*Targeted Outcomes*
Structural	The provisions, rules, and organisational architecture are encoded into the system	Define information models, ensure interoperability, and establish semantic clarity (terminology, signifiers, representations)	Reliability, robustness, and interpretive accuracy
Behavioural	The emergent actions and interaction patterns are produced when stakeholders engage with structural provisions	Anticipate workflows, support decision-making under constraints, minimise redundancy, and design for resilience in mandatory contexts	Efficiency, adaptability, reduced cognitive load, and sustainable practices
Experiential	The felt qualities of interaction that shape how stakeholders interpret and evaluate the system	Align timing, framing, and presentation with goals and contexts, striking a balance between utilitarian and hedonic qualities	Trust, confidence, clarity, legitimacy, systemic transparency

constructs, but also in terms of what they offer, when they occur, the context in which they are deployed, and ultimately how they are framed by the affected stakeholders' ways of making sense and meaning of them:

1. **Goal definition:** Establish the outcomes the artefact should enable (grounded in stakeholder contexts and constraints).
2. **Intention support:** Form a plan of possible stakeholders' artefact interactions (cues, pathways, and decisions).
3. **Artefact operation:** Confirm artefact provides appropriate affordances and interactions that enable reliable interactions with diverse stakeholder groups.
4. **Artefact responsiveness:** Ensure artefact interaction sequences are applicable, attainable, and accessible for the specific context.
5. **Interpretation facilitation:** Warrant that artefact supporting sense-making through semantic clarity, framing, and signifiers.
6. **Evaluation enablement:** Allow stakeholders to assess outcomes and determine appropriate next steps.

With the actionable stages of SBE, we aim to illustrate where and when the Structures, Behaviours, and Experientials of our framework might occur within an Interactive Information Artefact, and how they can benefit different stakeholder perspectives. In the presented format, the stages serve two purposes: 1) as a planning tool to ensure consistent delivery throughout the design process of Interactive Information Artefacts, and 2) as a diagnostic lens to identify and improve existing Interactive Information Artefacts, supporting the creation of design audits to analyse stakeholders, traits, context, or semantic clarity.

Although the stages are described herein as a numbered sequence (1– 6), they should be viewed as iterative loops that can overlap or be skipped, with multiple entry points depending on the purpose of their application (Figure 2).

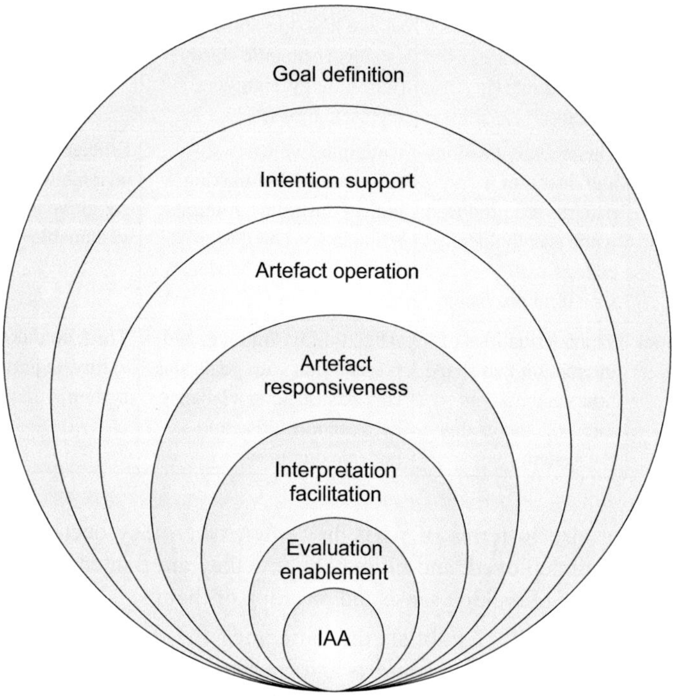

Fig. 2 The six-stage SBE model serves as an instrument for both planning and analysing the design of Interactive Information Artefacts.

Concluding Remarks

To conclude by combining the three suggested adaptations of MDA's 1) hedonic/utilitarian balance, 2) multi-stakeholder perspectives, and 3) meaning-making/semantic/contextual framing, Interactive Information Artefacts are positioned within a structured yet adaptable design frame SBE.

SBE offers a practical synthesis of the chapter's theoretical argument: that design should be evaluated not only based on the internal coherence of mechanics, dynamics, and aesthetics, but also on their alignment with temporal processes, semantic clarity, and the varied perspectives of stakeholders in real-world context situations. In this chapter, we have argued that directly applying the MDA framework to Interactive Information Artefacts risks missing vital design aspects. We have proposed three essential adaptations: integrating both utilitarian and hedonic traits, including multiple stakeholder perspectives, and prioritising meaning-making, semantics, and contextual/temporal framing as key design responsibilities. Collectively, these adaptations reposition MDA as a framework that not only connects mechanics, dynamics, and aesthetics but also analyses how design provisions develop over time, within different contexts, and across stakeholder viewpoints. The proposed Analysis Matrix puts these ideas into practice, providing both a planning framework and a diagnostic tool. By placing Interactive Information Artefacts within this adapted MDA perspective, designers and researchers can adopt a structured yet flexible approach to align structural provisions, behavioural patterns, and experiential outcomes in complex, information-rich systems.'

Using the lenses described in the SBE will assist in designing and evaluating Interactive Information Artefacts. However, additional factors surrounding the Interactive Information Artefact need to be considered, which are outside the scope of our framework. First and foremost, it is vital to understand the actual design problem or opportunity; this requires investigation and understanding before deciding whether to design any Interactive Information Artefact at all. Furthermore, differences in expertise, knowledge, and motivation among stakeholders must be considered. Just as different types of users (e.g., patients, doctors, administrators, etc.) have varying goals and needs, the expertise and knowledge within a user group influence both interaction with and information needs in the Interactive Information Artefact.

In addition to the users' expertise, resources, and skillsets, these aspects impact how and why the design (and maintenance) of the Interactive Information Artefact is carried out. Designers need to understand the boundaries of the Interactive Information Artefact's context and what lies beyond it. For example, a digital healthcare system is never used in isolation; there is a surrounding context that must be considered, but it cannot be fully understood without the Interactive Information Artefact. Factors such as the time needed to interact with the Interactive Information Artefact, cybersecurity, laws and regulations, and staff workload are relevant. The Interactive Information Artefact can be adapted to help mitigate these factors, but it is unlikely to eliminate their negative implications.

Another aspect not covered in this chapter is the emergence of AI tools and agents that utilise information, with which interaction designers can collaborate. Regardless of whether AI tools are used to develop elements within the structural, behavioural, or experiential lenses, or if AI-agents are considered new stakeholders altogether, we must ask whether this framework will remain useful for designers in the future, irrespective of how AI is adapted and employed.

With this chapter, we have introduced *Interactive Information Artefacts* as a constructed category that illuminates the interdependence of informational content and interactive form. By adapting the MDA framework into the SBE dimensions, we have articulated a lens that captures the distinctive demands of Interactive Information Artefacts: the need to balance utilitarian and hedonic qualities, to align diverse stakeholder perspectives, and to embed meaning-making, semantics, and contextual framing as primary design responsibilities. The SBE dimensions not only extend the analytical and design utility of MDA but also provide a practical scaffold for diagnosing and planning the design of information-rich, socially embedded systems. While future developments, such as the increasing integration of AI agents, will inevitably reshape the contexts in which Interactive Information Artefacts operate, the SBE framework provides a durable foundation for analysing and guiding design in complex domains where clarity, trust, and engagement must be achieved simultaneously.

References

Ashby, W.R. 1956. *An Introduction to Cybernetics*. Chapman & Hall, London

Benyon, D. 2019. *Designing User Experience: A Guide to HCI, UX, and Interaction Design* (4th Edn.). Pearson.

Bødker, S. 1991. *Through the Interface: A Human Activity Approach to User Interface Design*. Lawrence Erlbaum Associates.

Bødker, S. 2006. When second wave HCI meets third wave challenges. In: *Proceedings of the 4th Nordic Conference on Human–Computer Interaction: Changing Roles* (pp. 1–8). ACM. https://doi.org/10.1145/1182475.1182476.

Buchanan, R. 1992. . Wicked problems in design thinking. *Design Issues, 8*(2).

Dourish, P. 2001. *Where the Action is: The Foundations of Embodied Interaction*. MIT Press.

Evans, J.S.B. 2003. In two minds: Dual-process accounts of reasoning. *Trends in Cognitive Sciences, 7*(10): 454–459.

Evans, J.S.B. and Stanovich, K.E. 2013. Dual-process theories of higher cognition: Advancing the debate. *Perspectives on Psychological Science*, *8*(3): 223–241.

Forlizzi, J. and Battarbee, K. 2004. . Understanding experience in interactive systems. In: *Proceedings of the Fifth Conference on Designing Interactive Systems: Processes, Practices, Methods, and Techniques* (pp. 261–268). ACM. https://doi.org/10.1145/1013115.1013152.

Galy, E., Cariou, M. and Mélan, C. 2012. What is the relationship between mental workload factors and cognitive load types?. *International Journal of Psychophysiology*, *83*(3): 269–275.

García-Sánchez, E., García-Morales, V.J., and Martín-Rojas, R. 2018. Analysis of the influence of the environment, stakeholder integration capability, absorptive capacity, and technological skills on organizational performance through corporate entrepreneurship. *International Entrepreneurship and Management Journal*, *14*(2): 345–377. https://doi.org/10.1007/s11365-017-0440-6

Gigerenzer, G. and Selten, R. (Eds.). 2001. *Bounded Rationality: The Adaptive Toolbox*. MIT Press.

Haselager, P., van Dijk, J. and van Rooij, I. 2008. A lazy brain? Embodied embedded cognition and cognitive neuroscience. In: Calvo, P., and Gomila, A. (Eds.), *Handbook of Cognitive Science: Perspectives on Cognitive Science* (pp. 273–290). San Diego, CA: Elsevier.

Hassenzahl, M. 2003. The thing and I: Understanding the relationship between user and product. In: M.A. Blythe, A.F. Monk, K. Overbeeke, and P. C. Wright (Eds.), *Funology: From Usability to Enjoyment* (pp. 31–42). Kluwer Academic Publishers.

Hassenzahl, M. 2008. User experience (UX): Towards an experiential perspective on product quality. In: *Proceedings of the 20th International Conference of the Association Francophone d'Interaction Homme-Machine* (pp. 11–15). ACM. https://doi.org/10.1145/1512714.1512717.

Hassenzahl, M. and Tractinsky, N. (2006). User experience: A research agenda. *Behaviour and Information Technology*, *25*(2): 91–97. https://doi.org/10.1080/01449290500330331.

Horn, R.E. 1999. Information design: Emergence of a new profession. In: R. Jacobson (Ed.), *Information Design* (pp. 15–33). MIT Press.

Hunicke, R., LeBlanc, M. and Zubek, R. 2004.). MDA: A formal approach to game design and game research. In: *Proceedings of the AAAI Workshop on Challenges in Game AI* (pp. 1–5). AAAI Press.

Jordan, P.W. 2000. *Designing Pleasurable Products: An Introduction to the New Human Factors*. Taylor & Francis.

Junior, R. and Silva, F. (2021). Redefining the MDA Framework—The Pursuit of a Game Design Ontology. Information, 12(10), 395. https://doi.org/10.3390/info12100395

Juul, J. 2010. *A Casual Revolution: Reinventing Video Games and Their Players*. MIT Press.

Kahneman, D. (2011). Thinking, fast and slow. Farrar, Straus and Giroux.

Krippendorff, K. 2006. *The Semantic Turn: A New Foundation for Design*. CRC Press.

Kusuma, G.P., Wigati, E.K., Utomo, Y. and Suryapranata, L.K.P. 2018. Analysis of gamification models in education using MDA framework. *Procedia Computer Science*, *135*: 385–392.

Löwgren, J. and Stolterman, E. 2004. *Thoughtful Interaction Design: A Design Perspective on Information Technology*. MIT Press.

McCarthy, J., and Wright, P. 2004. *Technology as Experience*. MIT Press.

Nancy, P., Currie, W. and Whitley, E.A. 2016. Entangled stakeholder roles and perceptions in health information systems: A longitudinal study of the UK NHS N3 network. *Journal of the Association for Information Systems*, *17*(2): 107–161. https://doi.org/10.17705/1jais.00421.

Nelson, H.G. and Stolterman, E. 2014. *The Design Way: Intentional Change in an Unpredictable World*. MIT press.

Nielsen, J. (1993) Usability Engineering. Academic Press, Inc., Harcourt Brace & Company, San Diego, USA.

Norman, D.A. 1986. Cognitive engineering. In: D.A. Norman and S.W. Draper (Eds.). *User-centred System Design: New Perspectives on Human-computer Interaction* (pp. 31–61). Lawrence Erlbaum Associates.

Norman, D.A. 2004. *Emotional Design: Why We Love (or Hate) Everyday Things*. Basic Books.

Norman, D.A. 2013. *The Design of Everyday Things: Revised and Expanded Edition*. Basic Books.

Palmquist, A. 2023a. Stakeholders' design preferences for instructional gamification. *Behaviour and Information Technology*, *43*(12): 2717–2749. https://doi.org/10.1080/0144929X.2023.2255905.

Palmquist, A. 2023b. Design elements of conflict: A design study of a gamified smartphone application for employee onboarding. *Technology, Knowledge and Learning*, *28*(3): 1133–1173. https://doi.org/10.1007/s10758-023-09657-7.

Palmquist, A. 2023c. *Plug & Play? Stakeholders' Co-meaningmaking of Gamification Implementations in Workplace Learning Environments* [Doctoral thesis, University of Gothenburg, IT Faculty, Department of Applied Information Technology]. University of Gothenburg. ISBN 978-91-8009-8069-114-7.

Pettersson, R. (2002). Information design: an introduction. Amsterdam: John Benjamins Pub.

Pistono, A.M.A.D.A., dos Santos, A.M.P., Baptista, R.J.V. and Mamede, H.S. 2024. Framework for adaptive serious games. *Computer Applications in Engineering Education*, *32*(4): e22731.

Pontis, S. and Babwahsingh, M. 2023. *Information Design Unbound: Key Concepts and Skills for Making Sense in a Changing World*. Bloomsbury Publishing.

Rogers, Y., Sharp, H. and Preece, J. 2019. *Interaction Design: Beyond Human–computer Interaction* (5th Edn.). Wiley.

Salen, K. and Zimmerman, E. 2004. *Rules of Play: Game Design Fundamentals*. MIT Press.

Shannon, C.E., and Weaver, W. (1998.). *The Mathematical Theory of Communication*. University of Illinois Press.

Shedroff, N. (1994.). Information interaction design: A unified field theory of design. In: *Proceedings of the Fourth International Conference on Human–Computer Interaction* (pp. 23–27). Elsevier.

Silver, N. 2012. *The Signal and the Noise: The Art and Science of Prediction*. Penguin UK.

Simon, H.A. 1955. A behavioral model of rational choice. *The Quarterly Journal of Economics*, 99–118.

Simon, H.A. (1996.). *The Sciences of the Artificial* (3rd Edn.). MIT Press. (Original work published 1969).

Stolterman, E. and Wiberg, M. 2010. Concept-driven interaction design research. Human–Computer Interaction, *25*(2): 95–118. https://doi.org/10.1080/07370021003676623.

Suchman, L. 2007. *Human–machine Reconfigurations: Plans and Situated Actions* (2nd Edn.). Cambridge University Press.

Sweller, J., Ayres, P. and Kalyuga, S. 2011. *Cognitive Load Theory*. Springer.

Tufte, E.R. 2001. *The Visual Display of Quantitative Information* (2nd Edn.). Graphics Press.

Van der Heijden, H. 2004. User acceptance of hedonic information systems. *MIS Quarterly*, 695–704.

Walk, W., Görlich, D. and Barrett, M. 2017. Design, dynamics, experience (DDE): An advancement of the MDA framework for game design. In: *Game Dynamics: Best Practices in Procedural and Dynamic Game Content Generation* (pp. 27–45). Springer International Publishing.

Winograd, T. (Ed.). 1996. *Bringing Design to Software*. New York, NY: ACM Press.

Winograd, T. (1997) From Computing Machinery to Interaction Design. In P. Denning and R. Metcalfe (eds). Beyond Calculation: The Next Fifty Years of Computing. Springer-Verlag, Amsterdam, pp. 149–162.

Zubek, R. 2020. *Elements of Game Design*. MIT Press.

13 Information Design in Virtual Environments

Learn to Move–Move to Learn!

Helena Tobiasson[1*]

Introduction

> *"Considering the fact that the term experience in UX was more aspirational than literal when originally introduced, it now seems an almost prophetic manifestation for the true destination of designers, to design an immersive spatial dimension that previously never existed in UX, where physical movement, distance, and body, hand, and head movement are essential parts of the interaction."* (Hillman, 2021)

This citation from Cornell Hillman, an experienced designer, in his book *UX for XR* (2021) highlights the central aim of this chapter: to present, propose, and discuss perspectives drawn from embodiment, physical literacy, and methods from ergonomics and kinaesthetic learning. Together, these approaches are introduced as a proposal for information design and interaction in the context of creating immersive spaces.

Imagine information that moves with you, changes as you reach for it, and teaches through the movements your body makes. Designing for immersive spaces is not simply about adding another dimension to a flat page; it is a fundamentally different design problem that asks you to think in terms of bodies, rhythms, and situated action. When information lives inside three-dimensional space or mixed reality, it becomes an actor in a system where perception and movement co-create meaning.

This chapter will present and discuss design for engagement that respects bodies, cultivates skill, and deepens understanding in three-dimensional, multimodal spaces. It offers cross-disciplinary perspectives from Embodiment, Physical Literacy, Ergonomics, and Kinaesthetic Learning. It applies these theories and methods to the rapidly evolving landscapes of XR. The aim is to equip designers with an embodied lens through which immersive technology can be approached.

Theoretical perspectives from embodiment and physical literacy and methodological perspectives from Ergonomics and Kinaesthetic Learning will

[1] Ph.D. in Human Computer Interaction, Senior lecturer in Information Design, Department of Health Science, Innovation and Design, Mälardalen University, Box 325, 631 05 Eskilstuna, Sweden.

* Corresponding author: helena.tobiasson@mdu.se

DOI: 10.1201/9781032642598-13

be discussed and presented as parts that can strengthen information design as it migrates from screens into immersive spaces or Extended Reality (XR), a term used as an umbrella term to encompass the different digitally designed systems and spaces such as Augmented Reality (AR) where the physical world is seen through the head-set/the glasses and overlaid of elements that are digitally designed, Mixed Reality (MR) that is similar to AR but mixed in the sense that some physical objects can interact with digitally designed objects, Virtual Reality (VR), a digitally designed environment without a visual see-through into the physical world or reality. Design and technology will not stop here—it will most certainly continue to develop new ways of acting and being in relation with digital technology.

Transitioning from 2D to 3D is often framed as a technical challenge, but the deeper shift is conceptual. On a page, a monitor, or a screen information is framed, fixed, and consumed through a relatively constrained set of interactions. In XR, information is dynamic and relational; it unfolds as users move, turn, and change focus. The same dataset can feel different at ankle height compared with eye level, and a legend that makes sense when viewed statically may become distracting or even harmful when users must crane their necks to read it repeatedly. Designing for these conditions requires sensitivity to scale, density, and timing. It demands that designers decide not only what information to present but where, when, and how it should respond to bodily action.

Embodiment is the first perspective presented in this chapter. Human bodies are not passive receivers of visual signals; they are active multisensorial whose proprioception, touch, balance, and internal cues inform how we understand space and content. In XR, designers can no longer treat the body as incidental. Instead, information should be mapped to sensorimotor contingencies so that gestures, gait, and posture become legitimate input channels. When feedback aligns predictably with action, users may experience stronger agency and deeper presence. Misalignments, by contrast, can cause confusion, disorientation, and a breakdown of trust. Learning to design with embodiment in mind means practicing techniques that calibrate how virtual elements respond to real movement, how audio and haptic cues reinforce visual information, and how the representation of the self in space supports rather than undermines understanding. Physical literacy treats embodied competence as a design goal: users should not only be able to perform actions inside XR but develop durable movement skills that transfer beyond a single scenario.

Ergonomics provides the safety-focused complement to embodiment's experiential emphasis. Ergonomics asks how to fit environments to bodies so that meaningful interactions are sustainable across time. In immersive contexts, the stakes are different and often higher: prolonged neck flexion, repetitive reaches, or awkward wrist angles can create real physical discomfort and long-term strain. Ergonomic design in XR requires attention to reach envelopes, neutral postures, pacing, and the sequencing of tasks to avoid cognitive and musculoskeletal overload. It also requires designing onboarding flows that let users calibrate for their own bodies, adjusting height, reach, and input mappings so the experience becomes accessible to a wide range of sizes and abilities. Thinking ergonomically is not about removing challenges; it is about preserving health while enabling effective, prolonged engagement.

Kinaesthetic learning moves the focus from safety and perception to skill and transfer. Kinaesthetic learning leverages doing as a way of knowing. When learners assemble objects, trace spatial relationships, or coordinate movements with others, they are building internal models that support memory, problem solving, and creativity in ways that passive viewing does not. Designing for kinaesthetic learning requires scaffolding practice, offering progressive challenges, and creating feedback systems that reward quality of movement—smoothness, efficiency, and adaptiveness—rather than simple completion. It also invites designers to create collaborative tasks where timing and shared motion produce meaning collectively rather than individually. Bringing these perspectives together hope to strengthen Information design when designing for or in XR.

The move to XR also raises methodological questions. How do we build evaluation methods that capture long-term strain as well as short-term comprehension? How do we create prototyping techniques that allow rapid exploration of embodied interactions without expensive hardware? How do we reconcile the competing demands of accessibility and pedagogical challenge? These questions invite research, experimentation, and critical reflection, and Information design in XR can offer experiences that are compelling, inclusive, and educationally rich.

Imagine a training simulation where a novice's inefficient movements are smoothly scaffolded into muscle memory through gentle resistance, rhythmic prompts, and progressive challenges. Imagine data visualisations that reveal their structure as you walk through them, translating abstract patterns into embodied narratives that users can feel as well as see.

Whether we are designing for VR, MR, or AR, the body is never absent. It is the medium through which environments are perceived, tasks are enacted, and meaning is made. Designing for XR in virtual environments or evaluating user-experience, user-studies in these environments has an opportunity to invite the body back into action for work and play. It is about shaping the conditions under which human bodies move, feel, and learn. Next, a brief section on the technological development in XR.

XR: A Short Retrospect

Virtual environments—have matured from experimental labs into practical systems used in education, training, design, and collaboration. Their development combines advances in sensing and rendering hardware, spatial computing architectures, and new interaction paradigms that move beyond keyboard and mouse into gaze, gesture, voice, and haptics (Vohra, 2025).

A short retrospect starting off as recently as 2019 when Oculus Quest (from August 2022 Meta Quest) was launched and hand tracking matured to be included as an interaction modality and described as a natural interface for virtual environments. The same year (2019) the AR system, the Pokémon Go was very successful. It managed to bring users together from different generations and it designed a layer on real-world interactions to be discovered.

In the years 2019–2020, standalone headsets and hand tracking (wireless standalone devices) make 6DOF VR mainstream and camera-based hand tracking becomes viable, enabling natural pinch and grab metaphors and quick passthrough between virtual and real views. Then during 2021–2022, spatial computing and multimodal inputs has matured and eye tracking, foveated rendering (the technology that when the user is watching renders images in high resolution and renders images that appears in the peripheral sight in low resolution—computer capacity is economised), improved optics, and cloud/edge offload support higher fidelity visuals and multiuser persistence are gaining interests. Gaze, hands, voice, and body tracking begin to converge in production hardware. By 2023 the development of lightweight AR glasses makes consumer-oriented prototypes, optical see-through devices and better micro displays enable glanceable overlays and contextual AR workflows, shifting design toward minimal, always-available information layers. Then the following years, 2024–2025 sensor fusion and integrated XR ecosystems make possible combined sensors measuring and comprehend movement, cameras, depth sensors, eye trackers, and NPUs (Neural Processing Units, a fast microprocessor) produce robust spatial meshes and semantic scene understanding. Seamless mode switching between VR, AR, and mixed passthrough becomes standard and rich multimodal interactions are normalised.

By 2025 and onwards, XR has moved from isolated demos to integrated, multimodal ecosystems. For Information and Interaction design this means shifting from static screens to designing complex spaces that put more focus on attention models, temporal UI, multi sensory feedback, and ethical policies. These are examples of focus areas that now sit alongside visual design and information architecture as core responsibilities (Hillman, 2021).

The different terms VR, AR and MR will appear in the text, although the term XR will be used more frequently as topics or perspectives that will be discussed can be applied to the different strands, VR, AR or MR. Let's continue with a theoretical foundation.

Theoretical Foundation

To provide a backbone to the coming sections on Ergonomics and Kinaesthetic Learning in this section are presented theoretical perspectives on Embodiment and Physical literacy, starting out by acknowledging that embodiment is at focus in several other domains among others in cognitive science. Cognitive science, that emerged alongside the rise of digital computers and information-processing theory and at that time treated thinking mainly as information processing, has in some parts developed to extend the focus of interest from the brain into a broader theory including focus on the context and looks at cognitive processes as intertwined in the context (situated cognition). Situated cognition can be seen from four focus areas, one being *embedded*—cognition is related to both physical and social environments, the second, *embodied*—describing that in cognition, perception, and action, the body and the sensorimotor system is central for experience, the third focus area, *enacted* cognition—pointing towards interaction that occur between a body or an organism and the world as it unfolds through actions involving the sensorimotor system and skills. Possibilities for actions that are offered in an environment or affordances in that environment are used in enacted cognition. The fourth area of focus, *extended* cognition—here cognitive processes are seen as possible to incorporate physical objects that extend the processes beyond the body.

While situated cognition—understood as embodied and enacted—broadens the theoretical lens to include sensorimotor perspectives, this chapter takes a different approach. The focus here is on embodiment and physical movement from the perspective of the moving body itself.

Although cognitive theories offer fascinating insights, they will not be the primary framework for discussing movement in XRs. This is a deliberate choice: cognitive approaches often steer the conversation toward brain-based processing and functionality. Instead, the aim is to keep things simple and grounded—placing movement and the moving body at the centre of attention.

To rediscover and explore movement in XR, this chapter uses embodiment, physical literacy, ergonomics, and kinaesthetic learning as the bases for discussion. These perspectives allow us to understand the body not just as a processor of information, but as an active, expressive, and responsive participant in immersive environments. Next, the concept of embodiment will be presented and discussed.

Embodiment

Embodiment is widely discussed and researched from several perspectives among others philosophy, cognition, socio cultural perspective, in domains such as education, architecture, and healthcare (Kersting, Haglund, and Steier, 2021; Wang, et al., 2022; Mattes and Lang, 2021). Embodiment rooted in phenomenology, a philosophical strand in which the French philosopher Maurice Merleau-Ponty has been given credit for developing a distinctive account that incorporated the body as a central focus for first person subjective perspective of embodied existence. Through the lived body the individual perceives and experiences the world in which the subjective body lives, feels, and moves. Phenomenology, as described by Merleau-Ponty, presents subjectivity as always embodied, situated, and intersubjective (Merleau-Ponty and Landes, 2014). Phenomenology discusses consciousness on a much more detailed level here the focus resides mainly on the perspectives on the body as the central medium for embodied experience and in relation to consciousness the part where perception is presented as grounded in the body's active engagement with its surroundings (Merleau-Ponty and Landes, 2014). From a phenomenological perspective, the body is not just an object we "have", but the primary medium through which we experience, understand, and act in the world. In 1979, the American psychologist J.J. Gibson received attention through his seminal work, "The Ecological Approach to Visual Perception", where he presented the concept of affordances—action possibilities in the environment relative to the actor's capabilities (Gieseking et al., 2014). The concept was related to the enactive approaches in phenomenology, "what can I do with what I see around me", an action-oriented and operative intentionality also explored in the domain of ecological psychology (Masoudi et al., 2019).

In digital design, embodiment extends beyond the philosophical into the pragmatic: our sensory and motor systems fundamentally constrain and enable the ways in which we engage with for example technology. By the early 2000s, with the move towards embodied interaction (Dourish, 2001), researchers in human computer interaction began explicitly framing design as an *embodied* practice, moving away from mainly symbolic or screen-based concepts toward tangible, gestural, and spatial interfaces. This shift was accelerated by the rise of ubiquitous computing, tangible user interfaces, and motion tracking technologies.

In design and evaluation of immersive environments, theoretical and methodological perspectives related to embodiment could play an even more central role. The alignment of virtual and physical movements—sometimes called sensorimotor contingency—shapes users' sense of presence and agency (Slater and Sanchez-Vives, 2016). Researchers have explored how avatar representation, motion tracking fidelity, and haptic feedback influence perceived embodiment in VR (Banakou et al., 2022; Huygelier et al., 2021). Within neuroscience, a definition of the sense of embodiment reads as the integrated experience of being inside, owning, and controlling a body, and explains its three dimensions: agency, body ownership, and self-location. Where the agency is related to control of actions, body ownership related to feeling that the body is yours and self-location related to the experience of where you feel located. How substituting a real body with an avatar in virtual reality alters perception and behaviour, affecting interaction quality (Guy et al., 2023). In XR, embodiment is more than an abstract philosophical stance. It can be used as an operational design parameter that supports users' sense of presence, agency, and co-location. Designers make decisions about avatar fidelity, motion tracking accuracy, field of view alignment, and haptic feedback precisely to support sensorimotor contingencies—the tight coupling between perception and action that underpins the feeling of "being there" (Slater and Sanchez-Vives, 2016).

Importantly, embodiment carries socio cultural and ethical dimensions. The body is not a neutral interface; it is gendered, aged, racialised, and differently abled. Design choices—from motion capture calibration to gesture vocabularies—tend to either reinforce or challenge existing bodily norms and assumptions. Embodiment also extends to social presence: how users perceive themselves and others in a shared immersive environment (Della Longa et al., 2022). Choices about gesture vocabularies, posture mirroring, and facial expression mapping, influence not only interaction fluency but also trust, empathy, and inclusion.

Despite, as one might think, the centrality of embodiment in immersive design, there is not a clear definition nor is embodiment a unifying concept where researchers agree on a description or a methodology how to apply the notion of embodiment in virtual environments. Researchers have examined how the term embodiment is defined and used in VR game design research, arguing that inconsistent operationalisations weaken its value for analysis and design (Palmquist et al., 2025).

It is not always that present in the embodiment literature how bodies can come with disabilities, different sensory baselines, or culturally distinct movement repertoires. Keep in mind that assuming an able-bodied person, in doing so, risks marginalising or omitting bodies with different sensory, motor, or cognitive profiles. Although technological development continues there are still challenges in XR technologies to fully capture the full richness of bodily experience (e.g., fine-grained touch, balance cues, proprioception). Current sensors and actuators cannot fully capture or reproduce the richness of bodily experience—leading to partial or distorted embodiment. Over-reliance on incomplete embodiment simulations can lead to mismatches that break the presence or cause discomfort. In some cases, embodiment can be reduced to a set of measurable parameters (e.g., tracking latency, joint angle accuracy), detaching it from its philosophical and socio cultural roots. In so doing, it risks turning embodiment into a technical checklist rather than a holistic design sensibility. That said, how can the concept of embodiment be translated into actionable design methods? The following section will start to line up some examples.

Methodological Approaches

How can the concept of embodiment be augmented in the sense that the, somewhat taken for granted, lived body gets the focus or a voice in design? Bodystorming (Chilufya and Arvola, 2021) where designers physically act out scenarios to explore interaction possibilities before committing to digital prototypes, is an example of methods incorporating perspectives from embodiment. Another is Somaesthetic Exercises as Guided practices to heighten bodily awareness, often used in participatory workshops (Höök, 2018). Somaesthetic design promotes a shift from symbolic, language-based approaches to experiential, felt engagement, encouraging designers to test and refine their work through direct bodily experience. This approach aims at fostering empathy, creativity, and deeper connection between people and their environments and bridging philosophy and practice, offering a framework to reclaim the body as central to perception, meaning, and aesthetic pleasure. Somaesthetic design approach has found to be of relevance in fields like design, human-computer interaction, and education.

Design researchers (McVeigh-Schultz et al., 2024) examined the hybrid, often uncanny nature of embodied experience in social XR and argues that designing for these "weird" interactions requires methods that foreground bodily knowing alongside digital mediation. Grounded in embodied interaction and somaesthetic design, they draw on lived prototyping in VR to surface oddities of presence, reciprocity, and sensory mismatch that arise when bodies and virtual systems co-

constitute social encounters. A set of pragmatic tactics for designers and researchers are proposed—such as embodied sketching, deliberately eccentric constraints, scaffolded learning moments, and reflective postures—that help surface tacit knowledge, support hybrid embodiment, and cultivate new social norms in XR. These tactics are framed as tools to both explore creative possibilities and to address practical concerns like learnability, safety, and social comfort, inviting ongoing practice-based inquiry to shape a robust, pluralistic design repertoire for social XR. For Information design, a framework in XR focusing on embodiment should recognise bodily diversity as a design asset, not a constraint. Integrate multi sensory feedback (visual, auditory, haptic, proprioceptive) to support richer engagement. Allow personalisation of embodiment parameters (avatar form, control mappings) and embed critical reflection on the socio cultural implications of bodily representation and interaction.

In the following section Physical Literacy, a perspective on movement and action originally developed within the domain of physical activity and sports education, will be presented and discussed in relation to Information design in XR.

Physical Literacy

The term physical literacy emerged primarily from sport science and physical education, with Margaret Whitehead, a scholar in physical education and phenomenology. In her seminal work (2010) the aim was to open and provide a broader perspective on movement within the domain of sports science, related educational programmes and to the general public. The aim was likewise to critically discuss the preconceived associations on the concept of physical activity as something solely related to sports and, thereby, broader its applicability. Whitehead situated physical literacy in embodied dimensions as described from phenomenology. Four motives for developing Physical Literacy were outlined: promoting embodiment in human existence; emphasizing early childhood movement (often overlooked); addressing rising sedentary lifestyles and low levels of physical activity on a global scale; and countering school physical education's focus on elite performance that excluded less athletic students. Whitehead grounded physical literacy in phenomenology, existentialism, and monism, stressing it is not merely bodily capacity. Instead, physical literacy is a holistic concept aimed at nurturing the whole person—mind and body as one (Young et al., 2020).

Although a complex concept, a definition was developed and reads as follows; the motivation, confidence, physical competence, knowledge, and understanding to value and take responsibility for engaging in physical activities for life. In its original context, it was intended as a lifelong developmental journey, much like reading or numeracy—not an innate trait, but a cultivated capability. The different parts building the definition of the concept can be described as the emotional readiness to participate, and belief in one's ability to succeed in movement tasks (Motivation and Confidence), Mastery of fundamental motor skills (e.g., locomotion, manipulation, balance) and more complex movement patterns (Physical Competence), awareness of how and why to move in particular ways, and how movement affects health, performance, and experience (Knowledge and Understanding) and finally, the orientation toward integrating movement into daily living (Engagement for Life). The concept has been criticised as originally defined, emerging from Western sport paradigms, and potentially undervaluing non-Western movement traditions, forms of play, and everyday physical practices. Physical literacy as a concept has moved into other domains, among others, rehabilitation, city planning, and education. In each, the emphasis remains on holistic physical competence: movements are not evaluated purely on mechanical efficiency but on the interplay between skill, motivation, knowledge, and confidence. Physical Literacy when used as an applied design approach proposes to treat spaces, activities, and people as an integrated system aiming at being designed as to enhance positive movement experiences that build competence, confidence, motivation, and connectedness with the place or the environment. The approach is communicated as Physical Literacy Enhanced Environmental Design. Core principles are to suggest and promote a design that provides a level of challenge for all abilities, a type of challenge for all interests, movement intersection and opportunity, affordance-rich contexts, a four-domain neutral- risk approach, and connection to people, places, objects, and groups.

Practitioners—coaches, educators, and designers—can act as architects of learning environments to enhance possibilities for building physical literacy. Drawing from ecological dynamics and nonlinear pedagogy—movement development emerges through interaction with richly designed environments. Affordances—opportunities for action—embedded in physical spaces, should be emphasised and how intentional manipulation of these elements can guide exploration, creativity, and skill acquisition. Rather than prescribing fixed routines, an informed environment should advocate for adaptable, context-sensitive designs that invite learners to engage meaningfully with their surroundings. This approach resonates beyond sport and education; designers in urban planning, architecture, and even digital interfaces can apply these principles to create spaces that encourage intuitive interaction and embodied engagement. By viewing design as a catalyst for emergent behaviour, the role of the practitioner can be reframed from instructor to environmental curator, shaping experiences that nurture agency, adaptability, and lifelong movement competence (Newcombe, Davids, and Roberts, 2021). For Information design, physical literacy can be used as a concept to add perspectives, to understand what motivates people to learn to navigate, manipulate, and inhabit any space that requires bodily engagement. If concerned with perspectives on movement, physical literacy suggests Information designers to consider:

How movement is *introduced*: Is it scaffolded for gradual competence building?

How success is *defined*: Is it purely task completion, or does it encompass movement quality, creativity, and confidence?

How motivation is *sustained*: Does the system reward exploration and playful engagement?

How physical literacy is *assessed*: Is it inclusive, multidimensional, and culturally sensitive?

In XR contexts, "the physical world" extends into simulated or blended environments, and the literacy required includes an ability to navigate, manipulate, and meaningfully interact within those spaces. Here are some suggestions for principles to practice:

- **Scaffold movement complexity:** Start with low-stakes, familiar gestures before introducing novel or demanding motion sequences.
- **Build movement confidence through feedback:** Provide subtle cues (controller vibration, visual confirmation) when motions are performed correctly.
- **Allow for exploratory "safe play":** Users should have unstructured time to test the movement vocabulary before critical tasks.

Why the emphasis on physical movement and movement skills, and why in XR? A possible answer will be presented in the following section.

Rediscovering the Moving Body in Virtual Reality

For years, the human body in digital interaction has been largely passive—either seated or functioning as a tripod for smartphones. Our everyday engagement with technology has mostly involved tapping, swiping, or flicking on two-dimensional screens, with eyes fixated and hands doing the work. The body, in essence, was sidelined. There were brief moments when movement returned—like during the exergaming craze, which had us jumping, waving, pushing, and throwing. But even then, we remained in front of flat screens. Another example was geocaching, which encouraged physical exploration but was still tethered to handheld devices.

Now, with the rise of virtual environments, the moving body is making a comeback. Interaction design is no longer confined to two dimensions—it's expanding into three-dimensional, spherical spaces. This shift opens up a broader range of movement: actions can occur behind, above, below, or beside us. The body is once again an active participant, exploring motions that were previously dormant in digital interaction. This resurgence of embodied interaction may be one reason why virtual environments are booming. In physical-world digital setups, the body was often left out. But in virtual spaces, we're rediscovering the body as a multimodal interface—capable of nuanced, collaborative, and expressive movement.

Collaboration in virtual environments also brings bodies into contact with one another. Whether synchronizing movements, avoiding collisions, or working together to grasp, lift, or build, these interactions challenge both users and technology. While the tech still has room to grow, it's evolving rapidly. Ironically, virtual environments—once seen as escapist or sedentary—are now helping to promote physical activity and physical literacy. It's almost humorous that the digital realm is what's helping us rediscover the power and potential of the moving body.

As virtual systems increasingly invite body movements—from gestural control to spatial authoring—physical literacy can be used as a lens for understanding onboarding, engagement, and skill transfer. Building on the discussion of Physical Literacy and its role in shaping meaningful XR experiences, it becomes clear that designing for embodied interaction requires more than an understanding of movement and holistic learning. To ensure that these experiences are not only engaging but also safe, efficient, and sustainable for the user, attention must also turn to Ergonomics.

The next section explores how ergonomic principles complement Physical Literacy by addressing the physical demands in XR design and supporting long-term, user-centred practice.

Ergonomics

Ergonomics is a discipline concerned with optimising the fit between people and the systems they use. It emerged in the mid-20th century from wartime efforts to improve the safety and efficiency of complex machinery operation—aircraft cockpits, control rooms, and industrial equipment. Early ergonomics focused heavily on anthropometry (measurement of human body dimensions) and biomechanics (forces and motions in the body), while other parts encompassed cognitive ergonomics (mental workload, decision-making, perception) and organisational ergonomics (teamwork, workflow, socio-technical systems). By the 1980s and 1990s, ergonomic principles were being applied to computing systems, participating in the development of standards for *usability engineering* and *user-centred design.* As interfaces moved from command lines to Graphical User Interfaces (GUIs) to touchscreens, ergonomics expanded to include Input device design (keyboard layouts, pointing devices, touch surfaces), Display ergonomics (legibility, colour use, glare reduction) and Posture and reach in workstations (seated, standing, or mixed movements in workstations). At the time measurements were aiming at finding the best postures for prolonged seating in front of desktop computers where one area was explored. Some concepts commonly used in Ergonomics are Range of Motion (ROM), the angular distance a joint can move safely in each direction. Examples are neck flexion/extension, shoulder abduction/adduction, elbow flexion, and wrist flexion/extension. A related concept to ROM is Reach envelope, the three-dimensional space a user can comfortably access without full-body repositioning; it varies by user height, arm length, and stance. Two more concepts are, Neutral posture, that is the joint position that minimises muscle activation and compressive loads; natural upright head and torso alignment with relaxed shoulders is the baseline for comfort. And finally, Muscle load and fatigue, related to sustained or repetitive muscle activation that increases fatigue and discomfort; small, sustained forces (e.g., holding controllers up) accumulate quickly. Ergonomics, partly rooted in engineering, has gained critiques for being ultimately focused on *optimising for efficiency and productivity*, missing out on aspects of creativity, personal expression, or emotional engagement. That an ergonomic approach risks treating the human as a mechanical component in a system, ignoring the social, aesthetic, and affective dimensions (Hasanain, 2024). Within the domain of ergonomics there is a critique discussing and pointing towards that several methods and theories need development both in terms of ease of use and in terms of unclear terminology to make a greater impact in research and development. A closer collaboration with computer science is proposed and adjusted to demands from practice, rapidly in need of less time-consuming protocols and methods (De Winter and Eisma, 2024). Methods in ergonomics are very linear, and new methods are needed to meet challenges in advanced contexts and with complex, dynamic technology. Additionally, selection of anthropometric data may overlook or reflect limited demographic diversity, leading to "default" designs that fit some bodies well and marginalise others (Salmon et al., 2025). Ergonomics are using XR and motion capture (MoCap) to simulate and assess methodological basis for design. In immersive environments, focus areas from an ergonomic perspective includes, minimising physical strain, optimising perceptual clarity, and designing interaction flows that align with human capabilities and limitations, as well as a focus on perspectives on reach *(*static and dynamic*),* visual fatigue, motion sickness triggers, and cognitive load. Ergonomics in XR is not static—headsets, controllers, and tracking systems impose different kinds of strain than traditional interfaces. As XR modalities for interaction using hands, head and eye-movements are depending on the function of the tracking technologies and almost any physical body movement may be the modality for interaction, perspectives related to Ergonomics should be used when designing XR and related hardware to avoid possible pitfalls that could have been foreseen (Souchet et al., 2023). The unifying goal: design systems that support human well-being and performance in balance, rather than sacrificing one for the other.

Ergonomic Challenges in XR

A systematic review found increased physical workload in VR/AR tasks, especially in neck and shoulder muscles. Discomfort was linked to, Poor interface layout, Gesture target placement, Device type (e.g., headsets), and Lack of reach envelope design. The recommendation is a design for comfortable reach zones, improvement of device wearability and proposed studies of long-term effects of repetitive motion (Kuber and Rashedi, 2023).

Assessing posture helps identify risks of musculoskeletal injury, especially in XR environments where mid-air interaction and prolonged use are common. Two examples of methods are Rapid Upper Limb Assessment (RULA) and Rapid Entire Body Assessment (REBA). RULA evaluates postures of the neck, trunk, shoulders, arms, and wrists, along with external load and muscle use. It's designed for quick screening of high-risk postures—such as repetitive movements, sustained strain, or extreme joint angles. It was developed by researchers in Occupational Health, McAtamney and Corlett (1993). REBA assesses the whole body, including legs, and accounts for dynamic and static postures, handling, lifting, twisting, and tool use. It's suited to unpredictable work settings. It was created by Hignett and McAtamney (2000). Originally developed for physical workspaces, both methods are used to evaluate movement patterns in XR. See Table 1 for scoring. Although REBA lacks recent updates and may need refinement for XR use, the method helps assess gesture frequency, duration, and strain—critical in mid-air interaction (Joshi and Deshpande (2020). RULA has been used to estimate "interaction cost" in 3D space, visualised as coloured voxels (red = high effort, blue = low). This helps designers place UI elements to reduce strain (Evangelista Belo et al. (2021). It has also been used to explore "elbow-anchored" interaction zones to reduce arm fatigue, showing how posture-aware design supports comfort, especially in relaxed positions like sitting or reclining (Veras et al., 2021).

Table 1 Levels of scoring in RULA and REBA.

RULA Scoring (1–7)	*REBA Scoring (1–15)*
1–2: Acceptable	1: No risk
3–4: Low risk; monitor	2–3: Minor adjustments
5–6: Investigation needed	4–7: Likely changes needed
7: High risk; immediate action	8–10: Urgent intervention

In XR, ergonomic best practice can clash with gameplay or narrative goals. The most comfortable posture isn't always the most immersive. Designers must balance health with engagement—allowing freedom of movement while guiding users toward safer interaction patterns. With the technological development and advancement making information and interaction into XR contexts, posture and environment are dynamic. Ergonomic recommendations can collide with gameplay, narrative, or exploratory goals—the most comfortable posture is not always the most engaging one.

The following section will share how perspectives and awareness related to Ergonomics can support design in XR in its early phases. It may also be used during evaluation.

Ergonomic Design Guidelines for XR

Here follows a selection of areas to keep at focus from the perspective of physical ergonomics when designing for XR. A condensed version of the suggested guidelines is presented in Table 2. Describe exactly what users must do—the Primary tasks (e.g., select floating menus, grab/throw, annotate). Identify high-frequency micro-actions like pointing, holding gaze, or wrist rotations. Note session length, standing versus seated, controller vs. hand tracking, and expected repetition. Provide alternative UI layouts for seated and standing users; seated layouts should lower vertical targets and shorten reach distances. Add micro-pauses or alternate tasks to break repetitive movements; temporal effects of repetition on fatigue and stability deserve attention. Keep focus on Contexts and constraints. Minimize sustained neck flexion, reduce shoulder elevation time, keep targets in comfortable reach—set Ergonomic goals. Avoid extreme neck angles, limit required head tilt to about ±20° for frequent tasks and allow occasional greater tilt only for brief interactions. Move frequently used targets into neutral gaze and primary reach; avoid prolonged overhead or far-lateral placement. Keep primary interactive elements within a comfortable reach zone: place most controls within mid-reach (approximately shoulder height ±15° vertically, within 60–75% of arm length horizontally) to minimise shoulder and torso strain. Reduce precision pointing at arm's length; prefer near-body gestures or short bursts with rest states. The general focus in this part is Target placement. Visualise how headset mass shifts neck moments; position heavy virtual tasks lower or shorter to reduce cumulative neck torque. Use counterweights or strap adjustments to improve headset balance; improving wearability is a recommended strategy to reduce upper-body load. In short, consider device weight and balance. Combine DHM with VR prototypes to catch posture/clearance/strength issues before user testing, then validate with RULA/REBA and user ratings. Then invite users and observe posture, count repetitions, ask about fatigue, and if possible, log controller elevation and session duration to identify risky patterns. Apply these rules early in wireframes and prototypes to make VR interactions comfortable, inclusive, and sustainable from a physical ergonomic perspective. A reminder—make use of simulation early, people later.

Embodiment, physical literacy and ergonomics have been presented, combining theoretical and, to some extent, philosophical perspectives with practical, hands-on, and biomechanical considerations within ergonomics. At the outset of a design project that aims to provide opportunities to experiment and build experience from a sensorimotor perspective, methods drawn from the domain of kinaesthetic learning can be particularly valuable. The following section presents this.

Table 2 Ergonomic Design guidelines for XR.

Focus Area	*Condensed Description*
Primary tasks	Define core user actions (e.g., menu selection, grabbing, annotating). Highlight frequent micro-movements such as pointing, gaze holding, or wrist rotation.
Contexts and constraints	Note session duration, posture (seated/standing), and input type (controller/hand-tracking). Adapt layouts for seated users, shorten reach distances, and add pauses or varied tasks to reduce fatigue.
Ergonomic goals	Limit neck flexion and shoulder elevation; keep targets within comfortable reach. Allow only brief extreme angles (±20° tilt).
Target placement	Position frequently used elements in neutral gaze and mid-reach zones (shoulder height ±15°, 60–75% arm length). Avoid overhead or far-lateral targets; prefer near-body gestures over extended pointing.

Contd...

Table 2 *Contd...*

Focus Area	*Condensed Description*
Device weight & balance	Account for headset mass and neck load. Place heavy tasks lower, use counterweights or strap adjustments to improve balance and reduce strain.
Simulation before testing	Use digital human models and VR prototypes to identify posture/strength issues early. Validate with ergonomic tools (RULA/REBA) and user feedback. Observe repetitions, fatigue, and controller elevation during testing to refine designs.

Kinaesthetic Learning

Why make a move when learning?

Kinaesthetic learning refers to acquiring knowledge and skills through bodily movement, tactile engagement, and proprioceptive feedback. It is often described as "learning by doing"—but in its richer sense, it is about the integration of sensory and motor systems in the construction of understanding. Kinaesthetic approaches leverage muscle memory and sensorimotor coupling. The concept draws from theoretical traditions, among other the Experiential Learning Theory (Kolb, Boyatzis, and Mainemelis, 2014), that emphasises concrete experience as a foundation for reflection, conceptualisation, and experimentation. Kolb builds on Dewey and his work on reflective thought and action (Miettinen, 2000). Kinaesthetic learning stands on a constructivist perspective related to learning that positions learners as active participants who build knowledge through interaction with their environment (Richardson, 2022). Human perception of space is always embodied and multisensory, challenging the tendency in design thinking to prioritise vision above other senses. Drawing on neurological and phenomenological research, researchers Kwon and Iedema (2022) show how two processes—kinaesthetic resonance (learning and reasoning through movement) and synesthetic translation (linking information across different senses)—can expand spatial design beyond traditional, visually driven principles. To put this into practice, three models for design thinking are proposed: Synesthetic Translation, Kinaesthetic Resonance, and Kinaesthetic Engagement. These models demonstrate how multisensory strategies can help designers both imagine and evaluate built environments. Integrating kinaesthetic and cross-sensory perspectives can create designs that better support embodied navigation, emotional response, and a sense of coherence, and methods that strengthen designers' bodily awareness and multisensory literacy, bridging abstract ideas with lived spatial experience can show the way (Kwon and Iedema, 2022).

In immersive design, kinaesthetic approaches are especially valuable for exploring spatial reasoning, prototyping through movement, and collaborative choreography—skills that are highly relevant in virtual environments where doing is just as important as seeing. XR can amplify bodily skills, redistribute opportunities for action, and create new contexts for learning through movement. Movement is not merely physical exercise; it is a mode of knowing that shapes spatial reasoning, motor control, memory, and social coordination. Kinaesthetic learning benefits when environments are designed to invite meaningful effort, varied affordances, and clear feedback. XR can expand the action space by simulating terrains, constraints, and interactions that are impractical, dangerous, or impossible in everyday settings, providing learners with novel ways to explore balance, timing, reach, and force while preserving transferable elements of physical strain and proprioception. While movement-based approaches are powerful for many situations, not all individuals find physical enactment the most effective route to "grasp" the situation at hand. On an Individual level the preferable method for learning may be mainly verbal, or purely visual modes. In immersive technologies, alternative pathways are to be carefully built in. Skills learned kinaesthetically in a mediated environment may not transfer directly to unmediated contexts, especially if virtual physics differ from real-world dynamics. Movement vocabularies are culturally situated; kinaesthetic designs should respect and incorporate diverse traditions. In Information design, hands-on thinking and tinkering activities such as—sketching, modelling, prototyping are common methods. Kinaesthetic learning in XR is a way of integrating movement into the very fabric of design thinking. In XR prototyping, walking around a life-sized virtual object yields insights about proportion, balance, and affordance. Using full-body motion capture or VR controllers to let learners enact processes (e.g., assembling components, performing medical procedures) rather than merely observing. Learners physically enact processes or concepts, often with motion capture or haptic augmentation. Many design challenges are solved in the action and reflection of physical collaboration—gestures, positioning, shared manipulation of artefacts. Offer multiple movement modalities (e.g., standing, seated, controller-based, gesture-based) for each learning task.

XR has a distinctive capacity to level uneven starting points in physical ability by creating a shared environment where different bodies can meet on equal footing. Not only emphasising raw strength, speed, or reach, XR can mediate and translate physical input into interactions that value intention, timing, strategy, and coordination. Movement becomes a communicative and creative modality accessible through adjustable mappings between bodily effort and virtual effect, enabling people with limited mobility, chronic conditions or different skill sets to participate in competing games and fully in collaborative tasks.

Design choices are central: developers can scale spatial demands, alter movement-to-action mappings, and introduce assistive affordances that preserve meaningful agency without privileging a single body type. For example, virtual locomotion can be offered as a continuum of input methods—subtle weight shifts, arm gestures, eye gaze, or controller input—so that the same collaborative task is reachable by those who walk, use wheelchairs, rely on prosthetics or have limited range of motion. Haptic cues, audio guidance and adaptive pacing can provide equivalent perceptual feedback across modalities so that timing and rhythm, key elements in many cooperative activities, remain salient for everyone.

Beyond technical accessibility, XR can reframe competition and cooperation by emphasising complementary roles. Games and collaborative applications can reward different kinds of movement—precision rather than power, pattern recognition rather than endurance—so teams benefit from diverse bodily strengths. Avatars and environment constraints can be tuned to normalise differences: a player's reach can be virtually extended, speed moderated, or physical tasks abstracted into decision-based challenges that maintain the same social dynamics without excluding participants. Such adjustments ensure that success depends on shared strategy and communication rather than who happens to possess superior physical capacity.

Social presence in XR also supports equitable engagement. Real-time positional awareness, gesture mirroring and proxemic cues allow collaborators to sense each other's intent and adapt their actions. When virtual spaces encourage turn-taking, synchronous movement, and co-located tasks, they cultivate mutual attunement and empathetic coordination. These social affordances make collaboration more than parallel play; they convert embodied differences into resources for novel forms of teamwork. Careful evaluation and co-design with diverse users are essential. Inclusive metrics should measure not only task performance but perceived agency, enjoyment, and social connection. Iterative testing with participants across a spectrum of physical abilities uncovers which mappings and mechanics genuinely equalise experiences and which inadvertently reproduce real-world hierarchies.

When XR systems are deliberately crafted to translate, amplify, and redistribute movement, they create an arena where varied bodies can collaborate, compete, and engage on equitable terms, turning physical difference into collective possibility.

Synthesizing Embodied Perspectives

Translating Embodiment, Physical Literacy, Ergonomics, and Kinaesthetic Learning into XR design and practice, is about curating a toolkit that reflects the needs of your target group or participants, the constraints of the medium, and the intended experience outcomes. Combining ergonomics data (e.g., joint angle analysis) with kinaesthetic learner feedback and physical literacy progression scores. Using embodiment-driven prototyping but with adjustable movement ranges and multimodal feedback to accommodate different mobility profiles, and iterating between embodied testing, ergonomic adjustment, and kinaesthetic learning assessment in short design sprints.

Immersive design in XR seeks to create experiences that feel coherent across the body–environment loop, where users' movements, perceptions, and decisions unfold in ways that support interaction, learning, and well-being. For information designers, this means translating principles of embodiment, ergonomics, physical literacy, and kinaesthetic learning into practical design strategies.

In XR, users are not passive observers but active participants whose bodily actions shape their experience. Design choices around movement mapping, feedback, and representation directly influence whether users feel present and in control. Predictable sensorimotor responses build trust and enable intuitive interaction. Multimodal feedback—combining visual, auditory, and haptic cues—reinforces embodied engagement. Inclusive representation matters: avatars, control schemes, and interaction zones should accommodate diverse bodies and abilities. Simple calibration options during onboarding can dramatically improve comfort and effectiveness.

Ergonomics ensures that virtual environments support the body rather than constrain it. Designers must consider posture, reach, and movement pacing to avoid strain and fatigue. Frequently used controls should be placed within neutral reach zones, while less common interactions can be positioned farther away. Information should be layered to match movement phases, reducing overload and error. Ergonomic design also includes managing session length and encouraging posture variation. Onboarding plays a key role in preventing early mismatches—headset fit, interpupillary distance, and control sensitivity all affect long-term comfort. Long-term engagement depends on physical and perceptual wellbeing. Designers must monitor movement intensity and duration to prevent strain. Systems that prompt breaks or adapt activity levels help manage cumulative load. Visual–vestibular alignment is critical—mismatches can cause motion sickness and disrupt immersion. Locomotion techniques should prioritize user control and minimize unnecessary camera motion. Sequencing and pacing are essential: dynamic object positioning and micro-break prompts reduce fatigue while maintaining productivity.

Movement is a powerful medium for understanding. In XR, users learn by doing—manipulating, assembling, and navigating objects to build deeper knowledge. This is especially effective for spatial reasoning, procedural skills, and embodied storytelling. Repetition should be varied to maintain engagement and support transfer across contexts. Collaborative motion adds complexity and richness, fostering social coordination and shared affordances. Feedback should focus on movement quality—smoothness, timing, and efficiency—rather than just task completion. Real-time cues help users refine technique and accelerate skill acquisition.

Physical literacy treats movement as a skill that supports autonomy and adaptability. Designers should scaffold motor learning through progressive challenges, allowing users to build confidence and competence. Transferability is key: practice scenarios should vary in context and constraints to develop flexible movement strategies. Exploratory sandboxes offer safe spaces for experimentation, while reflexive tools like replays or heatmaps support self-monitoring and deliberate practice.

In XR information design, embodiment, ergonomics, and kinaesthetic learning are not checklists—they are guiding principles. Prioritise predictable movement mappings, inclusive calibration, and layered feedback. Design learning paths that reward movement quality and manage physical load. Test with diverse bodies and iterate based on both short-term usability and long-term wellbeing. When the body is treated as a design partner, XR becomes a space for skill development, agency, and deeper understanding.

Information Design in the Future of XR

Immersive design will likely become more physically nuanced as motion tracking, haptics, and MR overlays become richer. For Information design the challenge is to: Integrate critical awareness into every design brief. Prototype inclusively from the start, rather than "retrofitting" accessibility. Test not just for usability, but for bodily wellbeing and movement joy. Advocate for ethical standards in data handling, consent, and representation.

The question is whether we will use that capability to simply reproduce existing interaction paradigms or to invent genuinely body-centred design languages that foster creativity, inclusivity, and support for a variety of human movements. Can we discuss what natural means in these designed settings?

There are arguments that the long-standing natural interaction paradigm—designing interfaces that mimic familiar human actions like gestures, voice, and gaze—no longer suffices for many emerging technologies and contexts (Vatavu, 2024). A complementary "non-natural" design perspective is proposed that intentionally departs from familiarity to exploit novel affordances, support new forms of expression, and address constraints of AR, embodied devices, wearables, and AI-embedded environments. Non-natural designs can create expressive, efficient, or safer interactions when natural mappings are impossible, ambiguous, or limiting; they also enable creative experiences that reshape user expectations and behaviours. Non-natural design is framed not as a rejection of natural interaction but as a deliberate complement that expands the designer's toolkit (Vatavu, 2024).

What is natural and what is designed, and how can we engage and interact within those settings? As computer technology has evolved from room-sized machines to desktops, laptops, handheld devices, body-worn computers, and XR, interaction has shifted from 2D screen-based interfaces to 3D, room-scale, spatial experiences. Both spatial and bodily perspectives have, therefore, received greater attention. Concepts such as spatial computing have expanded, and maybe Mark Weiser, the father of ubiquitous computing, would say that we have made it toward calm technology, where the computer itself disappears.

Conclusion

When we take off the headsets, let go of the controllers, and the shimmering polygons at the screen, what remains is the human body—breathing, sensing, moving. Every immersive interaction, whether in VR, MR, or AR, is mediated through that living substrate. The theories we have explored—embodiment, physical literacy, and methods from Ergonomics, and kinaesthetic learning—reminds us that design for these environments is designed for lived experience. The opportunities lie in deliberately blending these perspectives.

To see the body not just as a user interface, but as a partner in the making of the design. To scaffold movement and not always remove and un load guided by efficiency, but for joy and growth. To create ergonomics that care for the body without anaesthetising its capacity for challenge. To bring kinaesthetic richness into design in ways that include, rather than exclude.

The critiques threaded through some sections are not weaknesses to be avoided; they are invitations to question, adapt, and expand.

The environments you shape will influence how people inhabit their own bodies, how they relate to others, and how they learn, work, and play. Design work in XR moves Information design from flat screens to lived space. Combining spatial thinking, multimodal craft, ethical sensitivity, and practical prototyping will be at the core when designing the next generation of immersive experiences.

Design as if the body matters—because it always does. The tissues and genetic material in the human body look almost the same as 10,000 years ago. Most of the human bodily systems (skeletal, muscle, metabolic, and cardiovascular) do not function in a sustainable way unless they are stimulated by possible individual levels of movement. Let each prototype, test, and iteration be an act of respect for the human form in all its diversity. Let your designs invite presence, foster competence, safeguard well-being, and spark the sheer delight of moving through worlds—both real and virtual.

Learn to move–move to learn!

References

Banakou, D., Kishore, S. and Slater, M. 2022. Virtual embodiment of white people in a Black virtual body leads to a sustained reduction in their implicit racial bias. *ACM Transactions on Applied Perception (TAP)*, *19*(1): 1–15.

Chilufya, E.M. and Arvola, M. 2021, November. Conceptual designing of a virtual receptionist: Remote desktop walkthrough and bodystorming in vr. In: *Proceedings of the 9th International Conference on Human-Agent Interaction* (pp. 112–120).

De Winter, J.C.F. and Eisma, Y.B. (2024). Ergonomics and Human factors: Fade of a discipline. *Ergonomics*, 1–9.

Della Longa, L., Valori, I., and Farroni, T. (2022). Interpersonal affective touch in a virtual world: Feeling the social presence of others to overcome loneliness. *Frontiers in Psychology*, *12*: 795283.

Dourish, P. (2001). *Where the Action Is: The Foundations of Embodied Interaction*. MIT Press.

Evangelista Belo, J.M., Feit, A.M., Feuchtner, T. and Grønbæk, K. 2021, May. XRgonomics: Facilitating the creation of ergonomic 3D interfaces. In: *Proceedings of the 2021 CHI Conference on Human Factors in Computing Systems* (pp. 1–11).

Gieseking, J.J., Mangold, W., Katz, C., Low, S. and Saegert, S. (Eds.). 2014. *The People, Place, and Space Reader* (1st Edn.). Routledge. https://doi.org/10.4324/9781315816852.

Guy, M., Normand, J. M., Jeunet-Kelway, C. and Moreau, G. 2023. The sense of embodiment in virtual reality and its assessment methods. *Frontiers in Virtual Reality*, *4*, 1141683.

Hasanain, B. 2024. The role of ergonomic and human factors in sustainable manufacturing: A review. *Machines*, *12* (3): *159*.

Hignett, S., and McAtamney, L. 2000. Rapid entire body assessment (REBA). *Applied Ergonomics*, *31*(2): 201–205.

Hillmann, C. 2021. The History and Future of XR. In: *UX for XR. Design Thinking*. Apress, Berkeley, CA. https://doi.org/10.1007/978-1-4842-7020-2_2.

Höök, K. 2018. *Designing with the Body: Somaesthetic Interaction Design*. MIT Press.

Huygelier, H., Mattheus, E., Vanden Abeele, V., Van Ee, R. and Gillebert, C.R. 2021. The use of the term virtual reality in post-stroke rehabilitation: a scoping review and commentary. *Psychologica Belgica*, *61*(1): 145.

Joshi, M. and Deshpande, V. 2020. Investigative study and sensitivity analysis of Rapid Entire Body Assessment (REBA). *International Journal of Industrial Ergonomics*, *79*, 103004.

Kersting, M., Haglund, J. and Steier, R. 2021. A growing body of knowledge: On four different senses of embodiment in science education. *Science and Education*, *30*(5): 1183–1210.

Kolb, D.A., Boyatzis, R.E., and Mainemelis, C. 2014. Experiential learning theory: Previous research and new directions. In: *Perspectives on Thinking, Learning, and Cognitive Styles* (pp. 227–247). Routledge.

Kuber, P. M., and Rashedi, E. 2023. Alterations in physical demands during virtual/augmented reality-based tasks: A systematic review. *Annals of Biomedical Engineering*, *51*(9): 1910–1932.

Kwon, J., and Iedema, A. 2022. Body and the senses in spatial experience: The implications of kinaesthetic and synesthetic perceptions for design thinking. *Frontiers in Psychology*, *13*, 864009.

Masoudi, N., Fadel, G.M., Pagano, C.C. and Elena, M.V. 2019. A Review of Affordances and Affordance-Based Design to Address Usability. in: *Proceedings of the 22nd International Conference on Engineering Design* (*ICED19*), Delft, The Netherlands, 5–8 August 2019. doi:10.1017/dsi.2019.141.

Mattes, D. and Lang, C. 2021. Embodied belonging: In/exclusion, health care, and well-being in a world in motion. *Culture, Medicine, and Psychiatry*, *45*(1): 2–21.

McAtamney, L. and Corlett, E. N. 1993. RULA: A survey method for the investigation of work-related upper limb disorders. *Applied ergonomics*, *24*(2): 91–99.

McVeigh-Schultz, J., Márquez Segura, E., and Isbister, K. 2024, October. Designing our Weird Social XR Future: Tactics to support hybrid ways of embodied knowing. In: *Proceedings of the Halfway to the Future Symposium* (pp. 1–11).

Merleau-Ponty, M. and Landes, D. A. 2014. *Phenomenology of Perception*. Routledge.

Miettinen, R. 2000. The concept of experiential learning and John Dewey's theory of reflective thought and action. *International Journal of Lifelong Education*, *19*(1): 54–72.

Newcombe, D., Davids, K. and Roberts, W. 2021. Practitioners as Architects of the Environment: How We Can Use Environmental Design Principles to Support Physical Literacy. In: *Nonlinear Pedagogy and the Athletic Skills Model* (pp. 91–101). Routledge.

Palmquist, A., Olsson, E., Soellaart, A., Johansen, N. and Goethe, O. 2025, May. Embodied Without Matter: An Analysis of Embodiment Definitions in VR Game Design Research. In: *International Conference on Human-Computer Interaction* (pp. 220–236). Cham: Springer Nature Switzerland.

Richardson, V. 2022. Constructivist Pedagogy. *Teachers College Record*, *105*(9): 1623–1640. https://doi.org/10.1177/016146810310500906 (Original work published 2003).

Salmon, P.M., Burns, C., Broadbent, S., Chari, S., Clay-Williams, R., Hancock, P.A., ... and Young, M.S. 2025. The Chartered Institute of Ergonomics and Human Factors at 75: Perspectives on contemporary challenges and future directions for Ergonomics and Human Factors. *Ergonomics*, *68*(6): 759–775.

Slater, M., and Sanchez-Vives, M.V. 2016. Enhancing our lives with immersive virtual reality. *Frontiers in Robotics and AI, 3*: 74.

Souchet, A.D., Lourdeaux, D., Pagani, A. and Rebenitsch, L. 2023. A narrative review of immersive virtual reality's ergonomics and risks at the workplace: Cybersickness, visual fatigue, muscular fatigue, acute stress, and mental overload. *Virtual Reality*, *27*(1): 19–50.

Vatavu, R.D. 2024, June. Embracing Non-Natural Design in Human-Computer Interaction. In: *Proceedings of the 2024 International Conference on Advanced Visual Interfaces* (pp. 1).

Veras, R., Singh, G., Farhadi-Niaki, F., Udhani, R., Patekar, P.P., Zhou, W., ... & Li, W. 2021, May. Elbow-anchored interaction: Designing restful mid-air input. In: *Proceedings of the 2021 CHI Conference on Human Factors in Computing Systems* (pp. 1–15).

Vohra, M. (Ed.). 2025. *Introduction to Extended Reality (XR) Technologies*. John Wiley & Sons.

Wang, S., Sanches de Oliveira, G., Djebbara, Z. and Gramann, K. 2022. The embodiment of architectural experience: A methodological perspective on neuro-architecture. *Frontiers in Human Neuroscience*, *16*, 833528.

Whitehead, M. (Ed.). 2010. *Physical Literacy: Throughout the Lifecourse*. Routledge.

Young, L., O'Connor, J. and Alfrey, L. 2020. Physical literacy: A concept analysis. *Sport, Education, and Society*, *25*(8): 946–959.

14 Information and Human-centred Design

Marie Sjölinder[1*] and Yvonne Eriksson[2]

Introduction

The chapter will discuss users and the perspective of different user groups and their relations, and different roles they have in different contexts and how situations affect users as aspects of complex systems. The chapter wants to broaden the concept of Information Design and discusses communication between various groups (actors) as designed information sharing. Context could also include a larger system that involves individuals, groups, equipment of various kinds, and different organisations. Even though we deal with small units they are parts of a larger system that could be of various levels of complexity. In a user situation, the involved individuals belong to different contexts. The focus of the chapter is healthcare systems in which various units and aspects are discussed. From a human-centred design perspective, methods used in this process, and how it is communicated and the relations between users in different health care settings are discussed. The chapter also deals with aspects such as context, situation, and human interaction, and perception of the context and the environment. The need of Human Centred Design (HCD) in the development of future health care systems are elaborated upon in terms of integration of technology, new ways of working, and knowledge in conducting this transformation in a way that is based on the HCD perspective.

Central for information design is human-centred design. The advantage of HCD is stated in ISO 9241-210:2019 Ergonomics of human-system interaction -- Part 210: Human-centred design for interactive systems. It is explained as followed:

> Using a human-centred approach to design and development has substantial economic and social benefits for users, employers and suppliers. Highly usable systems and products tend to be more successful both technically and commercially. /.../ The complete benefit of human-centred design

[1] RISE, Swedish Research Institute, Department of Health Science, Innovation and Design, Mälardalen University, Isafjordsgatan 22, Box 1263, 164 29 Kista.

[2] Professor in Information Design, Department of Health Science, Innovation and Design, Mälardalen University, Box 325, 631 05 Eskilstuna, Sweden.
Email: yvonne.eriksson@mdu.se[2]

* Corresponding author: marie.sjöinder@ri.se

DOI: 10.1201/9781032642598-14

> can be determined by taking into account the total life cycle cost of product, system or service, including conception, design, implementation, support, use, maintenance, and, finally, disposition. /…/ Using appropriate human-centred design methods could reduce the risk that product failing to meet stakeholder requirement or being rejected by its user (p. 11)

HCD is not a well-defined single method or set of principles that can be easily applied; it is rather a group of methods combined with a certain mind-set where humans are placed in focus. Another way of defining "Human-centred design (HCD) is a group of methods and principles that are aimed at designing useful, usable, pleasurable, and meaningful information, products, or services for people. The main principle of these methods is that they describe how to use insights about human beings—users, customers, or other stakeholders—to design products or services that meet their needs and aspirations" (2016, (van der Bijl-Brouwer, 2016).

HCD starts from the perspective of the individual needs and places these needs in a holistic context regarding information, products, and services. That means, for example, a design of home care for people with special needs must be regarded from a socio cultural aspect as well as from cognitive and behavioural perspectives, both in relation to care givers and care takers and their relatives and friends. According to Krippendorff (2004: 48), "Human-centredness takes seriously the premise that human understanding and behaviour goes hand-in-glove; that what artifacts are is inseparably linked to how their users perceive them, can imagine interfacing with them, use them and talk about their stake in them with others. Human-centred design is concerned less with assuring that artifacts work as intended (by their producers, designers, or other cultural authorities) than with enabling many individual or cultural conceptions to unfold into uninterrupted interfaces with technology."

The context of the public sector and the field of health care are very different from other design contexts (van der Bijl-Brouwer, 2016). When designing software systems or technology to be used within health care several aspects must be considered beside the actual functionality of the technology. The entire context must be considered, and the design is an iterative process rather than solving one stand-alone problem. The solutions must work within a complex organisational structure with its own eco system consisting of a variety of users with different roles and needs. The different stakeholders might view the need for the solution from different perspectives and there is a risk that the needs of stronger user groups will become more emphasised. This makes a human-centred approach even more important within the health context that consists of users with very different powers and the possibility to gain control over the situation. Here, the process of HCD can contribute to the understanding of different users and user perspectives. It can contribute to sorting out who is the main user, and which other user groups' needs to be considered. HCD can also in this context contribute to usage of the technology and make sure that the main user group(s) have access to the technology and that they can use it in an easy way. Because of this and in the long run this could also contribute to sustain trust in complex health systems (Adam et al., 2020).

Human Centred Design: Different Approaches

This section will give a short history of HCD and how it has been defined over the years. HCD has been developed over the last decades and has been a part of participatory design, user-centred design, and co-design. The focus on user and involvement of users is crucial for HCD and it has had different approaches over the years. Various kinds of user involvement and the focus on humans have been specified in different ways.

HCD and User centred design (UCD) is often used synonymously, but there are differences between the concepts. Don Norman has explained that HCD is a practice where designers focus on four key aspects: focus on people and their context; trying to understand and solve the right problems; taking a complex system with interconnected parts into consideration; doing small interventions (https://www.interaction-design.org/literature/topics/human-centered-design). The differences between HCD and UCD is mainly that you recognise the user as a human being, that is as a person.

The term HCD has its origins from around 1960 and its roots in fields such as engineering, computer science, and ergonomics. During the 1960s it consisted of a set of design techniques supporting creativity. Later, it became referred to as "design thinking". There is an ISO standard that covers aspects related to HCD, ISO 9241-210. It recommends six characteristics: 1) The adoption of multidisciplinary skills and perspectives, 2) Explicit understanding of users, tasks, and environments, 3) User-centred evaluation driven/refined design, 4) Consideration of the whole user experience, 5) Involvement of users throughout design and development, 6) Iterative process. (Giacomin, 2014). This ISO document provides recommendations based on HCD principles and they are related to the development of computer-based interactive systems, and they cover both hardware and software components.

Another way to take users' perspective into consideration is participatory design. Participatory design has its roots in the Scandinavian countries in the 1960s and 1970s. It is described as an approach that aims to involve all stakeholders in the design process and make sure that the outcome meets the needs and that it is usable (Greenbaum and Kyng, 1991; Schuler and Namioka, 1993). Participatory design is a process that can be applied in many settings. It could be used for involving users in product development, but it could also have a political dimension and be applied to involve citizens in different societal tasks (Bannon and Ehn, 2016). In 1971, Victor Papanek's book, *Design for a Real World* was published in which he stated that design was an object or system that works as a political tool and that it affects people's life and living conditions.

The term UCD appeared for the first time in the 1980s. It is an approach that focuses on the users in the planning, design, and development of a product.

Within this framework, components are added in terms of increasing the emphasis on design aspects related to the user. It places a large focus on early and continuous contact with possible user groups, and on setting up usability criteria and conduct evaluations (Keinonen, 2010).

In the beginning of the 1990s, the term co-creation was started to be used in terms of creating value by placing the customer in a role where he or she is a part of the design process and contributes with his/her ideas and solutions. Shifting from UCD to co-creation has large implications on methods and design practices since it demands a focus on collective creativity (Sanders and Stappers, 2008). New trends in society, digitalisation and demands from the public and the patients has also led to new ways of delivering care. Both new health care services and the way they are designed could be considered as a part of a co-creation process where the patients are about to increase their empowerment.

The role of design has changed over time and today organisations such as the Swedish foundation of Industrial design see their mission to contribute to societal challenges. They want to contribute to the solution of major challenges such as social, environmental, and economic sustainability, a globalised market, rapid technological development, and an ageing population that require new approaches and solutions.

The journey of HCD during the decades can be described as going from function-focused to human-focused. The aim has shifted from developing products and services that fulfil basic human needs towards supporting humans in social contexts and in fulfilling psychological needs and self-actualisation (Zhang and Dong H., 2008). From the perspective of developing systems within health care, this will have implications on how the patients (and other stakeholders) must be perceived by health care organizations. It is no longer an issue of just solving a problem or providing medical care—these newer HCD approaches also imply that it must be done in a way where social needs and psychological aspects among the different users and stakeholders are taken into consideration.

The motivation of HCD and Information Design in the health care setting is not only because of the focus on the patients' needs, it also has to do with health care systems power of their working situation. By that it means to have a professional influence on the system with focus on health care. Presently, there is a tendency that the health care systems are governed by politicians with no user involvement, either personnel or patients.

Methods within HCD

In a design process, it is common to talk about it as a journey since it is a kind of move from one state to another. Definition of the term journey is an act of travelling from one place to another. When a journey starts, it is often to a specific destination that is chosen beforehand for a particular purpose. The kind of vehicle that is used is commonly based on the distance and what is the most suitable for the time available for the trip, etc. The journey could be straight between A and

B, but it could contain several stops and potholes along the way. Especially if the journey is the purpose itself.

A journey is a common metaphor for life as such, as it reminds us that the destination is not our only goal. Like with any form of a journey, there are times when the roads are straight and times when they are winding. There are ups and downs and potholes along the way. A design journey is about a future stage. Therefore, one needs to describe and define the current stage. It is also necessary to be aware that there often exist different perspectives and different understanding of the existing/present. The different perspective could be explained by people's drivers' experiences, different positions, and roles in the design process.

To gain a common understanding of current conditions/situations, models are often used. Models are simplifications of reality—a way for us to make sense of what we see, hear, and read. They are used to help us understand and gain knowledge about a complex surrounding. The difficulty in getting an understanding of the surrounding depends on its complexity and jumble of processes on issues such as various times and length scales. In design, models are often used that are fundamental in order to isolate, simplify, and discuss a phenomenon. They facilitate practical management of some questions. Models could also be an idealised and simplified reality that ignores complexity, and has a holistic perspective. But they can lead us into blind alleys because of preconceptions built on norms that we are unaware of.

The aim of design is to change something that already exists, develop something new, fulfil users' needs, etc. The vision of design is to improve a current product, information artefact, or service system or environment. To be able to evaluate the result of the design, it is necessary to understand how the start looks, that is, the current state. If you don't know how it looks or works today, it is not possible to evaluate the result of your design. One needs to be aware of your one's role in the analyses of the current state.

- What does it mean to start a design journey?
- Where to start and where to go?
- How to be sure that it has been a movement from one condition or situation to a desired one?

The design journey includes several methods such as traditional ethnographic methods like interviews and observations. Explicit formulated design methods often start with ethnographical methods. The intended target group for a design artefact, an information campaign, or design of a workplace systems are interviewed, either in structured or semi-structured interviews. Semi-structured interviews are flexible and allow you to ask follow-up questions which could be useful for the result. While structured interviews make it possible to compare answers more rigorously. Observations from an ethnographic perspective are situated in the context where an artefact or a system is used. It requires that the artefact or system already exists, and that the intention is to make a change. Interviews and observations of the existing needs then to be carefully analysed with focus on what is the challenge/problem since it will work as a description of the current state.

Example of Design Methods

Design methods consist of different approaches and are often iterative. It could start with formulation of the challenge/problem or with a vision of a goal. By starting with a definition problem is often more demanding than one expects. The iterative process often looks like the following: it starts in a context where the purpose of the design project is defined that includes diagnosing the current state or situation. Followed by planning of the project, then taking action, and evaluating the result. The result could be a mock-up or a developed prototype. From that a new diagnosing taking place, and again followed by planning for a developed mock-up, prototype, or suggestion for a system. That needs to be evaluated. This process could take up to 3–5 rounds. For example, in a recent project that took place during the pandemic COVID-19, an HCD perspective was used to design video instructions for how to put on and take off personal protection equipment (PPE) in home care settings of old people (Eriksson et al., 2023). In contrast to existing videos made in a hospital context, with no relevance for home care, this video was created in an entrance of an apartment. The project involved personnel such as nurses responsible for hygiene routines, care givers, representatives from the unions and responsible management in the region. Using an iterative process with simple video prototypes, the project managed to make incremental improvements of the instructions. By involving personnel, people that should use the PPE the project was able to create instruction videos that were meaningful for the users (Eriksson et al., 2023). The motivation of taking an HCD perspective is to create a usable system for all involved. In the project 7 (seven) prototype videos were produced and shown for the personnel until we got all the details correct from users' perspective. The video prototypes made it possible for the personnel to recognize and identify new details that were missing in the instruction. The closer a final version came, the finer details were detected by the user. The user's engagement in the process became storing since we implemented the comments from the user in every new version of the prototypes. The final version of the instruction video was evaluated by 42 users. This resulted in additional valuable comments, from which we learn for the future (Eriksson et al., 2023).

A design journey is not necessarily what is expected. For example, in a project where a tool for social interaction between older adults and their friends and relatives were developed. As part of the design process the tool was tested in some older adults' apartments. After a while it turned out that the devices were almost not used at all. When this was investigated further it became obvious that the participating older adults either already had a computer or that they had too small social networks for using the device for communication with people outside the nursing home. However, the older adults did appreciate the functionality of being able to receive and share photos. Based on this, the care personnel tried to engage friends and relatives by asking them to be more active in sending photos and short messages. The usage increased to some extent, but still the device was not used on a frequent basis. Based on an idea from the care personnel, instead the devices were placed in the dining areas of the nursing home. When this was done, the residents started to use the devices as something to gather around and talk about friends and family members. It turned out that the device and the context around it provided a meaningful tool for social interaction between the residents, and not as expected, interaction with people outside the nursing home (Sjölinder and Scandurra, 2015; Sjölinder, 2022).

It is important to plan the design journey and to not get lost on the way. However, a design journey needs to involve individuals, often described as users. A simple definition of a user is someone who uses a product, machine, or services. It is a human being that is a part of the social, political, and economic system. This part will problematise the concept of the user from different perspectives and its relation to primary and secondary users of the product or services and how it is related/as a part of a system.

A user is a human being with specific living conditions such as socio economic background, education, and social identity. From a health care perspective, the humans, the user of the health care system, is defined as a patient. This anonymous definition "un-dress" the social identity of a person, both from patient perspective and from healthcare perspective. From the perspective of Erwin Goffman (1956), it is possible to talk about the role people perform as a patient. Goffman discussed that we perform in different ways depending on the situation, how we want to be perceived by others (Goffman, 1956). Of course, different people play various roles as a patient since the concept is not homogeneous. Nevertheless, in contemporary democratic Western countries, we are expected to be treated equally in the healthcare system, even though it is well known that socio economic conditions have an effect on the treatment.

It is a fine or blurred border between a user of a healthcare system and a victim of the same. To be a user one needs to be able to interact, in health care or nursing situations it is mainly what is offered regarding nursing or treatment that one must accept. In nursing homes for elderly care, the care takers are defined as users. It must be challenged what it means to be a user with dementia and by having problems defining and expressing wills and even needs. Taking the perspective of individuals or groups, one needs to have a deeper understanding of what it means to orient in the world, that is how perception works.

A design journey in a health care setting often involves the health care system, while a patient journey is about how to treat a disease or planning for an old person's care. As in a design journey one needs to describe the current state, which is often based on the description by the patient or client and interpreted and defined by professionals. There does not always exist an agreement between the patient and the professionals regarding the treatment process, even though the patient wants to get rid of their problems.

We will here describe a patient journey by starting with problematising the concepts user/patient- centred, and person- centred. A patient is a specific role that a person has in a health care system, while a person centred perspective includes the whole individual with all his/her roles. When one is regarded as a patient it is often with focus on a specific disease the person is diagnosed with or is checked over for. If the check over does not match the expected disease the patient will be left with their problems or hand over to other medical units. From a person-centred perspective, the patient will be actively involved in the care process. (Ahmed et al., 2020) discuss the power relation between patients and care professionals, where shared decision-making can be seen as a shift in power where patients are let into their domain. Self-management is a part of person-centred care, and from that perspective the patient allows the professionals into their domain. It has to do with mutual trust and respect. The physician must understand the patient's context and support self-management based on each patient's situation and his/her possibilities. For example, with respect to self-monitoring of physical parameters, to which extent this will be successful depends on motivation and interest from the patient. The care system must be flexible and use digital solutions when the prerequisites are in place, and when a patient is reluctant other solutions must be found. Measurements taken by the patient must be placed in a meaningful context so that the patient understands why these measurements should be taken. Finally, it is important with feedback so that the patient knows that someone actually follows the progression and acts upon deteriorations.

Communication

Crucial for taking a person-centred perspective that includes self-management is the communication between an actual person and professionals. Communication could not be discussed separately, it takes place in relation to individuals, groups, objects, environments and are situated in situations. Contexts are created in relation between people and places, such as healthcare situations. However, the three concepts/terms will be separate to be able to dig deeper into its meanings. We start with communication.

The etymological explanation of communication comes from the Latin word communicate, which means to have in common, to share the same understanding of something. The Swedish scholar, Peder Hård af Segerstad (2009), defines the communication process as follows: Knowledge of the carrier—information transmitted—the information interacts with the recipient's information system (the mental image)—new information is created—knowledge of the recipient—the recipient spreads the information further. We can conclude that communicating information is a complex process. However, the process recommended offers a structure that is neither mechanical nor ignores the directional nature of information in communication (someone is informing someone else). If it also takes a design thinking perspective into consideration, and each stage contains an iterative process—especially if such as an information designer takes the process seriously (Eriksson and Carlsson, 2022).

So, what does it mean to communicate in a healthcare or nursing situation? It is not enough to present facts for a patient, or how to take care of oneself after or under a treatment. In the communication act between a caregiver and a caretaker it is much that affects the result. For the doctor or nurse routine situation could for a patient be dramatic, for they the situation might be unique (a first-time experience). Taking a design thinking perspective, the doctor needs to empathise with the patient to come up with the best solution to take the treatment or change of lifestyle into the next step. When a patient gets informed by a doctor or nurse about things such as Diabetes 2, it is necessary to discuss the living situation before telling them what to change in diet and start exercising. For a patient to experience such information meaningfully, it needs to "talk to them". What are their current eating habits? And what causes these habits? Is it the diet that is the problem, or could it be something else? Probably most doctors or nurses ask those questions, but the tricky part here is to focus on the communication act/process as such. By listening to what kind of questions or comments the patient makes, it could work as a "guide" for how the message is received, that is to take a person-centred perspective and to start the journey.

It could be that a question that is as irrelevant for the doctor or nurse that gives information indicates that the patient does have totally different references or perspectives on what is communicated. This should not be interpreted as a lack of understanding in traditional meaning. It is often that people are heavily shaped by their environment, what is defined as the socio-semiotic perspective (Halliday, 1984). That is often much stronger than new facts that are presented for them, this is especially strong when it comes to lifestyle and politics. But it could also be that expectations from the nursing situation work as a grid between the patient and the doctor or nurse. If the patient expects that a treatment should include only pills it could be hard to perceive that such as exercise and diet is equally important. Even though the patient has that understanding, it could be undoable to follow because of the living situation. A doctor expert in diabetes said that he had found in his practice that women who got Diabetes 2 found it harder to follow the diet that is recommended than men, because they are the ones cooking and their husbands do not accept the diet. While men who were diagnosed had wives that followed the recommendations for their husbands. Here, the communication is about the life situation and it is important to involve not only the patient but also their partner.

There are also situations when patients have difficulties in communicating. When it comes to patients with dementia, communication includes multiple challenges. The loss of memory does not mean that a person cannot communicate, but the communication act is different from what most of us think of as communication. It is common that people with dementia talk their opinion and then forget about it shortly afterwards, but it does not mean what they communicated is of less importance.

Ilpo Koskinen (2023) argues that all design tasks, however simple or complex, begin with understanding the way humans ascribe meaning, both as individuals and as actors in complex societies. Thus, all design research must be interpretive at its core. USD is also an anchored experience and as such is contextual and situated (Koskinen, 2023). That means, from an HCD perspective, the design process needs to be situated in the context where the artefact or system will be used.

Relations between Users in Different Health care Settings

We have previously problematised the concept user, but in this section we will use the term since we will discuss relations between individuals, groups, and technology. The users are humans that act, they are actors. Users are part of a system built on relations, but they are people. Every system has various kinds of actors, who have different abilities to influence the system or part of the system. The relations in a system could be of different kinds, as mutual dependency, conflict interest, or as power relations. No matter which kind of relationship we are dealing with, they do have challenges. Dealing with the concept user, it is necessary not only to discuss relations both as a system where individuals and groups are a part of, but also as internal relations such as within a group or between groups in a specific context. The focus needs to be on people in relation to a system.

Why?

In most cases there is a clear intention to why something should be changed or created, either it is a part of a system, a small detail or creation of something new. However, it is also common that the reason for the change or the new changes that are going to be created will be blurred by the process itself. Therefore, it is important to keep the "why" vital during the process and do not forget it. For example, the aim is to make it easier for the patient to get access to information about different treatments. The focus must be on how to facilitate usability of the system that provides information. That means to reduce needless information but not make the information too condensed.

What?

Based on the example above it is the information that should be accessible. That will probably include both the content and the interface. The "what" activity needs to be focused on why something should be changed.

How?

The "How" is related to why and what. It is not always obvious how one should solve a problem or come up with a solution that satisfies the needs. If one knows why something needs to be changed and when the "what" is defined, it is time to start to choose method(s). From an HCD perspective, it is crucial to be aware of this power balance and make sure that all user groups have the possibility to affect the design of the systems. Even when the target user group is clear, it is important to take into consideration the needs of other secondary user groups. However, it is not that simple, when the designer comes into the picture, he or she is a further actor with his/her own perspective and background knowledge. Therefore, it is important to understand that the design process is a process that takes time, and that it must be a collaborative multi disciplinary work (Sjölinder et al., 2014).

Since a system consists of a variety of users/actors it is necessary to focus on why, what, and how the relations between them need to be discussed. The health care setting consists of numerous actors, for example, patients, relatives, doctors, nurses, specialists, administration, finance, IT, procurement, managers at different levels, decision- makers, consultants, and representatives from different companies (providing products or services). All these actors have their own perspective of what to contribute with, and about what to get out of the health care system. These actors have varying possibilities to affect the situation depending on context and authority within each specific context. The power and the authority in the relationship between the patient and the doctor has been investigated in several studies (Ha, 2010). The patient has the disadvantage of both being in an unfamiliar arena and possibly exposed to a condition or a diagnosis, and therefore dependent on the doctor or the health care system. On the other hand, this relationship has started to change in recent years, with more educated patients and a more symmetrical balance between the doctor and the patient (Kaba and Sooriakumaran, 2007). With cohesive and connected health care, the care of one patient could also involve further actors such as nurses, relatives, home health care, and care givers, all of them with their interpersonal balance of power.

Knowledge and trust must be built up, and different perspectives must be discussed. Within the health care setting some of the actors are responsible for other people's lives, and crucial systems must be designed in a way where medical staff feel that they can provide care without risks of conducting mistakes. On the other hand, there is the patient which everything is about, that must feel that he/she has control over his/her own situation. A further aspect related to the design of health care systems is the handling of sensitive data, this also places challenges related to how systems could be designed to best serve the different user groups' needs and at the same time meet privacy needs and regulations.

Context, Situation, and Human Interaction

A situation from a user perspective is when an individual or group is expected to use a service or product. It is not only the context that has an effect on the usability but also the situation. A situation that is experienced as stressful will influence the ability to use a product or services. This section will discuss from examples how well-designed products or services could be experienced differently because of the situation. The situation is often regarded as something isolated where communication takes place, but it is the situation in which the context is created.

A context could be limited but also include a larger system that involves individual groups, equipment of various kinds and different organisations. Even though we deal with small units as context they are parts of a larger system that could be more or less complex. The section will discuss the impact of context on how we take an HCD perspective and vice versa.

Context is often discussed or mentioned in singularise and as it is something fixed, while context is something that is created in relation to individuals, groups, and situations. But in addition, we need to take several contexts into consideration at the same time. A person grows up in a context, and is active during life in several contexts, in school, free time, work, new family, friends, etc. When they get older and maybe end up in a nursing home, they will be a part of a new context, but still, they belong to the previous context since they are talking about the actual individual in the context of friends, neighbours, and family. This is something that is embedded in the individual's body and by that is a part of the body schema. Therefore, it could/should not be ignored while talking about the nursing context.

Consequently, one has to deal with several contexts at the same time, the different context that belongs to the patient/the old person with dementia, and the contexts of the staff at the nursing home but also family and friends of the actual person. These contexts need to be regarded as the socio cultural background and belonging to the individuals.

Another aspect of context is the difference between planned or wanted situation and actual situation. For example, health care and care are exposed to increased numbers of patients/care takers and/or shortage of personnel. In the social care context, the care giver and the caretaker could be from different nationalities and there could be challenges with respect to communication. An HCD perspective could not change actual unexpected situations, but by considering principles from HCD when setting up the context and conducting plans the way the situation is perceived could be improved.

Perception of the Context and the Environment

Design moves in the field between the person who designs a system, information artefact or product and the person who is to perceive, interpret, and use it in a way that is relevant in relation to its purpose. In what way can phenomenology help us understand how such a process works? How can we benefit from the concept of a living body? How is it related to body schedule? What does design for a living body mean?

Perception needs to be understood on a deeper level, not only what we see, hear, and touch, but what it means to be a living body in the environment. Phenomenology gives the opportunity to look deeper into the meaning of perception and what affects our perception, and in which ways we perceive, analyse, and interpret various kinds of information. This goes in line with situatedness, whose theoretical position is that the ontological and functional are intertwined within the environmental, social, and cultural factors. Situated design is crucial for designing usable services or products.

Phenomenology can help us understand how we perceive the environment, and how it appears for us, either embedded or explicit. However, perception is a foundational, always-present quality of human experience and meaning, but difficult to grasp and articulate intellectually (Merleau-Ponty, 1962: 58). This is for two reasons: firstly, its presence and significance typically lie beneath conscious cerebral awareness; secondly, perception places itself in the background as it draws us out into the happenings of our world: "Perception hides itself from itself… it is of the essence of consciousness to forget its own phenomena thus enabling 'things' to be constituted…" (Merleau-Ponty, 1962: 58). Merleau-Ponty's central aim is to reconsider perception phenomenologically by "reawakening the basic experience of the world…" (1962: viii). That means we experience the world through our bodies, either we are aware of it or not. Because of that everything we perceive is related to the conditions of the body (fit, disabled, tired, painful, etc.)

To be aware of the meaning of the living body, we can better understand what it means to perceive and use information in specific contexts and in various situations. Specifically in relation to a healthcare situation. We all experience stress and feel vulnerable when we get information about healthcare issues. This is something that goes for all ages and living conditions. Even the people who are well-educated or well-prepared with questions could, after consulting a medical doctor, experience uncertainty. From a phenomenology perspective, the role of the living body and environmental embodiment are integral aspects of place and place experience (Seamon, 2018). So, when taking a user's perspective, it is also important to take the environment into consideration. For instance, it has been stressed by a researcher that "physical environment plays an important role in how everyday life is shaped and experienced for persons living in nursing homes as well as for the resident's family and friends" (Palmgren et al., 2022).

In relation to place experience, body schema is understood in relation to larger-scaled corporeal movements in relation to rooms, buildings, streets, public open spaces, neighbourhoods, and the like (Seamon, 2018). Merleau-Ponty explains everyday bodily mobility in terms of a body schema. Body-schema can be defined as a pre-cognitive, bodily intelligence and intentionality manifested through action and intertwining with the world at hand. As Merleau-Ponty explains, "Consciousness is being towards the thing through the intermediary of the body" (1962: 138–139). The understanding of body schema teaches us to understand that the body can act by its own memory and steering our intellect and not always vice versa. If we as designers take these phenomenological aspects into consideration, it will help us when walking beside another human being to understand that we walk beside each other with different bodily experiences, but with capability to exchange lived-body experiences.

The perception of the environment is often based on individuals' identity. The identity of individuals creates interaction between self-perception and outer input/influences. That means that the identities defined by others could be something that individuals often fulfil. This is something that needs to be taken into consideration when defining users from such as age, disabilities, aphasia, or dementia. For practical reasons we often group people by such characteristics, and from that we search information based on research in disciplines such as social science and medicine. It might fit a specific design situation, but it could also be misleading since the research conducted does often have other purposes and are not relevant for specific design projects. People with different kinds of disabilities, or dementia, have other qualities than the actual disorder, qualities, and interests that do not necessarily disappear because of the disease.

How a milieu is experienced also depends on how it is spatially organized and how the ability to orient oneself in the environment. The self-perception and by that the identity is not shaped in vacuum, it is created in interaction between the environment (culture) and self-perception. A build milieu that is designed with care and well suited for people supports their ability to orient and use the environment with embedded security. That could be such as how to find the way in a hospital, or the orient at the nursing home, or in grocery stores. Design that supports orientation in the environment or places makes people feel more comfortable, which is especially important for old people. Another aspect is the appearance of the environment. If the interior is designed with care and respect for the users (the visitor or the people living at an institution, such as a nursing home) it is embedded in the milieu (Bates). The same goes for the opposite, a less careful design creates a feeling among the users that they are not valuable human beings (Bates). An environment (interior and exterior) that is designed with respect for the users will contribute to their self-perception.

Within healthcare different stakeholders perceive the context and the involvement in the HCD process in different ways, as well as the system itself as the build environment. The patients are the actors around everything centres since it is him or her that needs or that is ill. On the other hand, to be able to meet the needs of the patients a complex medical system is needed that has its own needs to be able to operate in a safe and efficient manner. This complex machinery could, by the patients, be perceived by the patients as a system where they lose their control over their own situation. Within the healthcare organisations there are, among others, doctors, nurses, managers, and administrative staff. When applying HCD methods and involving these different groups, all of them have their own perspective of what is needed and of how it should work. It is crucial to incorporate this within the design process, and when designing for the healthcare context be aware of primary and secondary user groups and for whom and to which extent value must be created.

Another category of actors are the developers and the companies that design services and products for health care. They have their perspectives of the process related to budget, profit, and time they can spend. These perspectives are to some extent a prerequisite since without service providers and companies that design, deliver, and maintain services and products for health care many of the needs of the patients could not be met. However, of course, there is competition among the service providers, which leads us to the perspective of high-level managers and the policy makers. Their perspectives or perceptions of the HCD process are likely to be affected by budget and other aspects related to their role. How the HCD design processes are perceived by these groups, and how important they perceive the cost for it will have an impact on the outcome and on how well the user groups perceive that their needs will be met.

The Need of HCD in the Development of Future Health care Systems

It is well known that the number of elderly adults is increasing and that we will face challenges related to resources (Waite, 2004). Therefore, the need for people to take a larger responsibility for their own health will increase. Patient empowerment and taking control over one's own health demands both motivation and gaining of knowledge. This goes for both prevention and management of diseases. The health care system and the digital tools used must be designed in a way that they support the individuals in this process. The design of the systems must also convey trust—that the health care system trusts the individual and his/her efforts in managing his/her own health. For example, the use of sensor data provided by the individual must be balanced with the demand of accurate data from health care sensors. Of course, crucial medical data must come from healthcare certified equipment, but there is a large area where data from personal sensors could be used in a meaningful way, in the same way as applications for training and exercise are used today. Many values do not have to be certified or precise, the important thing might be to gain the motivation and the feeling of participation in one's own health. With respect to this, HCD can provide usage scenarios and design solutions that finds a balance between the safety aspects within health care and the motivating aspects of being able to share sensor data from one's own equipment with personnel in the health care system.

The increasing use of digital solutions in health care will lead to new ways of working and to changes in work tasks. It could change the field and nature of work, and affect workload and pace (Kaihlanen, 2023). In the transition, there is a risk that employees in several professions will feel that their professional pride is at risk. One important role of HCD in health care systems design will be to guide this transition in a way that employees see the benefits and find professional pride in their new and changed tasks, and to support the design of digital solutions so that these systems are easy to use and support the new tasks that have evolved.

Use of digital medical records and possibilities to share patient information between hospitals, regions, and counties will make it easier for patients to get care based on their medical history regardless of where they are. Extended possibilities to gather and share data has and will create new opportunities within health care. However, if patients should be willing to share their data both with the health care system and other people, for example, relatives, it is important that the usage is perceived as meaningful and that the users/patients feel that they get something out of data sharing. It is also important to design systems that make it possible for the users to decide if and with whom they are willing to share data. The users/patients must also have a feeling of trust in the health care system with respect to how the data are stored and used. Even if the systems are secure, the feeling of trust must also be conveyed to the users and the patients.

The MedTech industry and the development of implants are growing rapidly and within a few years many people will walk around with different kinds of medical equipment within their bodies. This will raise issues related to both security and trust. The equipment must meet certain safety standards, but it must also be possible to replace parts even if a company no longer exists. The monitoring aspect of the usage has both safety aspects in terms of being able to find the person and demands trust since it will always be possible to know where the person is by the positioning of the technology.

More advanced data analyses and usage of data have and will contribute to a better and more efficient care. Within this field, HCD and research within HCD could play an important role. AI (artificial intelligence) and decision support systems might affect several aspects such as trust, both in systems and in humans, human communication, and professional roles. An increasing use of AI-solutions are also going to face challenges related to ethics and the content that the models are based upon. For example, the models that the systems are based upon could be affected by the material that was used during their training. If this data is skewed towards a direction, it will have an impact on how the models and the AI solutions will behave. In this case, certain groups of people or certain aspects could be approached and treated in a different way. With respect to information design, the content must come from verified sources so that the users can be sure that the information given is correct. From a human perspective the professionals could feel that their knowledge and experience are set aside, and the patients could feel that they are less seen as individuals. On the other hand, AI solutions might contribute to a more individualised and targeted care for each patient since it could provide support in setting up patient specific treatments. However, the HCD perspective will be crucial in this transition towards a sustainable digitalised healthcare that contributes to reliable and usable systems. Therefore, for the development of future health care systems, including technology and new ways of working,

HCD will be needed. In this transition the HCD perspective will be crucial and closely related to success of the implementation of the healthcare systems. This will also be a challenge for information designers in the future.

References

Ahmed, F., Bari A.S.M.H. and Gavrilova. M.L. 2020. Emotion Recognition From Body Movement. *IEEE Access*, 8: 11761–11781. doi: 10.1109/ACCESS.2019.2963113.

Bannon, Liam J. and Ehn, Pelle. 2012. Design matters in participatory design. In: *Handbook of Participatory Design* (pp. 37–63). Routledge.

Eriksson, Y. and Carlsson, A-L. 2022. The challenges in designing meaningful information. In: Y. Eriksson (Ed.), *Different Perspectives in Design Thinking*. CRC Press. Francis and Taylor.

Eriksson, Y., Skagert, K. and Ekwall, P-E. 2023. Design process of live-action video instructions. In: G. Goldschmidt and E. Tarazi (Eds.), *Expanding the Frontiers of Design*. Francis & Taylor. Routledge.

Giacomin, J. 2014. What Is Human Centred Design? *An International Journal for All Aspects of Design, 17* (4): 606–623. Published online: 28 Apr. 2015. https://doi.org/10.2752/175630614X14056185480186.

Goffman, E.I. 1956. *The Presentation of Self Everyday Life*. University of Edingburgh. https://monoskop.org/images/1/19/Goffman_Erving_The_Presentation_of_Self_in_Everyday_Life.pdf.

Greenbaum, J. and Kyng, M. 1991. *Design at Work: Cooperative Design of Computer Systems* (pp. 3–24). Lawrence Erlbaum Associates, Inc., Hillsdale.

Ha, J.F. and Longnecker, N. 2010, Spring. Doctor-patient communication: A review. *Ochsner J., 10*(1): 38–43. PMID: 21603354; PMCID: PMC3096184.

Halliday, M, (1984). *Exploration in the Functions of Language*. Elsevier.

Kaba, R. and Sooriakumaran, P. 2007, February. The evolution of the doctor-patient relationship. *International Journal of Surgery, 5*: 57–65. https://doi.org/10.1016/j.ijsu.2006.01.005.

Kaihlanen, A.M., Laukka, E., Nadav, J., et al. 2023. The effects of digitalisation on health and social care work: A qualitative descriptive study of the perceptions of professionals and managers. *BMC Health Serv. Res, 23*: 714. https://doi.org/10.1186/s12913-023-09730-y.

Keinonen, T. 2010. Protect and Appreciate: –Notes on the Justification of User-Centered Design. *International Journal of Design, 4* (1).

Krippendorff, K. 2004. *Content Analysis: An Introduction to Its Methodology* (2nd Edn.). Thousand Oaks, CA: Sage Publications.

Mary B., Adam, Joy Minyenya-Njuguna, Wilson Karuri Kamiru, Simon Mbugua, Naomi Wambui Makobu, and Angela J. Donelson. 2020, November. *Health Policy and Planning, 35* (Supplement_2), ii150–ii162. https://doi.org/10.1093/heapol/czaa129.

Merleau-Ponty. M. 1962/1992. *Phenomenology of Perception*. London: Routledge.

Palmgren M, Rosenberg L, Gaber SN, Johansson K. Family members' reasoning in relation to pleasant environments in nursing homes. Dementia (London). 2023 Jan;22(1):235–251. DOI: 10.1177/14713012221142474. Epub 2022 Nov 25.

Sanders, Elizabeth B.-N., and Pieter Jan Stappers. 2008. Co-creation and the new landscapes of design. *Co-Design, 4* (1): 5–18. DOI: 10.1080/15710880701875068.

Schuler, D. and Namioka, A. 1993. *Participatory Design Principles and Practices*. Lawrence Erlbaum Associates, Inc., London.

Seamon, D. (2018). Life takes place: Phenomenology, lifeworlds, and place making. Routledge.

Sjölinder, M., Ehn, M., Boman, I-L., Folke, M., Hansson, P., Sommerfeld, D.K., Nylander, S. and Borg, J. 2014. A multidisciplinary approach in the development of a stroke rehabilitation tool. In: *Proceedings of HCI International 2014*, 22–27 June, Crete, Greece.

Sjölinder, M., and Scandurra I. 2015. Effects of Using Care Professionals in the Development of Social Technology for Elderly. In: *First International Conference, ITAP 2015, Held as Part of HCI International 2015 on Human Aspects of IT for the Aged Population. Design for Everyday Life*. Los Angeles, CA, USA, August 2–7 . *LNCS, 9194*: 181–192.

Sjölinder, M. 2022. Raising Users Confi dence in their own Technology Literacy as part of the Design process. In: Y. Eriksson (Ed.), *Different Perspectives in Design Thinking*. CRC Press. Francis and Taylor.

Van der Bijl-Brouwer, M. 2016. The Challenges of Human-Centred Design in a Public Sector Innovation Context. In: *Design Research Society Anniversary Conference*, 27–30 June, Brighton, UK.

Waite, L.J. 2004. The Demographic Faces of the Elderly. *Popul. Dev. Rev., 30*(Supplement):3–16. PMID: 19129925; PMCID: PMC2614322.

Zhang, T. and Dong, H. 2008. Human-centred design: An emergent conceptual model. Include 2009, Royal College of Art, April 8–10, 2009, London; Include 2009 proceedings (ISBN: 978-1-905000-80-7) Available in: http://www.hhc.rca.ac.uk/2084/all/1/proceedings.aspx.

Index

For Product Safety Concerns and Information please contact our EU representative GPSR@taylorandfrancis.com Taylor & Francis Verlag GmbH, Kaufingerstraße 24, 80331 München, Germany